This book comes with access to more content online.

Quiz yourself, track your progress,
and improve your grade!

Register your book or ebook at
www.dummies.com/go/getaccess.

Select your product, and then follow the prompts
to validate your purchase.

You'll receive an email with your PIN and instructions.

D1296320

Chemistry

ALL-IN-ONE

by Christopher R. Hren, John T. Moore, EdD, and Peter J. Mikulecky, PhD

A Wiley Brand

Chemistry All-in-One For Dummies®

Published by: **John Wiley & Sons, Inc.,** 111 River Street, Hoboken, NJ 07030-5774, www.wiley.com

Contents at a Glance

Contents at a Glance

Table of Contents

Introduction

Chemistry is at once practical and wondrous, humble and majestic. And for someone studying it for the first time, chemistry can be tricky and rather challenging in some spots.

That's why we wrote this book. It is designed to be an all-encompassing companion for you as you journey through the wonderful yet sometimes confusing world of chemistry. It is going to help you work through anything you might need in your class or whatever you want to investigate further in the wide chemical world.

Chemistry is sometimes called the central science (mostly by chemists), because in order to have a good understanding of biology or geology or even physics, you must have a good understanding of chemistry. We live and work in a world of chemistry, and after your journey is complete, hopefully you won't find the word *chemistry* so frightening.

About This Book

This book is a one-stop chemistry shop. Each chapter explains key concepts covered in any high school or introductory college chemistry class, along with example problems and opportunities for practice. You'll find the absolute basics that you need to succeed in a chemistry course, all the way up to some rather complicated material.

One thing that will stand out to you is all of the calculations throughout the book. Chemistry has a lot of math in it, and this book doesn't shy away from that. The beginning chapters of this book, though, are going to help you make sure you are comfortable working through the calculations you'll see throughout the rest of it, so make sure you don't skip over those if you think you need a little review. Once you are past those you're going to see every major chemistry topic covered that you're likely to encounter:

Each new topic provides

>> Example problems with answers and solutions

>> Practice problems with answers and solutions

Each chapter provides

>> An end-of-chapter quiz with problems representing the topics covered

>> Solutions to those quiz questions

Online quizzes are also available for even more practice and confidence-building.

Foolish Assumptions

Because you're interested in this book, we assume you probably fall into one of a few categories:

>> You're a student taking a high school chemistry course or a student in college taking an introductory chemistry class and are just not even sure where to start.

>> You're a parent of a student taking chemistry in high school and are trying to help your kid out with their chemistry but need some brushing up.

>> You're just naturally curious about science and mathematics and you want to get a little more acquainted with chemistry.

We also assume that you can add, subtract, multiply, and divide numbers without any real issue, but that you will still use a calculator for all this stuff. Calculators are there for a reason, and they are helpful, so please use them!

Icons Used in This Book

In this book, I use these five icons to signal what's most important along the way:

EXAMPLE

Each example is an algebra question based on the discussion and explanation, followed by a step-by-step solution. Work through these examples, and then refer to them to help you solve the practice test problems at the end of the chapter.

REMEMBER

This icon points out important information that you need to focus on. Make sure you understand this information fully before moving on. You can skim through these icons when reading a chapter to make sure you remember the highlights.

TIP

Tips are hints that can help speed you along when answering a question. See whether you find them useful when working on practice problems.

WARNING

This icon flags common mistakes that students make if they're not careful. Take note and proceed with caution!

YOUR TURN

When you see this icon, it's time to put on your thinking cap and work out a few practice problems on your own. The answers and detailed solutions are available so you can feel confident about your progress.

Beyond the Book

In addition to what you're reading right now, this book comes with a Cheat Sheet that provides quick access to some formulas, rules, and processes that are frequently used. To get this Cheat Sheet, simply go to www.dummies.com and type **Chemistry All-in-One For Dummies Cheat Sheet** in the Search box.

You'll also have access to online quizzes related to each chapter. These quizzes provide a whole new set of problems for practice and confidence-building. To access the quizzes, follow these simple steps:

1. **Register your book or ebook at Dummies.com to get your PIN.** Go to www.dummies.com/go/getaccess.

2. **Select your product from the drop-down list on that page.**

3. **Follow the prompts to validate your product, and then check your email for a confirmation message that includes your PIN and instructions for logging in.**

If you do not receive this email within two hours, please check your spam folder before contacting us through our Technical Support website at http://support.wiley.com or by phone at 877-762-2974.

Now you're ready to go! You can come back to the practice material as often as you want — simply log on with the username and password you created during your initial login. No need to enter the access code a second time.

Your registration is good for one year from the day you activate your PIN.

Where to Go from Here

This book is organized so that you can safely move from whichever chapter you choose to start with and in whatever order you like. You can strengthen skills you feel less confident in or work on those that need some attention.

If you need some help with scientific notation, unit conversions, or otherwise feel you could use a little practice on the math side of things then we recommend strongly that you look over Chapters 1 and 2. Those are going to get you ready for all of the other material you'll see throughout the book. After that Chapters 3, 4, and 5 are going to give you a solid grounding in matter, atoms, and the world-renowned periodic table. Those are probably where you're going to be starting in almost any chemistry class you encounter, so those are likely going to be a great place to begin your journey.

Beyond that, check out whatever chapters you might need help with. Each chapter is designed to be self-sufficient and will walk you through whatever material you need to understand a particular topic. Do keep in mind, though, that different aspects of chemistry are very much interrelated. Even though a chapter might be set up to be self-contained, there are likely going to be things in that chapter you are assumed to understand from previous chapters. The concepts you learn in chemistry rarely go away; they pop up again and again when you're learning new material so don't hesitate to look back at whatever you might need to review.

You can use the table of contents at the beginning of the book and the index in the back to navigate your way to the topic that you need to brush up on. Regardless of your motivation or what technique you use to jump into the book, you won't get lost because you can go in any direction from there.

Enjoy!

1
Getting Started with Chemistry

In This Unit . . .

Chapter **1**

Looking at Numbers Scientifically

Like any other kind of scientist, a chemist tests hypotheses by doing experiments. Better tests require more reliable measurements, and better measurements are those that have more accuracy and precision. Accurate and precise calculations are essential to successful experiments, so a large chunk of chemistry centers on ways to report and describe measurements.

How do chemists report their precious measurements? What's the difference between accuracy and precision? And how do chemists do math with measurements? These questions may not keep you awake at night, but knowing the answers to them will keep you from making mistakes in chemistry.

Using Exponential and Scientific Notation to Report Measurements

Because chemistry concerns itself with ridiculously tiny things like atoms and molecules, chemists often find themselves dealing with extraordinarily small or extraordinarily large numbers. Numbers describing the distance between two atoms joined by a bond, for example, run in the ten-billionths of a meter. Numbers describing how many water molecules populate a drop of water run into the trillions of trillions.

To make working with such extreme numbers easier, chemists turn to scientific notation, which is a special kind of exponential notation. In *exponential notation*, a number is represented as a value raised to a power of 10. The decimal point can be located anywhere within the number as long as the power of 10 is correct.

Suppose that you have an object that's 0.00125 meters in length. Express it in a variety of exponential forms:

$$0.00125 \text{ m} = 0.0125 \times 10^{-1} \text{ m, or}$$
$$0.125 \times 10^{-2} \text{ m, or}$$
$$1.25 \times 10^{-3} \text{ m, or}$$
$$12.5 \times 10^{-4} \text{ m, and so on}$$

All these forms are mathematically correct as numbers expressed in exponential notation. But in scientific notation the decimal point is placed so that only one digit other than zero is to the left of the decimal point. In the preceding example, the number expressed in scientific notation is 1.25×10^{-3} m. Most scientists express numbers in scientific notation.

In scientific notation, every number is written as the product of two numbers, a coefficient and a power of 10. In plain old exponential notation, a coefficient can be any value of a number multiplied by a power with a base of 10 (such as 10⁴). But scientists have rules for coefficients in scientific notation. In *scientific notation*, the coefficient is always at least 1 and always less than 10. For example, the coefficient could be 7, 3.48, or 6.0001.

TIP

To convert a very large or very small number to scientific notation, move the decimal point so it falls between the first and second digits. Count how many places you moved the decimal point to the right or left, and that's the power of 10. If you moved the decimal point to the left, the exponent on the 10 is positive; to the right, it's negative. (Here's another easy way to remember the sign on the exponent: If the initial number value is greater than 1, the exponent will be positive; if the initial number value is between 0 and 1, the exponent will be negative.)

To convert a number written in scientific notation back into decimal form, just multiply the coefficient by the accompanying power of 10.

In many cases, chemistry teachers refer to powers of 10 using scientific notation instead of their decimal form. With that in mind, here's a quick chart showing you the most common powers of 10 used in chemistry, along with their corresponding scientific notation.

EXAMPLE

Q. Convert 47,000 to scientific notation.

A. $47,000 = 4.7 \times 10^4$. First, imagine the number as a decimal:

47,000.

Next, move the decimal point so it comes between the first two digits:

4.7000

Then count how many places to the left you moved the decimal (four, in this case) and write that as a power of 10: 4.7×10^4.

Q. Convert 0.007345 to scientific notation.

A. $0.007345 = 7.345 \times 10^{-3}$. First, put the decimal point between the first two nonzero digits:

7.345

Then count how many places to the right you moved the decimal (three, in this case) and write that as a power of 10: $0.007345 = 7.345 \times 10^{-3}$.

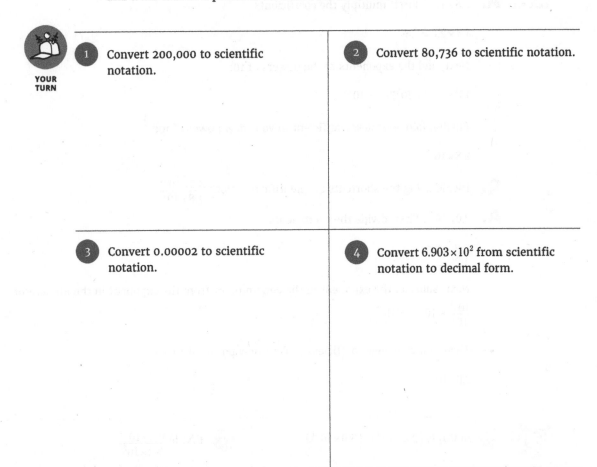

YOUR TURN

1   Convert 200,000 to scientific notation.

2   Convert 80,736 to scientific notation.

3   Convert 0.00002 to scientific notation.

4   Convert 6.903×10^2 from scientific notation to decimal form.

Multiplying and Dividing in Scientific Notation

A major benefit of presenting numbers in scientific notation is that it simplifies common arithmetic operations. The simplifying abilities of scientific notation are most evident in multiplication and division. (As we note in the next section, addition and subtraction benefit from exponential notation but not necessarily from strict scientific notation.)

REMEMBER To multiply two numbers written in scientific notation, multiply the coefficients and then add the exponents. To divide two numbers, simply divide the coefficients and then subtract the exponent of the *denominator* (the bottom number) from the exponent of the *numerator* (the top number).

Q. Multiply using the shortcuts of scientific notation: $(1.4 \times 10^2) \times (2.0 \times 10^{-5})$.

EXAMPLE **A.** 2.8×10^{-3}. First, multiply the coefficients:

$1.4 \times 2.0 = 2.8$

Next, add the exponents of the powers of 10:

$10^2 \times 10^{-5} = 10^{2+(-5)} = 10^{-3}$

Finally, join your new coefficient to your new power of 10:

2.8×10^{-3}

Q. Divide using the shortcuts of scientific notation: $\dfrac{3.6 \times 10^{-3}}{1.8 \times 10^4}$.

A. 2.0×10^{-7}. First, divide the coefficients:

$\dfrac{3.6}{1.8} = 2.0$

Next, subtract the exponent in the denominator from the exponent in the numerator:

$\dfrac{10^{-3}}{10^4} = 10^{-3-4} = 10^{-7}$

Then join your new coefficient to your new power of 10:

2.0×10^{-7}

YOUR TURN

⑤ Multiply $(2.2 \times 10^9) \times (5.0 \times 10^{-4})$.

 ⑥ Divide $\dfrac{9.3 \times 10^{-5}}{3.1 \times 10^2}$.

---

7 Using scientific notation, multiply 52×0.035.	**8** Using scientific notation, divide $\dfrac{0.00809}{20.3}$.

Using Scientific Notation to Add and Subtract

Addition or subtraction gets easier when you express your numbers as coefficients of identical powers of 10. To wrestle your numbers into this form, you may need to use coefficients less than 1 or greater than 10. So scientific notation is a bit too strict for addition and subtraction, but exponential notation still serves you well.

REMEMBER To add two numbers easily by using exponential notation, first express each number as a coefficient and a power of 10, making sure that 10 has the same exponent in each number. Then add the coefficients. To subtract numbers in exponential notation, follow the same steps but subtract the coefficients.

EXAMPLE **Q.** Use exponential notation to add these numbers: $3{,}710 + \left(2.4 \times 10^{2}\right)$.

A. 39.5×10^{2}. First, write both numbers with the same power of 10:

37.1×10^{2} and 2.4×10^{2}

Next, add the coefficients:

$37.1 + 2.4 = 39.5$

Finally, join your new coefficient to the shared power of 10:

39.5×10^{2}

Q. Use exponential notation to subtract: $0.0743 - 0.0022$.

A. 7.21×10^{-2}. First, convert both numbers to the same power of 10. We've chosen 10^{-2}:

7.43×10^{-2} and 0.22×10^{-2}

Next, subtract the coefficients:

$7.43 - 0.22 = 7.21$

Then join your new coefficient to the shared power of 10:

7.21×10^{-2}

YOUR TURN

9 Add $\left(398 \times 10^{-6}\right) + \left(147 \times 10^{-6}\right)$.

10 Subtract $\left(7.685 \times 10^{5}\right) - \left(1.283 \times 10^{5}\right)$.

11 Use exponential notation to add $0.00206 + 0.0381$.

12 Use exponential notation to subtract $9,352 - 431$.

Distinguishing between Accuracy and Precision

Whenever you make measurements, you must consider two factors, accuracy and precision. *Accuracy* is how well the measurement agrees with the accepted or true value. *Precision* is how well a set of measurements agree with each other. In chemistry, measurements should be *reproducible*; that is, they must have a high degree of precision. Most of the time chemists make several measurements and average them. The closer these measurements are to each other, the more confidence chemists have in their measurements. Of course, you also want the measurements to be accurate, very close to the correct answer. However, many times you don't know beforehand anything about the correct answer; therefore, you have to rely on precision as your guide.

Suppose you ask four lab students to make three measurements of the length of the same object. Their data follows.

	Student 1	Student 2	Student 3	Student 4
Trial 1	27.77 cm	27.30 cm	27.55 cm	27.30 cm
Trial 2	27.30 cm	27.60 cm	27.55 cm	27.29 cm
Trial 3	27.56 cm	27.97 cm	27.53 cm	27.31 cm
Average	27.54 cm	27.62 cm	27.54 cm	27.30 cm

The accepted length of the object is 27.55 cm. Which of these students deserves the higher lab grade? Both students 1 and 3 have values close to the accepted value, if you just consider their average values. (The average, found by summing the individual measurements and dividing by the number of measurements, is normally considered to be more useful than any individual value.) Both students 1 and 3 have made *accurate* determinations of the length of the object. The average values determined by students 2 and 4 are not very close to the accepted value, so their values are not considered to be accurate.

However, if you examine the individual determinations for students 1 and 3, you notice a great deal of variation in the measurements of student 1. The measurements don't agree with each other very well; their precision is low even though the accuracy is good. The measurements by student 3 agree well with each other; both precision and accuracy are good. Student 3 deserves a higher grade than student 1.

Neither student 2 nor student 4 has average values close to the accepted value; neither determination is very accurate. However, student 4 has values that agree closely with each other; the precision is good. This student probably had a consistent error in his or her measuring technique. Student 2 had neither good accuracy nor precision. The accuracy and precision of the four students is summarized below.

	Accuracy	Precision
Student 1	High	Low
Student 2	Low	Low
Student 3	High	High
Student 4	Low	High

Usually, measurements with a high degree of precision are also somewhat accurate. Because the scientists or students don't know the accepted value beforehand, they strive for high precision and hope that the accuracy will also be high. This was not the case for student 4.

So remember, accuracy and precision are not the same thing:

» **Accuracy:** Accuracy describes how closely a measurement approaches an actual, true value.

» **Precision:** Precision, which we discuss more in the next section, describes how close repeated measurements are to one another, regardless of how close those measurements are to the actual value. The bigger the difference between the largest and smallest values of a repeated measurement, the less precision you have.

The two most common measurements related to accuracy are *error* and *percent error*:

» **Error:** Error measures accuracy, the difference between a measured value and the actual value:

Actual value – Measured value = Error

» **Percent error:** Percent error compares error to the size of the thing being measured:

$$\frac{|\text{Error}|}{\text{Actual value}} = \text{Fraction error}$$

Fraction error × 100 = Percent error

Being off by 1 meter isn't such a big deal when measuring the altitude of a mountain, but it's a shameful amount of error when measuring the height of an individual mountain climber.

If you want a simpler all-in-one formula to help you remember percent error, here is all of the above put into one simple-to-use formula:

$$\frac{|\text{Actual value} - \text{Measured value}|}{\text{Actual value}} \times 100 = \text{Percent error}$$

EXAMPLE

Q. A police officer uses a radar gun to clock a passing Ferrari at 131 miles per hour (mph). The Ferrari was really speeding at 127 mph. Calculate the error in the officer's measurement.

A. –4 mph. First, determine which value is the actual value and which is the measured value:

- Actual value = 127 mph

- Measured value = 131 mph

Then calculate the error by subtracting the measured value from the actual value:

Error = 127 mph – 131 mph = –4 mph

Q. Calculate the percent error in the officer's measurement of the Ferrari's speed.

A. 3.15%. First, divide the error's absolute value (the size, as a positive number) by the actual value:

$$\frac{|-4 \text{ mph}|}{127 \text{ mph}} = \frac{4 \text{ mph}}{127 \text{ mph}} = 0.0315$$

Next, multiply the result by 100 to obtain the percent error:

$$\text{Percent error} = 0.0315 \times 100 = 3.15\%$$

YOUR TURN

13 Two people, Reginald and Dagmar, measure their weight in the morning by using typical bathroom scales, instruments that are famously unreliable. The scale reports that Reginald weighs 237 pounds, though he actually weighs 256 pounds. Dagmar's scale reports her weight as 117 pounds, though she really weighs 129 pounds. Whose measurement incurred the greater error? Who incurred a greater percent error?

---

14 Two jewelers were asked to measure the mass of a gold nugget. The true mass of the nugget is 0.856 grams (g). Each jeweler took three measurements. The average of the three measurements was reported as the "official" measurement with the following results:

- **Jeweler A:** 0.863 g, 0.869 g, 0.859 g
- **Jeweler B:** 0.875 g, 0.834 g, 0.858 g

Which jeweler's official measurement was more accurate? Which jeweler's measurements were more precise? In each case, what was the error and percent error in the official measurement?

Identifying Significant Figures

Significant figures are the number of digits that you report in the final answer of the mathematical problem you're calculating. If we told you that one student determined the density of an object to be 2.3 g/mL and another student figured the density of the same object to be 2.272589 g/mL, we bet that you'd believe that the second figure was the result of a more accurate experiment. You may be right, but then again, you may be wrong. You have no way of knowing whether the second student's experiment was more accurate unless both students obeyed the significant figure convention.

If we ask you to count the number of automobiles that you and your family own, you can do it without any guesswork involved. Your answer may be 0, 1, 2, or 10, but you know exactly how many autos you have. Those numbers are what are called *counted numbers*. If we ask you how many inches are in a foot, your answer will be 12. That number is an *exact number* — it's exact by definition. Another exact number is the number of centimeters per inch, 2.54. In both exact and counted numbers, you have no doubt what the answer is. When you work with these types of numbers, you don't have to worry about significant figures.

Now suppose that we ask you and four of your friends to individually measure the length of an object as accurately as you possibly can with a meter stick. You then report the results of your measurements: 2.67 meters, 2.65 meters, 2.68 meters, 2.61 meters, and 2.63 meters. Which of you is right? You are all within experimental error. These measurements are *measured numbers*, and measured values always have some error associated with them. You determine the number of significant figures in your answer by your least reliable measured number.

REMEMBER When you report a measurement, you should include digits only if you're really confident about their values. Including a lot of digits in a measurement means something — it means that you really know what you're talking about — so we call the included digits *significant figures*. The more significant figures (sig figs) in a measurement, the more accurate that measurement must be. The last significant figure in a measurement is the only figure that includes any uncertainty, because it's an estimate. Here are the rules for deciding what is and what isn't a significant figure:

>> **Any nonzero digit is significant.** So 6.42 contains three significant figures.

>> **Zeros sandwiched between nonzero digits are significant.** So 3.07 contains three significant figures.

>> **Zeros on the left side of the first nonzero digit are *not* significant.** So 0.0642 and 0.00307 each contain three significant figures.

>> **One or more *final zeros* (zeros that end the measurement) used after the decimal point are significant.** So 1.760 has four significant figures, and 1.7600 has five significant figures. The number 0.0001200 has only four significant figures because the first zeros are not final.

>> **When a number has no decimal point, any zeros after the last nonzero digit *may* or *may not* be significant.** So in a measurement reported as 1,370, you can't be certain whether the 0 is a certain value or is merely a placeholder.

Be a good chemist. Report your measurements in scientific notation to avoid such annoying ambiguities. (See the earlier section "Using Exponential and Scientific Notation to Report Measurements" for details on scientific notation.)

>> **If a number is already written in scientific notation, then all the digits in the coefficient are significant.** So the number 3.5200×10^{-6} has five significant figures due to the five digits in the coefficient.

Numbers from counting (for example, 1 kangaroo, 2 kangaroos, 3 kangaroos) or from defined quantities (say, 60 seconds per 1 minute) are understood to have an unlimited number of significant figures. In other words, these values are completely certain.

REMEMBER

The number of significant figures you use in a reported measurement should be consistent with your certainty about that measurement. If you know your speedometer is routinely off by 5 miles per hour, then you have no business protesting to a policeman that you were going only 63.2 mph in a 60 mph zone.

EXAMPLE

Q. How many significant figures are in the following three measurements?

(a) 20,175 yards

(b) 1.75 yards

(c) 1.750 yards

A. a) Five, b) three, and c) four significant figures. In the first measurement, all digits are nonzero, except for a 0 that's sandwiched between nonzero digits, which counts as significant. The coefficient in the second measurement contains only nonzero digits, so all three digits are significant. The coefficient in the third measurement contains a 0, but that 0 is the final digit and to the right of the decimal point, so it's significant.

YOUR TURN

15 Identify the number of significant figures in each measurement:

(a) 76.093×10^{-2} meters

(b) 0.000769 meters

(c) 769.3 meters

16 In chemistry, the potential error associated with a measurement is often reported alongside the measurement, as in 793.4 ± 0.2 grams. This report indicates that all digits are certain except the last, which may be off by as much as 0.2 grams in either direction. What, then, is wrong with the following reported measurements?

(a) 893.7 ± 1 grams

(b) 342 ± 0.01 grams

Doing Arithmetic with Significant Figures

Doing chemistry means making a lot of measurements. The point of spending a pile of money on cutting-edge instruments is to make really good, really precise measurements. After you've got yourself some measurements, you roll up your sleeves, hike up your pants, and do some math.

REMEMBER

When doing math in chemistry, you need to follow some rules to make sure that your sums, differences, products, and quotients honestly reflect the amount of precision present in the original measurements. You can be honest (and avoid the skeptical jeers of surly chemists) by taking things one calculation at a time, following a few simple rules. One rule applies to addition and subtraction, and another rule applies to multiplication and division.

Addition and subtraction

In addition and subtraction, round the sum or difference to the same number of decimal places as the measurement with the fewest decimal places. For example, suppose you're adding the following amounts:

$$2.675 \text{ g} + 3.25 \text{ g} + 8.872 \text{ g} + 4.5675 \text{ g}$$

Your calculator will show 19.3645, but you round off to the hundredths place based on the 3.25, which has the fewest number of decimal places. You round the figure off to 19.36. (See the later section "Rounding off numbers" for the rounding rules.)

Multiplication and division

In multiplication and division, you report the answer to the same number of significant figures as the number that has the *fewest* significant figures. Remember that counted and exact numbers don't count in the consideration of significant numbers. For example, suppose that you are calculating the density in grams per liter of an object that weighs 25.3573 (six sig figs) grams and has a volume of 10.50 milliliters (four sig figs). The setup looks like this:

$$\frac{25.3573 \text{ g}}{10.5 \text{ mL}} \times \frac{1,000 \text{ mL}}{1 \text{ L}}$$

Your calculator will read 2,414.981000. You have six significant figures in the first number and four in the second number (the 1,000 mL/L doesn't count because it's an exact conversion). You should have four significant figures in your final answer, so round the answer off to 2,415 g/L.

Notice the difference between the two rules. When you add or subtract, you assign significant figures in the answer based on the number of decimal places in each original measurement. When you multiply or divide, you assign significant figures in the answer based on the smallest number of significant figures from your original set of measurements.

TIP

Caught up in the breathless drama of arithmetic, you may sometimes perform multistep calculations that include addition, subtraction, multiplication, and division, all in one go. No problem. Follow the normal order of operations, doing multiplication and division first, followed by addition and subtraction. At each step, follow the simple significant-figure rules, and then move on to the next step.

Rounding off numbers

Sometimes you have to round numbers at the end of a measurement to account for significant figures. Here are a couple of very simple rules to follow and remember:

>> **Rule 1:** If the first number to be dropped is 5 or greater, drop it and all the numbers that follow it, and increase the last retained number by 1.

For example, suppose that you want to round off 237.768 to four significant figures. You drop the 6 and the 8. The 6, the first dropped number, is greater than 5, so you increase the retained 7 to 8. Your final answer is 237.8.

>> **Rule 2:** If the first number to be dropped is less than 5, drop it and all the numbers that follow it, and leave the last retained number unchanged.

If you're rounding 2.35427 to three significant figures, you drop the 4, the 2, and the 7. The first number to be dropped is 4, which is less than 5. The 5, the last retained number, stays the same. So you report your answer as 2.35.

EXAMPLE

Q. Express the following sum with the proper number of significant figures:

35.7 miles + 634.38 miles + 0.97 miles = ?

A. 671.1 miles. Adding the three values yields a raw sum of 671.05 miles. However, the 35.7 miles measurement extends only to the tenths place. Therefore, you round the answer to the tenths place, from 671.05 to 671.1 miles.

Q. Express the following product with the proper number of significant figures:

27 feet × 13.45 feet = ?

A. 3.6×10^2 feet². Of the two measurements, one has two significant figures (27 feet) and the other has four significant figures (13.45 feet). The answer is therefore limited to two significant figures. You need to round the raw product, 363.15 feet². You could write 360 feet², but doing so may imply that the final 0 is significant and not just a placeholder. For clarity, express the product in scientific notation, as 3.6×10^2 feet².

17 Express the answer to this calculation using the appropriate number of significant figures:

127.379 seconds − 13.14 seconds + $\left(1.2 \times 10^{-1} \text{ seconds}\right)$ = ?

18 Express the answer to this calculation using the appropriate number of significant figures:

$345.6 \text{ feet} \times \left(\dfrac{12 \text{ inches}}{1 \text{ foot}}\right) = ?$

19 Report the difference using the appropriate number of significant figures:

$\left(3.7 \times 10^{-4} \text{ minutes}\right) - 0.009 \text{ minutes} = ?$

20 Express the answer to this multistep calculation using the appropriate number of significant figures:

$\dfrac{87.95 \text{ feet} \times 0.277 \text{ feet} + 5.02 \text{ feet} - 1.348 \text{ feet}}{10.0 \text{ feet}} = ?$

Qualitative and Quantitative Observations

Observations are an essential part of any scientific discipline. In chemistry you are regularly required to make observations about experiments that you do in the lab. In most cases these observations are going to be made in the form of gathering data or taking measurements based on your experiment. Usually when you see the term *data* you likely assume a number is going to be the result of that data, but data in a chemistry lab can take the form of a numerical measurement or a descriptive observation. Both are completely valid and simply depend on what you are asked to do. It is important that you understand the distinction between these two types of observations.

>> **Qualitative Data**: If there is data gathered during an experiment that is based on your observations this data is called qualitative data. This goes for anything that you might be observing during an experiment or anything else you might record based on what you see, feel, or hear. In short, qualitative data does not involve numbers in any way. To help you remember this, simply think of qualitative data as measuring the "quality" of something. It is a subjective observation that you make based on your observations.

>> **Quantitative Data**: Any type of data that you gather through numerical measurement is considered quantitative. These measurements are exact and are not subjective in any way; they are absolute and based on a measuring device. In short, anything that is numerically based would be considered quantitative data. A helpful way to remember this is to think of the idea of "quantity" in quantitative. If it has a number in it, it is quantitative data.

EXAMPLE

Q. Identify the qualitative pieces of data described in the following statement:

You mix 50 ml of two solutions together in a test tube. Upon mixing the solution changes color and you feel the test tube getting warmer.

A. Your observation of a color change and heat being given off by the reaction are examples of qualitative data. The 50 ml is not qualitative in nature, it is quantitative data.

Q. Identify the quantitative observations that can be made from the following statement:

You take the mass of a sample of iron and find that it is 56 grams. The iron is shinny and feels smooth to the touch.

A. 56 grams of iron is quantitative data. The shiny appearance of iron and the way it feels are qualitative.

YOUR TURN

21 Identify the quantitative and qualitative data described in the passage below:

You are doing an experiment where you react lead nitrate with potassium iodide. You carefully measure out 1.5 grams of lead nitrate and 3.5 grams of potassium iodide and record the mass of each. You add 100 ml of water to a beaker and then add both substances into the beaker and mix them together. Upon mixing them the solution turns a bright yellow color.

Practice Questions Answers and Explanations

(1) 2×10^5. Move the decimal point immediately after the 2 to create a coefficient between 1 and 10. Because you're moving the decimal point five places to the left, multiply the coefficient, 2, by the power 10^5.

(2) 8.0736×10^4. Move the decimal point immediately after the 8 to create a coefficient between 1 and 10. You're moving the decimal point four places to the left, so multiply the coefficient, 8.0736, by the power 10^4.

(3) 2×10^{-5}. Move the decimal point immediately after the 2 to create a coefficient between 1 and 10. You're moving the decimal point five places to the right, so multiply the coefficient, 2, by the power 10^{-5}.

(4) **690.3.** You need to understand scientific notation to change the number back to regular decimal form. Because 10^2 equals 100, multiply the coefficient, 6.903, by 100. This moves the decimal point two places to the right.

(5) 1.1×10^6. First, multiply the coefficients: $2.2 \times 5.0 = 11$. Then multiply the powers of 10 by adding the exponents: $10^9 \times 10^{-4} = 10^{9+(-4)} = 10^5$. The raw calculation yields 11×10^5, which converts to the given answer when you express it in scientific notation.

(6) 3.0×10^{-7}. The ease of math with scientific notation shines through in this problem. Dividing the coefficients yields a coefficient quotient of $9.3/3.1 = 3.0$, and dividing the powers of 10 (by subtracting their exponents) yields a quotient of $10^{-5}/10^2 = 10^{-5-2} = 10^{-7}$. Marrying the two quotients produces the given answer, already in scientific notation.

(7) **1.82.** First, convert each number to scientific notation: 5.2×10^1 and 3.5×10^{-2}. Next, multiply the coefficients: $5.2 \times 3.5 = 18.2$. Then add the exponents on the powers of 10: $10^{1+(-2)} = 10^{-1}$. Finally, join the new coefficient with the new power: 18.2×10^{-1}. Expressed in scientific notation, this answer is $1.82 \times 10^0 = 1.82$. (*Note:* Looking back at the original numbers, you see that both factors have only two significant figures; therefore, you should round your answer to match that number of sig figs, making it 1.8. See the previous sections "Identifying Significant Figures" and "Doing Arithmetic with Significant Figures" for details.)

(8) 3.99×10^{-4}. First, convert each number to scientific notation: 8.09×10^{-3} and 2.03×10^1. Then divide the coefficients: $8.09/2.03 = 3.99$. Next, subtract the exponent in the denominator from the exponent in the numerator to get the new power of 10: $10^{-3-1} = 10^{-4}$. Join the new coefficient with the new power: 3.99×10^{-4}. Finally, express gratitude that the answer is already conveniently expressed in scientific notation.

(9) 545×10^{-6}. Because the numbers are each already expressed with identical powers of 10, you can simply add the coefficients: $398 + 147 = 545$. Then join the new coefficient with the original power of 10.

(10) 6.402×10^5. Because the numbers are each expressed with the same power of 10, you can simply subtract the coefficients: $7.685 - 1.283 = 6.402$. Then join the new coefficient with the original power of 10.

(11) 40.16×10^{-3} **(or an equivalent expression).** First, convert the numbers so they each use the same power of 10: 2.06×10^{-3} and 38.1×10^{-3}. Here, we use 10^{-3}, but you can use a different power as long as the power is the same for each number. Next, add the coefficients: $2.06 + 38.1 = 40.16$. Finally, join the new coefficient with the shared power of 10.

(12) **89.21×10^2 (or an equivalent expression).** First, convert the numbers so each uses the same power of 10: 93.52×10^2 and 4.31×10^2. Here, we've picked 10^2, but any power is fine as long as the two numbers have the same power. Then subtract the coefficients: $93.52 - 4.31 = 89.21$. Finally, join the new coefficient with the shared power of 10.

(13) **Reginald's measurement incurred the greater magnitude of error, and Dagmar's measurement incurred the greater percent error.** Reginald's scale reported with an error of 256 pounds $-$ 237 pounds $=$ 19 pounds, and Dagmar's scale reported with an error of 129 pounds $-$ 117 pounds $=$ 12 pounds. Comparing the *magnitudes* of error, you see that 19 pounds is greater than 12 pounds. However, Reginald's measurement had a percent error of $(19 \text{ pounds} / 256 \text{ pounds}) \times 100 = 7.42\%$, while Dagmar's measurement had a percent error of $(12 \text{ pounds} / 129 \text{ pounds}) \times 100 = 9.30\%$.

(14) Jeweler A's official average measurement was 0.864 grams, and Jeweler B's official measurement was 0.856 grams. You determine these averages by adding up each jeweler's measurements and then dividing by the total number of measurements, in this case three. Based on these averages, Jeweler B's official measurement is more accurate because it's closer to the actual value of 0.856 grams.

However, Jeweler A's measurements were more precise because the differences between A's measurements were much smaller than the differences between B's measurements. Despite the fact that Jeweler B's average measurement was closer to the actual value, the *range* of his measurements (that is, the difference between the largest and the smallest measurements) was 0.041 grams (0.875 g $-$ 0.834 g $=$ 0.041 g). The range of Jeweler A's measurements was 0.010 grams (0.869 g $-$ 0.859 g $=$ 0.010 g).

This example shows how low-precision measurements can yield highly accurate results through averaging of repeated measurements. In the case of Jeweler A, the error in the official measurement was 0.864 g $-$ 0.856 g $=$ 0.008 g. The corresponding percent error was $(0.008 \text{ g} / 0.856 \text{ g}) \times 100 = 0.9\%$. In the case of Jeweler B, the error in the official measurement was 0.856 g $-$ 0.856 g $=$ 0.000 g. Accordingly, the percent error was 0%.

(15) The correct number of significant figures is as follows for each measurement: **a) 5, b) 3,** and **c) 4.**

(16) The number of significant figures in a reported measurement should be consistent with your certainty about that measurement.

a. "893.7 $\pm$ 1 grams" is an improperly reported measurement because the reported value, 893.7, suggests that the measurement is certain to within a few tenths of a gram. The reported error is known to be greater, at $\pm$1 gram. The measurement should be reported as "894 $\pm$ 1 grams."

b. "342 $\pm$ 0.01 grams" is improperly reported because the reported value, 342, gives the impression that the measurement becomes uncertain at the level of grams. The reported error makes clear that uncertainty creeps into the measurement only at the level of hundredths of a gram. The measurement should be reported as "342.00 $\pm$ 0.01 grams."

(17) **114.36 seconds.** The trick here is remembering to convert all measurements to the same power of 10 before comparing decimal places for significant figures. Doing so reveals that 1.2×10^{-1} seconds goes to the hundredths of a second, despite the fact that the measurement contains only two significant figures. The raw calculation yields 114.359 seconds, which rounds properly to the hundredths place (taking significant figures into account) as 114.36 seconds, or 1.1436×10^2 seconds in scientific notation.

(18) **4.147×10³inches.** Here, you have to recall that defined quantities (1 foot is defined as 12 inches) have unlimited significant figures. So your calculation is limited only by the number of significant figures in the measurement 345.6 feet. When you multiply 345.6 feet by 12 inches per foot, the feet cancel, leaving units of inches:

$$\left(345.6 \ \cancel{ft}\right) \times \left(\frac{12 \text{ in.}}{1 \ \cancel{ft}}\right) = 4{,}147.2 \text{ in.}$$

The raw calculation yields 4,147.2 inches, which rounds properly to four significant figures as 4,147 inches, or 4.147×10^3 inches in scientific notation.

(19) **−0.009 minutes.** Here, it helps to convert all measurements to the same power of 10 so you can more easily compare decimal places in order to assign the proper number of significant figures. Doing so reveals that 3.7×10^{-4} minutes goes to the hundred-thousandths of a minute, and 0.009 minutes goes to the thousandths of a minute. The raw calculation yields −0.00863 minutes, which rounds properly to the thousandths place (taking significant figures into account) as −0.009 minutes, or -9×10^{-3} minutes in scientific notation.

(20) **2.80 feet.** Following standard order of operations, you can do this problem in two main steps.

Following the rules of significant-figure math, the first step yields 24.4 feet + 5.02 feet − 1.348 feet. Each product or quotient contains the same number of significant figures as the number in the calculation with the fewest number of significant figures.

After completing the first step, divide by 10.0 feet to finish the problem:

$$\frac{28.03 \text{ ft}}{10.0 \text{ ft}} = 2.803 \text{ ft} = 2.80 \text{ ft}$$

You write the answer with three sig figs because the measurement 10.0 feet contains three sig figs, which is the smallest available between the two numbers.

(21) The masses of each compound, 1.5 grams of lead nitrate and 3.5 grams of potassium iodide, are both quantitative data. The volume of water added, 100 ml, is quantitative data. The fact that these are measured quantities involving numbers make them quantitative pieces of data. The observation that the solution turned a bright yellow color upon adding the substances to the beaker and mixing is qualitative data. This is due to it being an observation you are making rather than a measured quantity you are determining.

If you're ready to test your skills a bit more, take the following chapter quiz that incorporates all the chapter topics.

Whaddya Know? Chapter 1 Quiz

Quiz time! Complete each problem to test your knowledge on the various topics covered in this chapter. You can then find the solutions and explanations in the next section.

1 Convert 56000 to scientific notation.

2 Convert 780 to scientific notation.

3 Convert 0.0032 to scientific notation

4 Convert 0.000000098 to scientific notation.

5 Solve the following: $(9.2 \times 10^6)(7.5 \times 10^{-2})$

6 Solve the following: $\dfrac{(3.10 \times 10^{-4})}{(1.33 \times 10^7)}$

7 Solve the following: $(2.5 \times 10^3) + (8.1 \times 10^2)$

8 Solve the following: $(8.55 \times 10^{-2}) - (4.2 \times 10^{-3})$

9 The actual mass of a rock is 5.6 g. A student decides to find the mass of this rock 5 times.

Upon completing this task the masses they recorded are: 5.6 g, 5.7 g, 5.3 g, 5.4 g, 5.8 g, 5.4 g, 5.6 g.

Classify their measurement results as accurate, precise, both, or neither.

10 A student does an experiment and determines the boiling point of water to is 104°C. The actual/known value of water's boiling point is 100°C. What is the student's percent error?

11 Calculate the correct number of significant figures in the following numbers:

  a. 0.0005

  b. 1000.44

  c. 100

  d. 100.0

12 Solve the following problem and write the answer with the correct number of significant figures:

$$456 \text{ cm} \times 2.0 \text{ cm}$$

13 Solve the following problem and write the answer with the correct number of significant figures:

$$83000 \text{ m} \div 4.550 \text{ m}$$

Answers to Chapter 1 Quiz

1. 5.6×10^4. You need to move the decimal point 4 places to the left to create a coefficient between 1 and 10. This means the exponent used will be positive.

2. 7.8×10^2. You need to move the decimal point 2 places to the left to create a coefficient between 1 and 10. This means the exponent used will be positive.

3. 3.2×10^{-3}. You need to move the decimal point 3 places to the right to create a coefficient between 1 and 10. This means the exponent used will be negative.

4. 9.8×10^{-8}. You need to move the decimal point 8 places to the right to create a coefficient between 1 and 10. This means the exponent used will be negative.

5. 6.9×10^5. To solve this problem you will do two steps. First multiply the coefficients: 9.2×7.5. This gives you 69. Next, add the exponents $6 + -2.0$. This gives you 4. Put it back together and you get 69×10^4. You then need to simplify that by moving the decimal one place over to the left, increasing the exponent by one, giving you 6.9×10^5 as the final answer.

6. 2.33×10^{-11} To solve this problem you will do two steps. First divide the coefficients: $3.10 \div 1.33$. This gives you 2.33. Next, subtract the exponents $-4 - 7$. This gives you -11. Put it back together and you get 2.33×10^{-11}.

7. 3.31×10^3. To solve this problem you need to convert one of the numbers so both have the same exponent. Convert 8.1×10^2 to $.81 \times 10^3$ and then add $2.5 + .81$. This gives you 3.31. Once you have done this, add the $\times 10^3$ back in and you have your correct answer.

8. 8.13×10^{-2}. To solve this problem you need to convert one of the numbers so both have the same exponent. Convert 4.2×10^{-3} to 0.42×10^{-2} and then subtract $8.55 - 0.42$. This gives you 8.13. Once you have done this add the $\times 10^{-2}$ back in and you have your correct answer.

9. Accurate and precise. These results can be considered both accurate and precise. The measurements from each trial are closely grouped together. No single result is an obvious outlier from the data set. This makes it precise. Since the results are also close to actual value of the measurement, the data is considered to be accurate.

10. 4% error. The actual value of the boiling point of water is given as 100°C and the measured value by the student is 104°C. Plug those numbers into the percent error formula as shown below to determine your answer.

$$\frac{|100°C - 104°C|}{100°C} \times 100 = 4\% \text{ error}$$

11. a. 1 significant figure. None of the zeros in the number are final zeroes so they are not significant. The 3 is the only significant figure in this number.

    b. 6 significant figures. All of the numbers are significant in this measurement. The nonzero numbers are significant. The zeros are also significant as they are between other significant figures.

c. 1 significant figure. The 1 is the only significant figure. The 2 trailing/final zeros are not significant because even though they are final zeros they are not after a decimal point as well.

d. 4 significant figures. All figures are significant. The one is a number so it is significant. The zero after the decimal point is significant because it is a final zero after the decimal point. The 2 middle zeroes are significant because they are between 2 other significant figures.

(12) 910 cm. If you solve this problem you will find the exact answer when calculated is 912 cm. Since you are multiplying you must round this answer to have the same number of significant figures as the lowest number of significant figures from the measurements you used when doing this calculation. 456 has 3 significant figures. 2.0 has 2 significant figures. This means your answer must have 2 significant figures.

(13) 18000 m. If you solve this problem you will find the exact answer when calculated is 18241.75824 m. That is quite a number. Since you are dividing you must round this answer to have the same number of significant figures as the lowest number of significant figures from the measurements you used when doing this calculation. 83000 has 2 significant figures. 4.550 has 4 significant figures. This means your answer must have 2 significant figures. This is a good problem to remember that you have to pay careful attention to sig figs. Even though 83000 has more digits, it has fewer significant figures than 4.550, so make sure when you are rounding problems to keep good track of significant figures.

Chapter 2

Using and Converting Units

Have you ever been asked for your height in centimeters, your weight in kilograms, or the speed limit in kilometers per hour? These measurements may seem a bit odd to those folks who are used to feet, pounds, and miles per hour, but the truth is that scientists sneer at feet, pounds, and miles. Because scientists around the globe constantly communicate numbers to each other, they prefer a highly systematic, standardized system. The *International System of Units*, abbreviated *SI* from the French term *Système International*, is the unit system of choice in the scientific community.

In this chapter, you find that the SI system offers a very logical and well-organized set of units. Scientists, despite what many of their hairstyles may imply, love logic and order, so SI is their system of choice.

TIP

As you work with SI units, try to develop a good sense for how big or small the various units are. That way, as you're doing problems, you'll have a sense for whether your answer is reasonable.

Familiarizing Yourself with Base Units and Metric System Prefixes

Much of the work chemists do involves measuring physical properties, such as the mass, volume, or length of a substance. Because chemists must be able to communicate their measurements to other chemists all over the world, they need to speak the same measurement language. This language is the SI system of measurement, related to the metric system, which you've hopefully used before. Minor differences exist between the SI and metric systems, but for the most part, they're very similar.

To correctly use the SI system, you need to have a firm understanding of what each prefix means. The good news: The SI system is a decimal system. In other words, it's easy to use as long as you know the prefixes.

SI has base units for mass, length, volume, and so on, and prefixes modify the base units. For example, *kilo–* means 1,000; a kilogram is 1,000 grams, and a kilometer is 1,000 meters. Use Table 2-1 as a handy reference for the abbreviations and meanings of some selected various SI prefixes.

TABLE 2-1   **SI (Metric) Prefixes**

Prefix	Abbreviation	Meaning
tera-	T	1,000,000,000,000 or 10^{12}
giga-	G	1,000,000,000 or 10^9
mega-	M	1,000,000 or 10^6
kilo-	k	1,000 or 10^3
hecto-	h	100 or 10^2
deka- or deca-	da	10 or 10^1
deci-	d	0.1 or 10^{-1}
centi-	c	0.01 or 10^{-2}
milli-	m	0.001 or 10^{-3}
micro-	μ	0.000001 or 10^{-6}
nano-	n	0.000000001 or 10^{-9}
pico-	p	0.000000000001 or 10^{-12}

The next step in mastering the SI system is to figure out all the possible units that you can run into when solving problems. Here's a quick explanation of the most common types of units you'll encounter.

Units of length

The base unit for length in the SI system is the *meter*. The exact definition of meter has changed over the years, but it's now defined as the distance that light travels in a vacuum in $1 \text{ cm} = 10^9/10^2$ nm of a second. Here are some SI units of length:

1 millimeter (mm) = 1,000 micrometers (μm)

1 centimeter (cm) = 10 millimeters (mm)

1 meter (m) = 100 centimeters (cm)

1 kilometer (km) = 1,000 meters (m)

Some common English-to-SI-system length conversions are

1 mile (mi) = 1.61 kilometers (km)

1 yard (yd) = 0.914 meters (m)

1 inch (in.) = 2.54 centimeters (cm)

Units of mass

The base unit for mass in the SI system is the *kilogram*. It's the weight of the standard platinum–iridium bar found at the International Bureau of Weights and Measures. Here are some SI units of mass:

1 milligram (mg) = 1,000 micrograms (μg)

1 gram (g) = 1,000 milligrams (mg)

1 kilogram (kg) = 1,000 grams (g)

Some common English-to-SI-system mass conversions are

1 pound (lb) = 454 grams (g)

1 ounce (oz) = 28.4 grams (g)

1 pound (lb) = 0.454 kilograms (kg)

1 grain (gr) = 0.0648 grams (g)

1 carat (car) = 200 milligrams (mg)

Units of volume

In the SI system, volume is measured in base units called *cubic meters*. However, chemists normally use the *liter*, 0.001 m³, to measure volume. Here are some SI units of volume:

1 milliliter (mL) = 1 cubic centimeter (cm³) = 1,000 microliters (μL)

1 liter (L) = 1,000 milliliters (mL)

Some common English-to-SI-system volume conversions are

1 quart (qt) = 0.946 liters (L)

1 pint (pt) = 0.473 liters (L)

1 fluid ounce (fl oz) = 29.6 milliliters (mL)

1 gallon (gal) = 3.78 liters (L)

Units of temperature

Kelvin is the base unit for temperature in the SI system. 0 K is called *absolute zero*, the temperature at which all atomic/molecular motion ceases. Water freezes at 273 K and boils at 373 K. Following are the three major temperature conversion formulas:

Celsius to Fahrenheit: $°F = (9/5)°C + 32$

Fahrenheit to Celsius: $°C = (5/9)(°F - 32)$

Celsius to Kelvin: $K = °C + 273$

Units of pressure

The SI unit for pressure is the *pascal*, where 1 pascal equals 1 newton per square meter. (A *newton* is a unit of force equal to $1 \text{ kg} \cdot \text{m/s}^2$.) But pressure can also be expressed in a number of ways, so here are some common pressure conversions:

1 millimeter of mercury (mm Hg) = 1 torr

1 atmosphere (atm) = 760 millimeters of mercury (mm Hg) = 760 torr

1 atmosphere (atm) = 29.9 inches of mercury (in. Hg)

1 atmosphere (atm) = 14.7 pounds per square inch (psi)

1 atmosphere (atm) = 101 kilopascals (kPa)

Units of energy

The SI unit for energy (heat being one form) is the *joule*, but most folks still use the metric unit of heat, the *calorie*. Here are some common energy conversions:

1 calorie (cal) = 4.184 joules (J)

1 nutritional (food) Calorie (Cal) = 1 kilocalorie (kcal) = 4,184 joules (J)

1 British thermal unit (BTU) = 252 calories (cal) = 1,053 joules (J)

SI base units

In case you're looking for a simple and handy way to keep track of the most common SI base units, look at Table 2-2 to see them all nicely arranged for your enjoyment.

TABLE 2-2   SI Base Units

Measurement	SI Unit	Symbol	Non-SI Unit
Amount of a substance	mole	mol	no non-SI unit
Length	meter	m	foot, inch, yard, mile
Mass	kilogram	kg	pound
Temperature	kelvin	K	degree Celsius, degree Fahrenheit
Time	second	s	minute, hour

EXAMPLE

Q. You measure a length to be 0.005 m. How can this be better expressed using a metric system prefix?

A. 0.005 is 5×10^{-3} m, or 5 mm.

Q. What metric base unit would be the most appropriate to use if you were going to measure the distance a runner covers during a marathon?

A. Meters are used to measure how long something is or how much distance is covered. When taking and reading measurements, remember which units you're using and understand which unit best applies.

YOUR TURN

 1 How many nanometers are in 1 cm?

2 Your lab partner has measured the mass of your sample to be 2,500 g. How can you record this more nicely (without scientific notation) in your lab notebook using a metric system prefix?

Looking at Density

Sometimes units combine two or more units of measurement. These kinds of quantities are called *derived units*, and they're built from combinations of different base units. Area, volume, and pressure, which we cover in detail later in the book, are considered to be derived units. There's one derived unit, however, that's almost universally discussed early in every chemistry class. It's quite important and quite useful. We are, of course, talking about density.

Getting to the bottom of density basics

Density (d) is the ratio of the mass (m) to the volume (V) of a substance. Mathematically, it looks like this:

$$d = \frac{m}{V}$$

Usually, mass is described in grams (g) and volume is described in milliliters (mL), so density is g/mL. Because the volumes of liquids vary somewhat with temperature, chemists also usually specify the temperature at which a density measurement is made. Most reference books report densities at 20 degrees Celsius, because it's close to room temperature and easy to measure without a lot of heating or cooling. The density of water at 20 degrees, for example, is 1 g/mL.

Another term you may hear is *specific gravity (sg)*, which is the ratio of the density of a substance to the density of water at the same temperature. Specific gravity is just another way for you to get around the problem of volumes of liquids varying with the temperature. Specific gravity is used with urinalysis in hospitals and to describe automobile battery fluid in auto repair shops. Note that specific gravity has no units of measurement associated with it, because the unit g/mL appears in both the numerator and the denominator and therefore cancels out. In most cases, the density and specific gravity are almost the same, so it's common to simply use the density.

TIP

You may see density reported as g/cm³ or g/cc. These units are the same as g/mL. A cube measuring 1 centimeter on each edge (written as 1 cm³) has a volume of 1 milliliter (1 mL). Because 1 mL = 1 cm³, g/mL and g/cm³ are interchangeable. And because a cubic centimeter (cm³) is commonly abbreviated cc, g/cc also means the same thing. (You hear cc a lot in the medical profession. When you receive a 10 cc injection, you're getting 10 milliliters of liquid. That's a lot. You better believe we're running the other way when we see a nurse coming with a 10 cc shot!)

Measuring density

Calculating density is pretty straightforward. You measure the mass of an object by using a balance or scale, determine the object's volume, and then divide the mass by the volume.

Determining the volume of liquids is easy, but solids can be tricky. If the object is a regular solid, like a cube, you can measure its three dimensions and calculate the volume by multiplying the length by the width by the height (volume = $l \times w \times h$). But if the object is an irregular solid, like a rock, determining the volume is more difficult. With irregular solids, you can measure the volume by using something called Archimedes' principle.

Archimedes' principle states that the volume of a solid is equal to the volume of water it displaces. The Greek mathematician Archimedes discovered this concept in the third century BC, greatly simplifying the process for finding an object's density. Say that you want to measure the volume of a small rock in order to determine its density. First, put some water into a graduated cylinder with markings for every mL and read the volume. (The example in Figure 2-1 shows 25 mL.) Next, put the rock in, making sure that it's totally submerged, and read the volume again (29 mL in Figure 2-1). The difference in volume (4 mL) is the volume of the rock.

TIP

Anything with a density lower than water's floats when put into water, and anything with a density greater than 1 g/mL sinks.

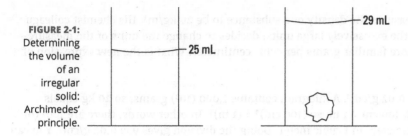

FIGURE 2-1:
Determining
the volume
of an
irregular
solid:
Archimedes'
principle.

29 mL

25 mL

For your pondering pleasure, Table 2-3 lists the density of some common materials.

TABLE 2-3 **Densities of Typical Solids and Liquids**

Substance	Density (in g/mL)
Gasoline	0.68
Ice	0.92
Water	1.00
Table salt	2.16
Iron	7.86
Lead	11.38
Mercury	13.55
Gold	19.3

Note that gold has a pretty high density. One of those gold bars that are stored in Fort Knox weighs over 30 pounds. Remember that fact when you see those burglars on TV tossing a bunch of gold bars into a bag, throwing it over their shoulders, and carrying them away.

If you know the density of a substance and either its mass or volume, you can calculate the other.

EXAMPLE

Q. Suppose you have a 25.0 mL sample of mercury (d = 13.55 g/mL). What would be the mass of that sample of mercury?

A. The mass of the sample would be 339 grams. To solve, follow these easy steps:

1. Start with the density formula.

$$d = \frac{m}{V}$$

2. Switch it around so you're solving for the mass.

$$m = dV$$

3. Put in your density and volume and solve for the mass (m):

$$m = \frac{13.55 \text{ g}}{1 \text{ mL}} \times \frac{25.0 \text{ mL}}{1} = 339 \text{ g}$$

That's about 3 / 4 of a pound!

Q. A physicist measures the density of a substance to be 20 kg/m³. His chemist colleague, appalled with the excessively large units, decides to change the units of the measurement to the more familiar grams per cubic centimeter. What is the new expression of the density?

A. The density is 0.02 g/cm³. A kilogram contains 1,000 (10^3) grams, so 20 kg equals 20,000 g. Well, 100 cm = 1 m, so (100 cm)³ = (1 m)³. In other words, there are 100³ (or 10^6) cubic centimeters in 1 cubic meter. Doing the division gives you 0.02 g/cm³. You can write out the conversion as follows:

$$\left(\frac{20\,\text{kg}}{10\,\text{m}^3}\right)\left(\frac{10^3\,\text{g}}{1\,\text{kg}}\right)\left(\frac{1\,\text{m}^3}{10^6\,\text{cm}^3}\right)=0.02\,\frac{\text{g}}{\text{cm}^3}$$

YOUR TURN

3 The pascal, a unit of pressure, is equivalent to 1 newton per square meter. If the newton, a unit of force, is equal to a kilogram-meter per second squared, what is the pascal expressed entirely in basic units?

4 A student measures the length, width, and height of a sample to be 10 mm, 15 mm, and 5 mm, respectively. If the sample has a mass of 0.9 dag, what is the sample's density in grams per milliliter?

5 Another student takes some measurements of an unknown solution. She determines that the mass of the solution is 89.5 g and that the volume of the solution is 145.2 mL. What is the density of the unknown solution?

Using Conversion Factors

So what happens if you need to convert between one set of units and another, perhaps from a non-SI unit to an SI unit or between SI units? Well, first, you need to understand what a conversion factor is. Second, you need to know how to set up a conversion problem and solve it.

A *conversion factor* is a ratio represented by a fraction that is equal to 1. It simply uses your knowledge of the relationships between units to convert from one unit to another. When using a conversion factor, you're not actually changing anything about the physical quantity a measurement represents. You're simply changing the units in which that quantity is reported. For example, if you know that there are 2.54 centimeters in every inch (or 2.2 pounds in every kilogram or 101.3 kilopascals in every atmosphere), then converting between those units becomes simple algebra.

REMEMBER

All the numbers and measures you encounter in chemistry represent physical quantities of matter. When using conversion factors, you're simply representing that physical quantity in another way, using a different unit. A meter stick is always going to be the same length, whether you say it's 1 m long or 100 cm long. Peruse Table 2-4 for some useful conversion factors. And remember that if you know the relationship between any two units, you can build your own conversion factor to move between those units.

TABLE 2-4 **Conversion Factors**

Unit	Equivalent To	Conversion Factors
Length		
1 meter	3.3 feet	$\dfrac{3.3\ \text{ft}}{1\ \text{m}}$ or $\dfrac{1\ \text{m}}{3.3\ \text{ft}}$
1 foot	12 inches	$\dfrac{1\ \text{ft}}{12\ \text{in.}}$ or $\dfrac{12\ \text{in.}}{1\ \text{ft}}$
1 inch	2.54 centimeters	$\dfrac{1\ \text{in.}}{2.54\ \text{cm}}$ or $\dfrac{2.54\ \text{cm}}{1\ \text{in.}}$
Volume		
1 gallon	16 cups	$\dfrac{16\ \text{c}}{1\ \text{gal}}$ or $\dfrac{1\ \text{gal}}{16\ \text{c}}$
1 cup	237 milliliters	$\dfrac{1\ \text{c}}{237\ \text{mL}}$ or $\dfrac{237\ \text{mL}}{1\ \text{c}}$
1 milliliter	1 cubic centimeter	$\dfrac{1\ \text{mL}}{1\ \text{cm}^3}$ or $\dfrac{1\ \text{cm}^3}{1\ \text{mL}}$
Mass		
1 kilogram	2.2 pounds	$\dfrac{1\ \text{kg}}{2.2\ \text{lb}}$ or $\dfrac{2.2\ \text{lb}}{1\ \text{kg}}$
Time		
1 hour	3,600 seconds	$\dfrac{1\ \text{hr}}{3,600\ \text{s}}$ or $\dfrac{3,600\ \text{s}}{1\ \text{hr}}$
Pressure		
1 atmosphere	101.3 kilopascals	$\dfrac{1\ \text{atm}}{101.3\ \text{kPa}}$ or $\dfrac{101.3\ \text{kPa}}{1\ \text{atm}}$
1 atmosphere	760 millimeters of mercury (mm Hg)	$\dfrac{1\ \text{atm}}{760\ \text{mm Hg}}$ or $\dfrac{760\ \text{mm Hg}}{1\ \text{atm}}$

* One of the more peculiar units you'll encounter in your study of chemistry is mm Hg, or millimeters of mercury, a unit of pressure. Unlike SI units, mm Hg doesn't fit neatly into the base-10 metric system, but it reflects the way in which certain devices like blood pressure cuffs and barometers use mercury to measure pressure.

WARNING

Chemistry teachers are sneaky. They often give you quantities in non-SI units and expect you to use one or more conversion factors to change them to SI units — all this before you even attempt the "hard part" of the problem! So with that in mind, expect to use conversion factors throughout the rest of this book.

Here is a simple example illustrating the use of the conversion factor.

EXAMPLE

Q. An absent-minded professor named Steve measures the mass of a sample to be 0.75 lb and records his measurement in his lab notebook. His astute lab assistant, who wants to save the prof some embarrassment, knows that there are 2.2 lb in every kilogram. The assistant quickly converts the doctor's measurement to SI units. What does she get?

A. The sample's mass is 0.34 kg.

$$\left(0.75 \text{ lb}\right)\left(\frac{1 \text{ kg}}{2.2 \text{ lb}}\right) = 0.34 \text{ kg}$$

Notice that something very convenient happens because of the way this calculation is set up. In algebra, whenever you find the same quantity in a numerator and in a denominator, you can cancel it out. Canceling out the pounds (lb) is a lovely bit of algebra because you don't want those units around, anyway. The whole point of the conversion factor is to get rid of an undesirable unit, transforming it into a desirable one — without breaking any rules. Always let the units be your guide when solving a problem. Ensure the right ones cancel out, and if they don't, go back and flip your conversion factor.

Remember an algebra rule: You can multiply any quantity by 1, and you'll always get back the original quantity. Now look closely at the conversion factors in the example: 2.2 lb and 1 kg are exactly the same thing! Multiplying by 2.2 lb / 1 kg or by 1 kg / 2.2 lb is really no different from multiplying by 1.

Q. A chemistry student, daydreaming during lab, suddenly looks down to find that he's measured the volume of his sample to be 1.5 cubic *inches*. What does he get when he converts this quantity to cubic centimeters?

A. The volume is 25 cm³.

$$\left(1.5 \text{ in.}^3\right)\left(\frac{2.54 \text{ cm}}{1 \text{ in.}}\right)^3 = \left(1.5 \text{ in.}^3\right)\left(\frac{16.39 \text{ cm}^3}{1 \text{ in.}^3}\right) = 25 \text{ cm}^3$$

Rookie chemists often mistakenly assume that if there are 2.54 centimeters in every inch, then there are 2.54 cubic centimeters in every cubic inch. No! Although this assumption seems logical at first glance, it leads to catastrophically wrong answers. Remember that cubic units are units of volume and that the formula for volume is *length × width × height*. Imagine 1 cubic inch as a cube with 1-inch sides. The cube's volume is $1 \text{ in.} \times 1 \text{ in.} \times 1 \text{ in.} = 1 \text{ in.}^3$.

Now consider the dimensions of the cube in centimeters: 2.54 cm × 2.54 cm × 2.54 cm. Calculate the volume using these measurements, and you get 2.54 cm × 2.54 cm × 2.54 cm = 16.39 cm^3! This volume is much greater than 2.54 cm³! To convert units of area or volume using length measurements, square or cube everything in your conversion factor, not just the units, and everything works out just fine.

6 A sprinter running the 100.0 m dash runs how many feet?

7 At the top of Mount Everest, the air pressure is approximately 0.330 atmospheres, or a third of the air pressure at sea level. A barometer placed at the peak would read how many millimeters of mercury?

8 A *league* is an obsolete unit of distance used by archaic (or nostalgic) sailors. A league is equivalent to 5.6 km. If the characters in Jules Verne's novel *20,000 Leagues Under the Sea* travel to a depth of 20,000 leagues, how many kilometers under the surface of the water are they? If the radius of the Earth is 6,378 km, is this a reasonable depth? Why or why not?

9 The slab of butter that Paul Bunyan slathered on his morning pancakes is 2.0 ft wide, 2.0 ft long, and 1.0 ft thick. How many cubic meters of butter does Paul consume each morning?

Working with the Factor Label Method

You may find that you're sometimes unclear on how to actually set up a particular chemistry problem in order to solve it. A scientific calculator handles the math, but it can't tell you what you need to multiply or what you need to divide.

That's why you need to know about the *factor label method,* which is sometimes called the *unit conversion method* or *dimensional analysis.* It can help you set up chemistry problems and solve them correctly. Two basic rules are associated with the factor label method:

>> **Rule 1:** Always write the unit and the number associated with the unit. Rarely in chemistry will you have a number alone. Units are your guide to determining how to solve the problem, so always include them.

>> **Rule 2:** Carry out mathematical operations *with* the units, canceling them until you end up with the unit you want in the final answer. In other words, treat a unit just like you would a number. For example, if you have a unit of g (grams) on top and g on the bottom of a fraction, as in $\frac{g}{g}$, they can cancel out just like the 2s in $\frac{2}{2}$. Both fractions end up equaling 1.

Here are some examples illustrating how to perform a conversion using the factor label method.

Q. A chemistry student measures a length of 423 mm, yet the lab she's working on requires that it be in kilometers. What is the length in kilometers?

EXAMPLE

A. The length is 4.23×10^{-4}. You can go about solving this problem in two ways. We first show you the slightly longer way involving two conversions and then shorten it to a nice, simple one-step problem.

This conversion requires you to move across the metric-system prefixes you find in Table 2-2. When you're working on a conversion that passes through a base unit, it may be helpful to treat the process as two steps, converting to and from the base unit. In this case, you can convert from millimeters to meters and then from meters to kilometers:

$$\left(423 \ \text{mm}\right)\left(\frac{1 \ \text{m}}{1{,}000 \ \text{mm}}\right)\left(\frac{1 \ \text{km}}{1{,}000 \ \text{m}}\right) = 4.23 \times 10^{-4} \ \text{km}$$

You can see how *millimeters* cancels out and you're left with *meters.* Then *meters* cancels out, and you're left with your desired unit, *kilometers.*

The second way you can approach this problem is to treat the conversion from milli- to kilo- as one big step:

$$\left(423 \ \text{mm}\right)\left(\frac{1 \ \text{km}}{10^{6} \ \text{mm}}\right) = 4.23 \times 10^{-4} \ \text{km}$$

Notice the answer doesn't change; the only difference is the number of steps required to convert the units. Based on Table 2-2 and the first approach we showed you, you can see that the total conversion from millimeters to kilometers requires converting 10^{6} mm to 1 km. You're simply combining the two denominators in the two-step conversion (1,000 mm and 1,000 m) into one. Rewriting each 1,000 as 10^{3} may help you see how the denominators combine to become 10^{6}.

Q. Suppose that you have an object traveling at 75 miles per hour. What is its speed in kilometers per second?

A. The speed is 0.034 km/s. To solve this problem using the factor label method, follow these steps:

1. **Write down what you start with.**

$$\frac{75 \text{ mi}}{\text{hr}}$$

Note that per Rule 1, the expression shows the unit and the number associated with it.

2. **Convert miles to feet, canceling the unit of miles per Rule 2.**

$$\frac{75 \cancel{\text{ mi}}}{\text{hr}} \times \frac{5{,}280 \text{ ft}}{1 \cancel{\text{ mi}}}$$

3. **Convert feet to inches.**

$$\frac{75 \cancel{\text{ mi}}}{\text{hr}} \times \frac{5{,}280 \cancel{\text{ ft}}}{\cancel{\text{ mi}}} \times \frac{12 \text{ in.}}{1 \cancel{\text{ ft}}}$$

4. **Convert inches to centimeters.**

$$\frac{75 \cancel{\text{ mi}}}{\text{hr}} \times \frac{5{,}280 \cancel{\text{ ft}}}{\cancel{\text{ mi}}} \times \frac{12 \cancel{\text{ in.}}}{1 \cancel{\text{ ft}}} \times \frac{2.54 \text{ cm}}{1 \cancel{\text{ in.}}}$$

5. **Convert centimeters to meters.**

$$\frac{75 \cancel{\text{ mi}}}{\text{hr}} \times \frac{5{,}280 \cancel{\text{ ft}}}{\cancel{\text{ mi}}} \times \frac{12 \cancel{\text{ in.}}}{1 \cancel{\text{ ft}}} \times \frac{2.54 \cancel{\text{ cm}}}{1 \cancel{\text{ in.}}} \times \frac{1 \text{ m}}{100 \cancel{\text{ cm}}}$$

6. **Convert meters to kilometers.**

$$\frac{75 \cancel{\text{ mi}}}{\text{hr}} \times \frac{5{,}280 \cancel{\text{ ft}}}{\cancel{\text{ mi}}} \times \frac{12 \cancel{\text{ in.}}}{1 \cancel{\text{ ft}}} \times \frac{2.54 \cancel{\text{ cm}}}{1 \cancel{\text{ in.}}} \times \frac{1 \cancel{\text{ m}}}{100 \cancel{\text{ cm}}} \times \frac{1 \text{ km}}{1{,}000 \cancel{\text{ m}}}$$

7. **Stop and stretch.**

8. **Now convert hours to minutes in the denominator of the original fraction.**

$$\frac{75 \cancel{\text{ mi}}}{\cancel{\text{hr}}} \times \frac{5{,}280 \cancel{\text{ ft}}}{\cancel{\text{ mi}}} \times \frac{12 \cancel{\text{ in.}}}{1 \cancel{\text{ ft}}} \times \frac{2.54 \cancel{\text{ cm}}}{1 \cancel{\text{ in.}}} \times \frac{1 \cancel{\text{ m}}}{100 \cancel{\text{ cm}}} \times \frac{1 \text{ km}}{1{,}000 \cancel{\text{ m}}} \times \frac{1 \cancel{\text{ hr}}}{60 \text{ min}}$$

9. **Convert minutes to seconds.**

$$\frac{75 \cancel{\text{ mi}}}{\cancel{\text{hr}}} \times \frac{5{,}280 \cancel{\text{ ft}}}{\cancel{\text{ mi}}} \times \frac{12 \cancel{\text{ in.}}}{1 \cancel{\text{ ft}}} \times \frac{2.54 \cancel{\text{ cm}}}{1 \cancel{\text{ in.}}} \times \frac{1 \cancel{\text{ m}}}{100 \cancel{\text{ cm}}} \times \frac{1 \text{ km}}{1{,}000 \cancel{\text{ m}}} \times \frac{1 \cancel{\text{ hr}}}{60 \cancel{\text{ min}}} \times \frac{1 \cancel{\text{ min}}}{60 \text{ s}}$$

10. **Do the math to get the answer now that you have the units of kilometers per second (km/s).**

The calculator gives you 0.033528 km/s. Round off that answer to the correct number of significant figures (see Chapter 1 for details on how to do so):

0.034 km/s

If you'd like to write the answer in scientific notation (again, see Chapter 1), the speed is

3.4×10^{-2} km/s

Tip: Note that although the setup of this example is correct, it's certainly not the only correct setup. Depending on what conversion factors you know and use, there may be many correct ways to set up a problem and get the correct answer.

Q. Suppose that you have an object with an area of 35 inches squared, and you want to figure out the area in meters squared.

A. The area is 0.023 m². Follow these easy steps to make this calculation.

1. **Write down what you start with.**

$$\frac{35 \text{ in.}^2}{1}$$

2. **Convert from inches to centimeters.**

Remember: You have to cancel inches *squared*. You must square the inches in the new fraction, and if you square the unit, you have to square the number also. And if you square the denominator, you have to square the numerator, too:

$$\frac{35 \text{ in.}^2}{1} \times \frac{(2.54 \text{ cm})^2}{(1 \text{ in.})^2}$$

3. **Convert from centimeters squared to meters squared in the same way.**

$$\frac{35 \text{ in.}^2}{1} \times \frac{(2.54 \text{ cm})^2}{(1 \text{ in.})^2} \times \frac{(1 \text{ m})^2}{(100 \text{ cm})^2}$$

4. **Now that you have the units of meters squared (m²), do the math to get your answer.**

The calculator gives you 0.0225806 m². Rounded off to the correct number of significant figures, the answer is

0.023 m², or 2.3×10^{-2} m²

With a little practice, you'll really like and appreciate the factor label method. It ends up making things dramatically easier. Don't give up on it. Pay attention to your units, and you'll be just fine.

YOUR TURN

10 How many meters are in 15 ft?

11 If Steve weighs 175 lb, what's his weight in grams?

12 How many liters are in 1 gal of water?

13 If the dimensions of a solid sample are 3 in. × 6 in. × 1 ft, what's the volume of that sample in cubic centimeters? Give your answer in scientific notation or use a metric prefix.

14 If there are 5.65 kg per every half liter of a particular substance, is that substance liquid mercury (density 13.5 g/cm³), lead (density 11.3 g/cm³), or tin (density 7.3 g/cm³)?

Practice Questions Answers and Explanations

(1) 1×10^7 **nm.** Both 10^2 centimeters and 10^9 nanometers equal 1 meter. Set the two measurements equal to one another (10^2 cm = 10^9 nm) and solve for centimeters by dividing. This conversion tells you that 1 cm = $10^9/10^2$ nm, or 1×10^7 nm.

(2) **2.5 kg.** Because 1,000 g are in 1 kg, simply divide 2,500 by 1,000 to get 2.5.

(3) $1\,Pa = 1\dfrac{kg}{m \cdot s^2}$. First, write out the equivalents of pascals and newtons given in the problem:

$$1\,Pa = \frac{1\,N}{m^2} \text{ and } 1\,N = \frac{1\,kg \cdot m}{s^2}$$

Now substitute *newtons* (expressed in fundamental units) into the equation for the *pascal*:

$$1\,Pa = \frac{\dfrac{1\,kg \cdot m}{s^2}}{m^2}$$

Simplify this equation to $1\,Pa = \dfrac{1\,kg \cdot m}{m^2 \cdot s^2}$ and cancel out the *meter*, which appears in both the top and the bottom, leaving $1\,Pa = 1\dfrac{kg}{m \cdot s^2}$.

(4) **12 g/mL.** Because a milliliter is equivalent to a cubic centimeter, the first thing to do is to convert all the length measurements to centimeters: 1 cm, 1.5 cm, and 0.5 cm. Then multiply the converted lengths to get the volume: $(1\,cm)(1.5\,cm)(0.5\,cm) = 0.75\,cm^3$, or 0.75 mL. The mass should be expressed in grams rather than decagrams; there are 10 grams in 1 decagram, so 0.9 dag = 9 g. Using the formula $d = m/V$, you calculate a density of 9 g per 0.75 mL, or 12 g/mL.

(5) **0.616 g/mL.** Simply plug the given values for volume and mass into the density formula and solve for density:

$$d = \frac{m}{V} = \frac{89.5\,g}{145.2\,mL} = 0.616\ g/mL$$

(6) **330 ft.** Set up the conversion factor as follows:

$$(100.0\,m)\left(\frac{3.3\,ft}{1\,m}\right) = 330\,ft$$

(7) **251 mm Hg.**

$$(0.330\,atm)\left(\frac{760\,mm\,Hg}{1\,atm}\right) = 251\,mm\,Hg$$

(8) 1.12×10^5 **km.**

$$(20{,}000\ \text{leagues})\left(\frac{5.6\,km}{1\,\text{league}}\right) = 1.12 \times 10^5\,km$$

The radius of the Earth is only 6,378 km, and 20,000 leagues is 17.5 times that radius! So the ship would've burrowed through the Earth and been halfway to the orbit of Mars if it had truly sunk to such a depth. Jules Verne's title refers to the distance the submarine travels through the sea, not its depth.

(9) **0.11 m³.** The volume of the butter in feet is 2.0 ft × 2.0 ft × 1.0 ft, or 4 ft³.

$$\left(4.0\ \text{ft}^3\right)\left(\frac{1\ \text{m}}{3.3\ \text{ft}}\right)^3 = \left(4.0\ \text{ft}^3\right)\left(\frac{1\ \text{m}^3}{35.937\ \text{ft}^3}\right) = 0.11\ \text{m}^3$$

(10) **4.6 m.** You have to convert all the way from feet to meters. Looking at the conversion factors in Table 2-4, you should see that you can convert feet into inches and then inches into centimeters. Then you can easily convert centimeters into meters.

$$\left(15\ \text{ft}\right)\left(\frac{12\ \text{in.}}{1\ \text{ft}}\right)\left(\frac{2.54\ \text{cm}}{1\ \text{in.}}\right)\left(\frac{1\ \text{m}}{100\ \text{cm}}\right) = 4.6\ \text{m}$$

(11) **7.95×10⁴ g.** There's no direct pound-to-gram conversion factor in Table 2-4, so you must determine the correct path to take. In this case, you can convert from pounds to kilograms and then from kilograms to grams:

$$\left(175\ \text{lb}\right)\left(\frac{1\ \text{kg}}{2.2\ \text{lb}}\right)\left(\frac{1{,}000\ \text{g}}{1\ \text{kg}}\right) = 7.95 \times 10^4\ \text{g}$$

(12) **3.8 L.** You must determine the correct pathway to get from gallons to liters using the conversions provided in Table 2-4. To do so, convert from gallons to cups, then to milliliters, and finally to liters:

$$\left(1\ \text{gal}\right)\left(\frac{16\ \text{c}}{1\ \text{gal}}\right)\left(\frac{237\ \text{mL}}{1\ \text{c}}\right)\left(\frac{1\ \text{L}}{1{,}000\ \text{mL}}\right) = 3.8\ \text{L}$$

(13) **3.54×10³ cm³.** First convert all the inch and foot measurements to centimeters:

$$\left(3\ \text{in.}\right)\left(\frac{2.54\ \text{cm}}{1\ \text{in.}}\right) = 7.62\ \text{cm}$$

$$\left(6\ \text{in.}\right)\left(\frac{2.54\ \text{cm}}{1\ \text{in.}}\right) = 15.24\ \text{cm}$$

$$\left(1\ \text{ft}\right)\left(\frac{12\ \text{in.}}{1\ \text{ft}}\right)\left(\frac{2.54\ \text{cm}}{1\ \text{in.}}\right) = 30.48\ \text{cm}$$

The volume is therefore 7.62 cm × 15.24 cm × 30.48 cm, or $3.54 \times 10^3\ \text{cm}^3$.

(14) **The substance is lead.** To set up this problem, be sure to begin with the correct initial amounts. The problem tells you there are 5.65 kg for every half liter of substance. This translates into 5.65 kg/0.5 L. After you've established the initial value, use conversion factors to find the density:

$$\left(\frac{5.65\ \text{kg}}{0.5\ \text{L}}\right)\left(\frac{1{,}000\ \text{g}}{1\ \text{kg}}\right)\left(\frac{1\ \text{L}}{1{,}000\ \text{mL}}\right)\left(\frac{1\ \text{mL}}{1\ \text{cm}^3}\right) = 11.3\ \frac{\text{g}}{\text{cm}^3}$$

This answer is exactly the density of lead.

If you're ready to test your skills a bit more, take the following chapter quiz that incorporates all the chapter topics.

Whaddya Know? Chapter 2 Quiz

Ready for a quiz? The 10 questions in this section will test the skills you learned in this chapter. When you're done, check out the section that follows for answers and explanations.

1. Convert 56 grams to milligrams.

2. Convert 3500 mL to L.

3. Convert 1.5×10^4 km to mm.

4. Determine the density of an object in grams/ml. The object in question has a mass of 5.40 kilograms and a volume of 400.0 cm³.

5. Convert 45 grams/liter to kg/ml.

6. Convert 400 kilograms to lb.

7. How many seconds does it take for 5 years to pass on earth? (Don't worry about leap years in this. Keep it simple. 365 days = 1 year.)

8. Convert 500 atm to kPa.

9. What is the length of 1 mile in kilometers?

10. Convert 2.2×10^{-3} centimeters to miles.

Answers to Chapter 2 Quiz

(1) **5.6×10^4 mg.** The conversion between grams and milligrams is 1000 mg in 1 gram, or in exponential form, as you should be solving these, it is 10^3 mg = 1 g. To solve this you will need to use the factor label method and perform a conversion:

$$\left(56 \ \cancel{g} \right)\left(\frac{10^3 \, \text{mg}}{1 \, \cancel{g}} \right) = 5.6 \times 10^4 \, \text{mg}$$

(2) **3.5×10^1 L.** The conversion between milliliters and liters is 1000 mL = 1 L or, written in exponential form, 10^3 mL = 1L. To solve this you will need to use the factor label method and perform a conversion:

$$\left(3500 \ \cancel{\text{mL}} \right)\left(\frac{1 \text{L}}{10^3 \, \cancel{\text{mL}}} \right) = 3.5 \times 10^1 \, \text{L}$$

(3) **1.5×10^{10} mm.** The conversion between kilometers and mm is 1 km = 10^6 mm. To solve this you will need to use the factor label method and perform a conversion:

$$\left(1.5 \times 10^4 \ \cancel{\text{km}} \right)\left(\frac{10^6 \, \text{mm}}{1 \, \cancel{\text{km}}} \right) = 1.5 \times 10^{10} \, \text{mm}$$

(4) **13.5 g/mL.** The formula for the density of an object is density = mass ÷ volume. However, before you can plug your numbers in, you need to convert your starting units to grams and milliliters as the question asks.

$$\left(5.4 \ \cancel{\text{kg}} \right)\left(\frac{10^3 \, \text{g}}{1 \, \cancel{\text{kg}}} \right) = 5.6 \times 10^3 \, \text{g}$$

$$\left(400.0 \ \cancel{\text{cm}^3} \right)\left(\frac{1 \, \text{mL}}{1 \, \cancel{\text{cm}^3}} \right) = 400.0 \ \text{mL}$$

Once you have converted your values to the correct units you then plug them into the density formula and solve:

$$\frac{m}{V} = \text{density} = \frac{5.6 \times 10^3 \, \text{g}}{400.0 \ \text{mL}} = 13.5 \ \frac{\text{g}}{\text{mL}}$$

(5) $4.5 \times 10^{-5} \ \frac{\text{kg}}{\text{mL}}$. To solve this problem you can break it up into two steps. First, convert from grams to kilograms:

$$\left(45 \ \frac{\cancel{g}}{\text{L}} \right)\left(\frac{1 \ \text{kg}}{10^3 \, \cancel{g}} \right)$$

then convert from liters to milliliters:

$$\left(45 \ \frac{\cancel{g}}{\cancel{\text{L}}} \right)\left(\frac{1 \ \text{kg}}{10^3 \, \cancel{g}} \right)\left(\frac{1 \cancel{\text{L}}}{10^3 \ \text{mL}} \right) = 4.5 \times 10^{-5} \ \frac{\text{kg}}{\text{mL}}$$

You'll notice these calculations are all on the same line. It is best not to split them up and instead solve them all on the same line. It dramatically reduces your chance for error.

You should simplify your answer to the correct scientific notation by moving the decimal over one more place to the left as is reflected in the final answer.

6 **880 lb.** The conversion factor between kilograms and pounds is 2.2 kg to 1 lb. Use the factor label method as shown here to solve:

$$\left(400 \ \cancel{kg}\right)\left(\frac{2.2 \ lb}{1 \cancel{kg}}\right) = 880 \ lb$$

7 **157,680,000 seconds.** To solve this problem you'll need to perform several conversions. You could always look up the direct conversion from years to seconds but for the sake of learning, here is a solution showing the basic path that you can take from years to seconds when converting:

$$\left(5 \ \cancel{years}\right)\left(\frac{365 \ \cancel{days}}{1 \ \cancel{years}}\right)\left(\frac{24 \ \cancel{hours}}{1 \ \cancel{days}}\right)\left(\frac{60 \ \cancel{minutes}}{1 \ \cancel{hours}}\right)\left(\frac{60 \ seconds}{1 \ \cancel{minutes}}\right) = 157,680,000 \ seconds$$

8 **50662.5 kPa.** The conversion from atm to kPa is 1 atm = 110.325 kPa. These are units of pressure. The conversion using the factor label method is shown below:

$$\left(500 \ \cancel{atm}\right)\left(\frac{101.325 \ kPa}{1 \ \cancel{atm}}\right) = 50662.5 \ kPa$$

9 **1.61 km.** 1 mile is equal to 1.61 km. Use this information to solve the problem with the factor label method:

$$\left(1 \ \cancel{mile}\right)\left(\frac{1.61 \ km}{1 \ \cancel{mile}}\right) = 1.61 \ km$$

10 **1.37×10⁻⁸ miles.** To solve this problem you need to first perform a conversion from the metric system to the imperial system of measurement going from centimeters to inches. Then you need to convert from inches to miles. The most common route for you is likely to be going from inches to feet and then to miles. This is shown here. The answer has been rounded to the tenths place.

$$\left(2.2 \times 10^{-3} \ \cancel{cm}\right)\left(\frac{1 \ \cancel{in}}{2.54 \ \cancel{cm}}\right)\left(\frac{1 \ \cancel{ft}}{12 \ \cancel{in}}\right)\left(\frac{1 \ mile}{5280 \ \cancel{ft}}\right) = 1.37 \times 10^{-8} \ miles$$

Chapter **3**

The Basic Properties of Matter

ook around you. All the stuff you see — your chair, the water you're drinking, the paper this book is printed on — is matter. Matter is anything that has mass and occupies space. It's the material part of the universe and it is the major focus of chemistry. Chemistry can be simply defined as the study of matter. There can be a bit more to it than that but the general idea of chemistry is that you are looking at what makes up matter and how that matter changes. This chapter is going to give you a basic introduction into what matter is in more depth, how matter is classified, and some of the different properties that can be observed in matter.

Describing the States of Matter

Matter is anything that has mass and occupies space. It's the material part of the universe. Whenever scientists contemplate matter, they do so in one of two ways — with a macroscopic viewpoint or with a microscopic one. Here's an overview of each one:

» **Macroscopic:** When scientists view matter in a macroscopic way, they're considering matter as the "stuff" they can physically observe — a lump of coal, 5 pounds of sugar, a pinch of salt. The macroscopic world is the world you can directly observe through your senses. Most people use this viewpoint when looking at the world around them.

>> **Microscopic:** The chemist can switch perspective to a microscopic one. The microscopic level isn't just what you can observe through a microscope. It goes far beyond that. It's the level of individual particles, such as carbon atoms, sugar molecules, and sodium and chloride ions. The microscopic view is the world of scientists' theories and models.

Scientists can switch back and forth between these viewpoints without even thinking, and the hope is that you can grow to feel more comfortable with viewing matter in both of these ways. Whichever perspective you use, matter can exist in one of three states: solid, liquid, and gas.

Solids

At the macroscopic level, the level at which you directly observe with your senses, a solid has a definite shape and occupies a definite volume. Think of an ice cube in a glass — it's a solid. You can easily weigh the ice cube and measure its volume. At the microscopic level (where items are so small that people can't directly observe them), the particles that make up the ice are very close together and aren't moving around very much (see Figure 3-1a).

FIGURE 3-1:
Solid (a),
liquid (b),
and gaseous
(c) states of
matter.

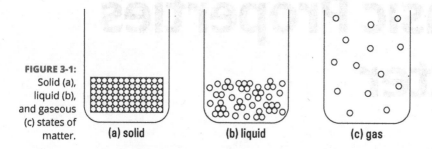

(a) solid   (b) liquid   (c) gas

The particles that make up the ice (also known as *water molecules*) are close together and have little movement because, as in many solids, the particles are pulled into a rigid, organized structure of repeating patterns called a *crystal lattice*. The particles that are contained in the crystal lattice are still moving, but barely — it's more of a slight vibration. Depending on the particles, this crystal lattice may be of different shapes.

Liquids

When an ice cube melts, it becomes a liquid. Unlike solids, liquids have no definite shape, but they do have a definite volume, just like solids do. For example, a cup of water in a tall, skinny glass has a different shape from a cup of water in a pie pan, but in both cases, the volume of water is the same — one cup. Why? The particles in liquids are not really much farther apart than the particles in solids, but they're moving around much more (refer to Figure 3-1b).

Some of the particles in liquids may be near each other, clumped together in small groups. Because the particles are moving much faster in liquids, the attractive forces among them aren't as strong as they are in solids — which is why liquids don't have a definite shape. However, these attractive forces are strong enough to keep the substance confined in one large mass — a liquid — instead of going all over the place.

Gases

If you heat water, you can convert it to steam, the gaseous form of water. A gas has no definite shape and no definite volume. In a gas, particles are much farther apart than they are in solids or liquids (refer to Figure 3-1c), and they're moving relatively independently of each other. Because of the distance between the particles and the independent motion of each of them, the gas expands to fill the area that contains it (and thus it has no definite shape).

Because a great deal of distance separates gas particles, you can easily compress a gas but not a solid — or, to a certain extent, a liquid where the particles are very close together. If you hold a balloon and squeeze, you can actually force those gas particles closer together because of all the empty space between the gas particles. Check out Chapter 17 for more information on gases.

EXAMPLE

Q. What is the difference in the spacing and arrangement of particles in a solid, a liquid, and a gas?

A. Solid particles are close together and organized into a rigid structure. They have a very low amount of motion. Liquid particles are still close together relative to gas particles but have more space between them than a solid particles. Liquid particles have the ability to move and flow over and around one another but they do not have total freedom of movement. Gas particles are very spaced apart. They have no organization in terms of their arrangement and move randomly.

YOUR TURN

1 For each of the following states of matter, identify which will take the shape of their container and which will fill that container (have the same volume as their container)?

(a) Solid

(b) Liquid

(c) Gas

Classifying Pure Substances and Mixtures

One of the basic concepts in science is classification. As we discuss in the preceding section, chemists can classify matter as solid, liquid, or gas. But matter can be classified in other ways as well. In this section, we discuss how all matter can be classified as either a pure substance or a mixture (see Figure 3-2).

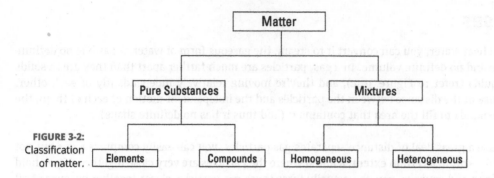

FIGURE 3-2:
Classification
of matter.

Keeping it simple with pure substances

A *pure substance*, such as salt or sugar, has a definite and constant composition or makeup. A pure substance can be either an element or a compound, but the composition of a pure substance doesn't vary.

Elementary, my dear reader

An *element* is composed of a single kind of atom. An *atom* is the smallest particle of an element that still has all the properties of the element. Here's an example: Gold is an element. If you slice and slice a chunk of gold until only one tiny particle is left that can't be chopped any more without losing the properties that make gold *gold*, then you've got an atom.

The atoms in an element all have the same number of protons. *Protons* are subatomic particles — particles of an atom. Subatomic particles come in three major kinds, which Chapter 4 covers in great, gory detail.

REMEMBER

The important thing to remember right now is that elements are the building blocks of matter. And they're represented in a strange table you've probably seen a time or two — the periodic table.

Compounding the problem

A *compound* is composed of two or more elements in a specific ratio. For example, water (H_2O) is a compound made up of two elements, hydrogen (H) and oxygen (O). These elements are combined in a very specific way — in a ratio of two hydrogen atoms to one oxygen atom (hence, H_2O). A lot of compounds contain hydrogen and oxygen, but only one has that special two-to-one ratio we call water. Even though water is made up of hydrogen and oxygen, the compound water doesn't have the physical and chemical properties of hydrogen or oxygen; water's properties are unique, entirely different from the two elements of which it's composed.

Chemists can't easily separate the components of a compound. They have to resort to some type of chemical reaction.

Throwing mixtures into the mix

Mixtures are physical combinations of pure substances (elements and/or compounds) that have no definite or constant composition; the composition of a mixture varies according to who prepares the mixture. Suppose we asked two people to prepare a salad. Unless these two people used the same recipe, these mixtures would vary somewhat in their relative amounts of lettuce, croutons, and so on. The mixtures would be slightly different. However, each component of a mixture — that is, each pure substance that makes up the mixture (in the salad example, each *ingredient*) — retains its own set of physical and chemical characteristics. Because of this, it's relatively easy to separate the various substances in a mixture.

Although chemists have a difficult time separating compounds into their specific elements, the different parts of a mixture can be easily separated by physical means, such as filtration. For example, suppose you have a mixture of salt and sand, and you want to purify the sand by removing the salt. You can do this by adding water, dissolving the salt, and then filtering the mixture. You then end up with pure sand.

Mixtures can come in two different forms:

>> **Homogeneous mixtures:** Sometimes called *solutions,* this type of mixture is relatively uniform in composition; every portion of the mixture is like every other portion. If you dissolve sugar in water and mix it really well, your mixture is basically the same no matter where you sample it.

>> **Heterogeneous mixtures:** If you put some sugar in a jar, add some sand, and then give the jar a couple of shakes, your mixture doesn't have the same composition throughout the jar. Because the sand is heavier, there's probably more sand at the bottom of the jar and more sugar at the top. In this case, you have a *heterogeneous mixture,* a mixture whose composition varies from position to position within the sample.

Q. Why is any type of salad a good example of a heterogeneous mixture?

EXAMPLE **A.** A salad is always composed of multiple parts, and these parts are never individual elements, nor are they combined in a perfectly set ratio every single time. In addition, a salad is never perfectly uniform throughout, making it a heterogeneous mixture.

YOUR TURN

2 Classify each of the following as an element, compound, heterogeneous mixture, or homogeneous mixture:

(a) Boron

(b) Italian salad dressing

(c) The air you breathe in every day

(d) Methane (CH_4) gas

(e) A bowl of cereal

3 Is the air we breathe a homogenous or heterogenous mixture? Why?

Nice Properties You've Got There

When chemists study chemical substances, they examine two types of properties:

>> **Chemical properties:** These properties enable a substance to change into a brand-new substance, and they describe how a substance reacts with other substances. Does a substance change into something completely new when water is added — like how sodium metal changes to sodium hydroxide? Does it burn in air?

>> **Physical properties:** These properties describe the physical characteristics of a substance. The color, luster, hardness, and so on of a substance are its physical properties, and so is its ability to conduct electricity.

Some physical properties are *extensive properties*, properties that depend on the amount of matter present. Mass and volume are extensive properties. A large chunk of gold has a larger mass and volume than a smaller chunk. *Intensive properties*, however, don't depend on the amount of matter present. Hardness is an intensive property. A large chunk of gold, for example, has the same hardness as a small chunk of gold. The mass and volume of these two chunks are different (extensive properties), but the hardness is the same. Intensive properties are especially useful to chemists because they can use intensive properties to identify a substance.

Q. Why are intensive properties used to identify substances, and why are extensive properties not used to identify substances?

EXAMPLE

A. *Intensive properties* are a unique set of properties that are completely independent of the amount of matter present. An individual property is not unique to a particular substance, but you can take intensive properties as a whole to identify substances with great accuracy. For example, although several substances may have the same melting point or boiling point, it's highly unlikely that two substances have the same melting point, boiling point, density, solubility, and more. In that way, intensive properties uniquely describe each substance.

Extensive properties, on the other hand, would serve very little purpose in identifying a substance. Any number of substances can assume the same mass or volume.

YOUR TURN

 Identify each of the following as an intensive or extensive property of matter:

(a) Density

(b) Length

(c) Color

(d) Melting point

(e) Mass

Practice Questions Answers and Explanations

(1)

   a. Solids will neither take the shape of their container nor fill the container you put them in. Solids have a fixed shape and a fixed volume.

   b. Liquids will take the shape of a container you put them but they will not fill the container. Liquids have a variable shape based on their container but they have a fixed volume and will not fill all available space.

   c. Gases will take the shape of their container and fill their container. Gases have no fixed shape or fixed volume. They will fill any volume you put them in.

(2) To solve these problems, refer to Figure 3-2. Start at the top of the chart for each one and trace your way down until you come to a stop.

   a. **Element.** Boron is an element on the periodic table.

   b. **Heterogeneous mixture.** Italian salad dressing is a mixture of multiple components arranged in a nonuniform way.

   c. **Homogeneous mixture.** Air is a uniform mixture of many gaseous elements.

   d. **Compound.** Methane is a combination of two elements in a definite ratio each time. In this case, one carbon will always combine with four hydrogen atoms to form methane gas.

   e. **Heterogeneous mixture.** Cereal is a mixture of components that are not uniform throughout.

(3) The air we breathe is a mixture of different gases, mainly nitrogen and oxygen along with smaller amounts of other gases that are uniformly distributed throughout. There is no specific part of the air that has more oxygen molecules over there or a greater concentration of nitrogen molecules in one spot. The gases that make up air are uniformly distributed making it a homogenous mixture.

(4) If a property is independent of the amount present, it's intensive; if the property depends on the amount, it's extensive.

   a. **Intensive.** Density depends on the ratio of mass/volume but not on a set amount of either.

   b. **Extensive.** Length obviously is a measure of how long something is, which is entirely dependent on the amount you have.

   c. **Intensive.** Color is not dependent on an amount. A substance will have the same color regardless of how much of the substance you have.

   d. **Intensive.** A substance will melt at the same temperature, no matter the amount of the substance. For example, 100 kg of ice and 1 mg of ice will both start melting at 0°C. One sample will take longer than the other to completely melt, but they'll both reach their melting point at the same temperature.

   e. **Extensive.** Mass is entirely dependent on how much matter a substance has.

Whaddya Know? Chapter 3 Quiz

Quiz time! Complete each problem to test your knowledge on the various topics covered in this chapter. You can then find the solutions and explanations in the next section.

1 What state of matter has the greatest potential for movement of its particles based on their spacing?

2 If a solid were to change to a liquid, how would the spacing and arrangement of the particles change?

3 Classify the following as being a pure substance or a mixture:

 a. Hydrogen

 b. Water

 c. Salt water

 d. Carbonated water

 e. NaCl

4 You shake up a bag of M&M candies of all different colors in hopes that the bag will become a homogenous mixture of candy. However, after shaking the bag of for 5 minutes, the mixture is still a heterogenous mixture of candy. Why is that? Can this mixture ever become homogenous?

5 Identify the following as being intensive or extensive properties of matter:

 a. Freezing point

 b. Malleability

 c. Volume

 d. Flammability

 e. The temperature of an object

6 Classify the following as being an element or a compound:

 a. Helium

 b. Iron

 c. Carbon Dioxide

 d. Glucose

 e. Water

7 Which phase of matter has a definite volume and a definite shape?

8 Which phase of matter has no definite volume or defined shape?

9 Which phase of matter has a defined volume and but no defined shape?

10 Are the particles found in a crystal lattice completely still or do they have some level of motion present?

Answers to Chapter 3 Quiz

(1) **Gases** have the great potential for movement based on the spacing of their particles. Since gas particles are the most spaced apart, they have the ability to move around far more than liquid or solid particles. Liquid particles do have the potential for freedom of movement, but that movement is limited by the somewhat close packing of liquid particles relative to gases. Solids are very tightly packed together and their particles have no freedom of movement because they are in in a rigid organized structure.

(2) When a solid changes state to a liquid the particles move further apart due to an increase in energy. As the state changes to a liquid, the particles begin moving more. As the matter goes from a solid to a liquid, the particles also become less organized. A solid is composed of a highly organized structure of particles. When it changes states to a liquid, that organization is greatly reduced because the particles are less structured and have more freedom of movement.

(3)

a. **Pure substance.** Hydrogen is an element. All elements are pure substances.

b. **Pure substance**. Water is a chemical compound. All chemical compounds are pure substances.

c. **Mixture.** Salt water is a solution of salt and water. All solutions are mixtures.

d. **Mixture.** Carbonated water is a solution of water and dissolved carbon dioxide. As stated above, all solutions are mixtures.

e. **Pure substance.** NaCl is sodium chloride, though it is more commonly known as salt. Salt is a chemical compound, making it a pure substance. If the salt were to be dissolved in water as in the question above, then it would become part of a mixture but on its own salt is a pure substance.

(4) A mixture of M&Ms cannot be a homogenous mixture at any point, no matter how much you shake the bag up and try to evenly disperse the particles. Even if you could somehow get a uniform distribution of all the colored M&Ms it would not be a homogenous mixture. If they were perfectly distributed and no area of the bag had a greater concentration of one color than another the mixture itself would still not be completely the same throughout. There would still be areas of different colors throughout the entire bag, making it a heterogeneous mixture. All parts of a homogenous mixture must be indistinguishable from one another.

(5)

a. **Intensive property.** The temperature that something freezes at does not depend on the amount of matter present; 100 liters of water will freeze at the same temperature as 1 liter of water.

b. **Intensive property.** Malleability is basically how easy a material is to bend or shape through force. This is another property that is the same for something regardless of whether it is a 5 gram sample of iron or a 500 gram sample of iron.

c. **Extensive property.** Volume is the amount of space something takes up, so that by definition depends on the amount of the object present. If something is physically bigger it takes up more space, making it have a greater volume than a smaller object.

d. **Intensive property.** Flammability is another property that is not impacted by the amount of the matter. Make sure you do not get the idea of flammability confused with how long or well something might burn. Obviously a greater amount of flammable material, like lighter fluid, will burn longer and better than a smaller quantity of the same material. That is not the question, however. Whether you have a large amount of material or a small amount of lighter fluid, the flammability — whether or not it will combust — does not change. It will combust, making it an intensive property.

e. **Intensive property.** Temperature can be measured in a large sample or a small sample of matter. It doesn't matter how much you have of that matter. How well something changes temperature definitely depends on the amount of matter you have, but simply taking a measurement of temperature does not making this an intensive property.

6

a. **Element.** Remember, if it can be found on the periodic table, it is an element.

b. **Element.** Iron is an element that can be found on the periodic table.

c. **Compound.** Carbon dioxide is composed of two elements, carbon and oxygen, making this a compound.

d. **Compound.** Glucose is composed of carbon, hydrogen, and oxygen, making this a compound.

e. **Compound.** Water is composed of hydrogen and oxygen, making this a compound.

7 **Solid.** Solids have a defined shape and a defined volume.

8 **Gas.** Gases have no defined shape or volume. They will expand to fill whatever area you put them in.

9 **Liquid.** Liquids have a defined volume but no definite shape. A liquid will not fill whatever container you put it in but it will take the shape of that container.

10 **Particles in a solid crystal lattice still have motion.** It is very slow and minimal but all particles in matter are moving constantly. There is no state of matter that exists where molecular motion completely and totally ceases.

Chapter 4

Breaking Down Atoms into Their Subatomic Particles

"Big stuff is built from smaller pieces of stuff. If you keep breaking stuff down into smaller and smaller pieces, eventually you'll reach the smallest possible bit of stuff. Let's call that bit an *atom*." This is how the Greek philosopher Democritus might have explained his budding concept of "atomism" to a buddy over a flask of Cretan wine. Like wine, the idea had legs.

For hundreds of years, scientists have operated under the idea that all matter is made up of smaller building blocks called *atoms.* So small, in fact, that until the invention of the electron microscope in 1931, the only way to find out anything about these tiny, mysterious particles was to design a very, very clever experiment. Chemists couldn't exactly corner a single atom in a back alley somewhere and study it alone — they had to study the properties of whole gangs of atoms and try to guess what individual ones might be like. Through remarkable ingenuity and incredible luck, chemists now understand a great deal about the atom. After reading this chapter, so will you.

The Atom: Protons, Electrons, and Neutrons

The *atom* is the smallest part of matter that represents a particular element. For quite a while, the atom was thought to be the smallest part of matter that could exist. Picture an atom as a microscopic LEGO. Atoms come in a variety of shapes and sizes, and you can build larger structures out of them. Like a LEGO, an atom is extremely hard to break. In fact, so much energy is stored inside atoms that breaking them in half results in a nuclear explosion. This section examines the parts of an atom.

Breaking an atom into its parts

In the latter part of the 19th century and early part of the 20th, scientists discovered that atoms are composed of certain subatomic particles and that, no matter what the element, the same *subatomic particles* make up the atom. The number of the various subatomic particles is the only thing that varies. The atom is considered the smallest possible unit of an element, because after you break an atom into subatomic particles, the pieces lose the unique properties of that element.

Scientists now recognize that there are many subatomic particles (this really makes physicists salivate). But in order to be successful in chemistry, you really need to be concerned with only the three major subatomic particles:

>> **Proton:** The subatomic particle that has a positive charge; it's found in the atom's dense central core.

>> **Neutron:** The subatomic particle that has no charge; like the proton, it's found in the atom's dense central core.

>> **Electron:** The subatomic particle that has a negative charge; it's found outside the atom's dense central core.

Table 4-1 summarizes the characteristics of these three subatomic particles.

TABLE 4-1   **The Three Major Subatomic Particles**

Name	Symbol	Charge	Mass (g)	Mass (amu)	Location
Proton	p^+	+1	1.673×10^{-24}	1	Nucleus
Neutron	n^0	0	1.675×10^{-24}	1	Nucleus
Electron	e^-	−1	9.109×10^{-28}	0.0005	Outside nucleus

In Table 4-1, the masses of the subatomic particles are listed in two ways: grams and *amu*, which stands for *atomic mass units*. Expressing mass in amu is much easier than using the gram equivalent.

Atomic mass units are based on something called the carbon-12 scale, a worldwide standard that's been adopted for atomic weights. By international agreement, a carbon atom that contains 6 protons and 6 neutrons has an atomic weight of exactly 12 amu, so 1 amu is $\frac{1}{12}$ of this carbon atom. (What do carbon atoms and the number 12 have to do with anything? Just trust us.) Because the masses of protons and neutrons are almost exactly the same in grams, both protons and neutrons are said to have a mass of 1 amu. Notice that the mass of an electron is much smaller than that of either a proton or a neutron. It takes almost 2,000 electrons to equal the mass of a single proton.

Table 4-1 also shows the electrical charge associated with each subatomic particle. Matter can be electrically charged in one of two ways: positively or negatively. The proton carries one unit of positive charge, the electron carries one unit of negative charge, and the neutron has no charge; it's neutral.

REMEMBER

Scientists have discovered through observation that objects with like charges, whether positive or negative, repel each other, and objects with unlike charges attract each other.

The atom itself has no charge. It's neutral, meaning the number of positive charges present is equal to the number of negative charges present. In other words, the number of protons in an atom equals the number of electrons in an atom. Sometimes, though, atoms can have differing numbers of electrons. When that occurs, they're called *ions*. We cover ions in Chapter 9.

Narrowing the focus to the nucleus

The last column in Table 4-1 lists the locations of the three subatomic particles. Protons and neutrons are located in the *nucleus*, a dense central core in the middle of the atom. Because they're always found in the nucleus, they're sometimes called *nuclear particles*. Electrons move around the nucleus in a cloud of many different energy levels and are discussed at great length in Chapter 6.

Now the nucleus is very, very small and very, very dense when compared to the rest of the atom. Typically, atoms have diameters that measure around 10^{-10} meters. (That's really small!) Nuclei are around 10^{-15} meters in diameter. (That's *really, really* small!) For example, if the Superdome in New Orleans represented a hydrogen atom, the nucleus would be about the size of a pea.

Not only is the nucleus very small, but it also contains most of the mass of the atom. In fact, for all practical purposes, the mass of the atom is the sum of the masses of the protons and neutrons. (In your chemistry class, you'll ignore the minute mass of the electrons when performing calculations. The electrons' mass is necessary only for very, very, very precise calculations.)

Q. If a gold atom contains 197 nuclear particles, 79 of which are protons, how many neutrons and how many electrons does the gold atom have?

EXAMPLE

A. An atom of gold contains 118 neutrons and 79 electrons. The *nucleus* contains all the protons and neutrons in an atom, so if 79 of the 197 particles in a gold nucleus are protons, the remaining 118 particles must be neutrons. All atoms are electrically neutral, so there must be a total of 79 electrons (in other words, 79 negative charges) to balance out the 79 positive charges of the protons.

This type of logic leads to a general formula that you can use to calculate proton or neutron counts. This formula is $M = P + N$, where M is the atomic mass, P is the number of protons, and N is the number of neutrons.

Q. If a carbon atom contains 14 nuclear particles and 8 of those are neutrons, how many protons and electrons are present?

A. Six protons and 6 electrons are present. The carbon atom has 14 total particles in its nucleus, and 8 of those are neutrons. To determine the number of protons, you simply need to subtract the neutrons from the total number ($14 - 8 = 6$), showing that 6 protons are present. Because the atom is neutral, the number of protons must equal the number of electrons, so there are also 6 electrons.

YOUR TURN

1 If an atom has 71 protons, 71 electrons, and 104 neutrons, how many particles reside in the nucleus, and how many are outside of the nucleus?

2 If an atom's nucleus has a mass of 31 amu and contains 15 protons, how many neutrons and electrons does the atom have?

A Brief History of the Atom

Now, how exactly do we know all this stuff about the atom, and where did this theoretical model of the atom come from? Well, generations of ingenious scientists have tackled this problem. The result of all the clever experimentation and tricky math has been a series of models, each a bit more refined than the one before. The result is the model of the atom that we present in this chapter and throughout the book. Knowing the history of a theory is always a good idea, so here's an introduction to three of the most influential scientists who addressed the question of what an atom looks like.

J. J. Thomson: Cooking up the "plum pudding" model

The first subatomic particle was discovered by J. J. Thomson in the late 1800s. Thomson performed a series of experiments using a device called a cathode-ray tube, a contraption that eventually evolved into the modern television. Thomson realized that a beam of electrons (or a cathode ray, in scientific speak) could be deflected in a magnetic field. This result and others got Thomson thinking that the whole "indivisible atom" model had its limitations; atoms were actually composed of other, subatomic particles. Thomson proposed a model of the atom, called the "plum pudding" model. You've probably never had the misfortune to taste plum pudding, a traditional English dessert consisting of dried plums mixed into a thick pudding. Thomson's model was as important to chemistry as the dessert is foul tasting.

Thomson, like all chemists of his day, knew that two negative charges repel each other, so he pictured the atom as a collection of evenly spaced, negatively charged particles. He called these charges *corpuscles*, but Thomson's corpuscles have subsequently been awarded the much nicer name *electrons.* Thomson also knew that the atom was electrically neutral (meaning it has zero overall charge), so he figured the atom must also contain an amount of positive charge equal to the negative charge of the electrons. Because the proton had not yet been discovered, Thomson imagined this positive charge as a soup in which all of the negative electrons were suspended, just like plums are suspended in plum pudding.

Ernest Rutherford: Shooting at gold

The next leap forward was made by Ernest Rutherford in 1909. Rutherford set out to test Thomson's plum pudding model of the atom. To do so, Rutherford made an extremely thin sheet of gold foil and shot alpha particles (helium nuclei) at it. This may sound a little crazy, but there was method to Rutherford's madness. If the plum pudding model were true, Rutherford expected that when alpha particles crashed into the gold foil, loosely bound electrons of the gold atoms would deflect the alpha particles by a few degrees at most. When Rutherford attempted it, however, he found that 1 in every 8,000 (or so) alpha particles deflected by 90 degrees or more! He famously compared the result to watching a shot bullet bounce off a piece of tissue paper, a result perplexing enough to drop the jaw of even a jaded chemist.

After much pondering, Rutherford eventually realized that most of an atom must be empty space and that most of the atom's mass — and one of its two kinds of charge — must be concentrated at the center. So 7,999 of every 8,000 particles Rutherford shot at the foil missed this tiny bundle of mass, passing straight through, but every so often a lucky shot smashed into the nucleus and was deflected at a large angle.

REMEMBER

We now know that the positive charge of an atom is concentrated at its center in the form of protons. The protons reside there along with all the atom's neutrons. There is something very strange and counterintuitive about this idea. Like charges repel each other, so having all the positive charge of an atom concentrated in one tiny, central area is truly bizarre. A very, very strong force must be holding together all those positive charges. This force, unimaginatively dubbed the *strong force,* is something we can take for granted; nuclear physicists can't take it for granted, but that's their headache. (Physicists call this force and others found in the nucleus *nuclear glue.* Sometimes this "glue" isn't strong enough, and the nucleus breaks apart in a process called *radioactivity.*)

Niels Bohr: Comparing the atom to the solar system

Enter Niels Bohr in the early 1900s. Dr. Bohr was not an experimentalist like Thomson or Rutherford. Bohr was a *theorist,* which basically means he sat around pondering things. Having pondered his way to an aha moment, his job became to prove his ideas mathematically.

Bohr was aware of Rutherford's gold foil experiment. It occurred to Bohr that the atom may operate very much like the solar system, with most of the mass concentrated in the center (at the sun), with smaller bodies (the planets) orbiting the center at specific distances. According to this model, low-mass electrons in an atom orbit the central nucleus, which contains all the massive protons and neutrons. This conceptual leap was neat, logical, and crazy. The understanding of electricity and magnetism at the time suggested that there was no way for a negative charge to orbit at a constant distance from a positive charge. Classic theories suggested that an orbiting electron would eventually spiral into a central nucleus.

Bohr bypassed this problem with a neat little trick called "quantization of angular momentum." Basically, Bohr invented a whole new set of rules for how electrons should behave in an atom. Strangely enough, the predictions of his mathematical model matched experiments so well that nobody could prove him wrong. Although now an entire branch of physics, called *quantum mechanics,* is much more accurate than Bohr's model, his predictions are so nearly right and so convenient that we gratefully leave quantum mechanics to math-happy physicists. We'll stick with the picture painted for us by Bohr.

Deciphering Chemical Symbols: Atomic and Mass Numbers

When looking at all the information available to describe an atom of any element, you can easily get lost. Thankfully, though, two very important numbers, the *atomic number* and the *mass number,* tell you much of what you need to know about an atom. Chemists tend to memorize these numbers like baseball fans memorize batting averages, but clever chemistry students like you need not resort to memorization. You have the ever-important periodic table of the elements at your disposal. We discuss the logical structure and organization of the periodic table in detail in Chapter 5, so for now we simply explain what the atomic and mass numbers mean without going into great detail about their consequences. At the end of this section, we've also included a table that lists the elements in alphabetical order and shows each element's chemical symbol, name, atomic mass, and atomic number (you'll very much love the alphabetical order part — trust us).

Atomic numbers are like name tags: They identify an element as carbon, nitrogen, beryllium, and so on by telling you the number of protons in the nucleus of that element. Atoms are known by their numbers of protons. Adding a proton or removing one from the nucleus of an atom changes the elemental identity of an atom.

In the periodic table, you can find the atomic number above the one- or two-letter abbreviation for an element. The short letter abbreviation is the element's chemical symbol. Notice that the elements of the periodic table are lined up in order of atomic number, as if they've responded to some sort of roll call. Atomic number increases by 1 each time you move to the right in the periodic table; when a row ends, the sequence of increasing atomic numbers begins again at the left side of the next row down. You can check out the periodic table for yourself in Chapter 5.

The second identifying number of an atom is its mass number. The mass number reports the mass of the atom's nucleus in atomic mass units (amu). Because protons and neutrons have a mass of 1 amu each (as you find out earlier in this chapter), the mass number equals the sum of the numbers of protons and neutrons:

Mass Number = Protons + Neutrons

An electron has only $1/1,836$ of the mass of a proton or neutron, so in most cases the mass of the electron is ignored.

Chemists commonly use the symbolization shown in Figure 4-1, called *isotope notation*, to represent the mass number and atomic number of an individual element.

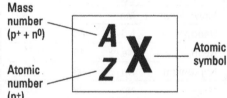
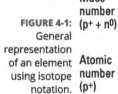

As shown in Figure 4-1, chemists use the placeholder X to represent the chemical symbol. You can find an element's chemical symbol on the periodic table or in a list of elements. The placeholder Z represents the atomic number, which is the number of protons in the nucleus. A represents the mass number, the sum of the number of protons plus neutrons. The mass number is listed in amu.

Table 4-2 shows the elements in alphabetical order. (Not all the known elements are included in the table — just all the ones you'll be using in your chemistry class.) When you're trying to quickly locate some information about one element or another, this table can prove to be far handier than the periodic table. Use it wisely.

TABLE 4-2 The Elements

Name	Symbol	Atomic Number	Mass Number	Name	Symbol	Atomic Number	Mass Number
Actinium	Ac	89	227.028	Cerium	Ce	58	140.115
Aluminum	Al	13	26.982	Cesium	Cs	55	132.905
Americium	Am	95	243	Chlorine	Cl	17	35.453
Antimony	Sb	51	121.76	Chromium	Cr	24	51.996
Argon	Ar	18	39.948	Cobalt	Co	27	58.933
Arsenic	As	33	74.922	Copper	Cu	29	63.546
Astatine	At	85	210	Curium	Cm	96	247
Barium	Ba	56	137.327	Dubnium	Db	105	262
Berkelium	Bk	97	247	Dysprosium	Dy	66	162.5
Beryllium	Be	4	9.012	Einsteinium	Es	99	252
Bismuth	Bi	83	208.980	Erbium	Er	68	167.26
Bohrium	Bh	107	262	Europium	Eu	63	151.964
Boron	B	5	10.811	Fermium	Fm	100	257
Bromine	Br	35	79.904	Fluorine	F	9	18.998
Cadmium	Cd	48	112.411	Francium	Fr	87	223
Calcium	Ca	20	40.078	Gadolinium	Gd	64	157.25
Californium	Cf	98	251	Gallium	Ga	31	69.723
Carbon	C	6	12.011	Germanium	Ge	32	72.61
Gold	Au	79	196.967	Mendelevium	Md	101	258
Hafnium	Hf	72	178.49	Mercury	Hg	80	200.59
Hassium	Hs	108	265	Molybdenum	Mo	42	95.94
Helium	He	2	4.003	Neodymium	Nd	60	144.24
Holmium	Ho	67	164.93	Neon	Ne	10	20.180
Hydrogen	H	1	1.0079	Neptunium	Np	93	237.048
Indium	In	49	114.82	Nickel	Ni	28	58.69
Iodine	I	53	126.905	Niobium	Nb	41	92.906
Iridium	Ir	77	192.22	Nitrogen	N	7	14.007
Iron	Fe	26	55.845	Nobelium	No	102	259
Krypton	Kr	36	83.8	Osmium	Os	76	190.23
Lanthanum	La	57	138.906	Oxygen	O	8	15.999
Lawrencium	Lr	103	262	Palladium	Pd	46	106.42
Lead	Pb	82	207.2	Phosphorus	P	15	30.974
Lithium	Li	3	6.941	Platinum	Pt	78	195.08
Lutetium	Lu	71	174.967	Plutonium	Pu	94	244
Magnesium	Mg	12	24.305	Polonium	Po	84	209
Manganese	Mn	25	54.938	Potassium	K	19	39.098
Meitnerium	Mt	109	266	Praseodymium	Pr	59	140.908
Promethium	Pm	61	145	Tantalum	Ta	73	180.948
Protactinium	Pa	91	231.036	Technetium	Tc	43	98

Name	Symbol	Atomic Number	Mass Number	Name	Symbol	Atomic Number	Mass Number
Radium	Ra	88	226.025	Tellurium	Te	52	127.60
Radon	Rn	86	222	Terbium	Tb	65	158.925
Rhenium	Re	75	186.207	Thallium	Tl	81	204.383
Rhodium	Rh	45	102.906	Thorium	Th	90	232.038
Rubidium	Rb	37	85.468	Thulium	Tm	69	168.934
Ruthenium	Ru	44	101.07	Tin	Sn	50	118.71
Rutherfordium	Rf	104	261	Titanium	Ti	22	47.88
Samarium	Sm	62	150.36	Tungsten	W	74	183.84
Scandium	Sc	21	44.956	Uranium	U	92	238.029
Seaborgium	Sg	106	263	Vanadium	V	23	50.942
Selenium	Se	34	78.96	Xenon	Xe	54	131.29
Silicon	Si	14	28.086	Ytterbium	Yb	70	173.04
Silver	Ag	47	107.868	Yttrium	Y	39	88.906
Sodium	Na	11	22.990	Zinc	Zn	30	65.39
Strontium	Sr	38	87.62	Zirconium	Zr	40	91.224
Sulfur	S	16	32.066				

EXAMPLE

Q. What are the name, atomic number, mass number, number of protons, number of electrons, and number of neutrons of each of the following four elements: $^{35}_{17}Cl$, $^{37}_{17}Cl$, $^{190}_{76}Os$, and $^{39}_{19}K$?

A. The answers to questions like these, favorites of chemistry teachers, are best organized in a table. First, look up the symbols Cl, Os, and K in Table 4-2 or the periodic table in Chapter 5 and find the names of these elements. Enter what you find in the first column.

To fill in the second and third columns (Atomic Number and Mass Number), read the atomic number and mass number from the lower left and upper left of the chemical symbols given in the question. The atomic number equals the number of protons; the number of electrons is the same as the number of protons, because elements have zero overall charge. So fill in the proton and electron columns with the same numbers you entered in column two.

Last, subtract the atomic number from the mass number to get the number of neutrons, and enter that value in column six. Voilà! The entire private life of each of these atoms is now laid before you. Your answer should look like the following table.

Element Name	Atomic Number	Mass Number	Number of Protons	Number of Electrons	Number of Neutrons
Chlorine	17	35	17	17	18
Chlorine	17	37	17	17	20
Osmium	76	190	76	76	114
Potassium	19	39	19	19	20

3 Write the proper chemical symbol for an atom of bismuth with a mass of 209 amu.

4 Fill in the following chart for $_1^1$H, $_{24}^{52}$Cr, $_{77}^{192}$Ir, and $_{42}^{96}$Mo.

Element Name	Atomic Number	Mass Number	Number of Protons	Number of Electrons	Number of Neutrons

5 Write the $_Z^A X$ form of the two elements shown in the following table.

Element Name	Atomic Number	Mass Number	Number of Protons	Number of Electrons	Number of Neutrons	
Tungsten	74	184	74		74	110
Lead	82	207	82	82	125	

6 Use the periodic table or Table 4-2 and your knowledge of atomic numbers and mass numbers to fill in the missing pieces in the following table.

Name	Atomic Number	Mass Number	Number of Protons	Number of Electrons	Number of Neutrons
Silver		108			
	16				16
		64	29		
				18	22

Keeping an Eye on Ions

Because an atom itself is neutral, it is usually said that the number of protons and electrons in atoms are equal. But in some cases an atom can acquire an electrical charge. For example, in the compound sodium chloride — table salt — the sodium atom has a positive charge and the chlorine atom has a negative charge. Atoms (or groups of atoms) in which there are unequal numbers of protons and electrons are called ions.

The neutral sodium atom has 11 protons and 11 electrons, which means it has 11 positive charges and 11 negative charges. Overall, the sodium atom is neutral, and it's represented as Na. But the sodium ion contains one more positive charge than negative charge, so it's represented as Na+ (the + represents its net positive electrical charge).

This unequal number of negative and positive charges can occur in one of two ways: An atom can gain a proton (a positive charge) or lose an electron (a negative charge). So which process is more likely to occur? Well, the rough guideline is that gaining or losing electrons is easy but gaining or losing protons is very difficult. So atoms become ions by gaining or losing electrons.

If an ion is formed by the loss of an electron, the ion has more protons than electrons, or more positive charges. Those positive ions are called *cations*. You represent the overall positive charge in cations with the little plus sign in the notation (like Na^+). If the atom loses two electrons instead of one, the result is still a cation, but it has a stronger positive charge (actually twice as strong as in the case in which only one electron was lost) and is represented with the number of electrons lost and a plus sign (like Mg^{2+} for a magnesium cation with two missing electrons, or Al^{3+} for aluminum with three electrons gone). If an ion is created by gaining an electron, the number of electrons exceeds the number of protons, so the ion acquires a negative charge. Negatively charged ions are called *anions*, and they're represented with a little negative sign (−). If chlorine (Cl) gains an electron, it becomes a chlorine ion because it has unequal numbers of protons and electrons, and as an anion (a negatively charged ion), it's represented as Cl^-.

EXAMPLE

Q. How many electrons does an oxygen ion, O^{2-}, gain or lose when it becomes an ion? What is the total number of protons and electrons found in the oxygen ion?

A. An oxygen ion has a negative 2 charge. Any ion with a negative charge indicates that the ion gained electrons. Since the charge is negative 2, it means it gained 2 electrons. All oxygen atoms, regardless of whether they are an ion or not, will always contain 8 protons. The number of electrons can be calculated by referencing the number of electrons in a neutral oxygen atom, 8 electrons, and then adding the 2 electrons gained when it became an ion. So 8 electrons + 2 electrons = 10 total electrons.

Q. An ion has 26 protons and 24 electrons. What element is this and what is its ionic charge?

A. To identify the element you need to look at the number of protons, 26, and find that iron has 26 protons on the periodic table. Next if this element has 26 protons and 24 electrons that means it has 26 positive charges and only 24 negative charges. This results in there being an overall net charge of 2+ on this ion. You would write the ion like this: Fe^{2+}.

7 An atom has 13 protons and 10 electrons. Write the correct ionic symbol of this ion.

 8 What is the number of protons and electrons in sulfur ion, S^{2-}?

Accounting for Isotopes Using Atomic Masses

It's Saturday night. The air is charged with possibility. Living in the moment, you peruse your personal copy of the periodic table. What's this? You notice that the numbers that appear below the atomic symbols seem to be related to the elements' mass numbers, but they're not nice whole numbers. You know that the mass number of an atom equals the sum of the numbers of protons and neutrons in the atom's nucleus, so how could a decimal value appear? There's no such thing as half a proton or a quarter of a neutron. What does it all mean?

As it turns out, most elements have several different configurations with differing numbers of neutrons in each one. *Isotopes* are atoms of the same element that have different mass numbers; the differences in mass number arise from different numbers of neutrons. The messy-looking numbers with all those decimal places are atomic masses. An *atomic mass* is a weighted average of the masses of all the naturally occurring isotopes of an element. Chemists have measured the percentage of each element that exists in different isotopic forms. In the weighted average of the atomic mass, the mass of each isotope contributes in proportion to how often that isotope occurs in nature. More–common isotopes contribute more to the atomic mass.

Consider the element carbon, for example. Carbon occurs naturally in three isotopes. These isotopes can be represented by using isotope notation (see Figure 4-1) or by writing the element name and then a hyphen followed by the mass number (for example, carbon-12):

>> **Carbon-12** ($^{12}_{6}C$, or carbon with six protons and six neutrons) is boring old run-of-the-mill carbon, accounting for 99 percent of all the carbon out there.

>> **Carbon-13** ($^{13}_{6}C$, or carbon with six protons and seven neutrons) is a slightly rarer (though still dull) isotope, accounting for most of the remaining 1 percent of carbon atoms. Taking on an extra neutron makes carbon-13 slightly heavier than carbon-12 but does little else to change its properties.

However, even this minor change has some useful consequences. Scientists compare the ratio of carbon-12 to carbon-13 within meteorites to help determine their origins.

>> **Carbon-14** ($^{14}_{6}C$, or carbon with six protons and eight neutrons) shows its interesting little face in only one out of every trillion or so carbon atoms. So if you're thinking you won't be working with large samples of carbon-14 in your chemistry lab, you're right!

These three isotopes are why you see carbon's atomic mass on the periodic table written as 12.01. If you do a quick bit of deductive reasoning, you can probably determine that carbon-12 is far and away the most common of the three isotopes due to the average atomic mass being closest to 12.

REMEMBER

Precise measurements of the amounts of different isotopes can be important. You need to know the exact measurements if you're asked to figure out an element's atomic mass. To calculate an atomic mass, you need to know the masses of the isotopes and the percentage of the element that occurs as each isotope (this is called the *relative abundance*). To calculate an average atomic mass, make a list of each isotope along with its mass and its percent relative abundance. Multiply the mass of each isotope by its relative abundance. Add the products. The resulting sum is the atomic mass based on the weighted average of all the naturally occurring isotopes of an element.

Certain elements, such as chlorine, occur in several very common isotopes, so their average atomic mass isn't close to a whole number. Other elements, such as carbon, occur in one very common isotope and several very rare ones, resulting in an average atomic mass that's very close to the whole-number mass of the most common isotope.

EXAMPLE

Q. Chlorine occurs in two common isotopes. It appears as $^{35}_{17}Cl$ 75.8% of the time and as $^{37}_{17}Cl$ 24.2% of the time. What is its average atomic mass?

A. The average atomic mass is 35.5 amu. First, multiply each atomic mass by its relative abundance, using the decimal form of your percentage (75.8% = 0.758; remember, to convert to decimal form, move the decimal two places to the left in your percentage):

$$(35 \text{ amu})(0.758) = 26.53 \text{ amu}$$
$$(37 \text{ amu})(0.242) = 8.95 \text{ amu}$$

Then add the two results together to get the average atomic mass:

$$26.53 \text{ amu} + 8.95 \text{ amu} = 35.5 \text{ amu}$$

Compare your answer to the value on your periodic table. If you've done the calculation correctly, the two values should match or at least be very, very similar.

YOUR TURN

9 Magnesium occurs in three fairly common isotopes, $^{24}_{12}Mg$, $^{25}_{12}Mg$, and $^{26}_{12}Mg$, which have percent abundances of 78.9%, 10.0%, and 11.1%, respectively. Calculate the average atomic mass of magnesium.

Practice Questions Answers and Explanations

(1) **175 inside, 71 outside.** The nucleus of an atom consists of protons and neutrons, so this atom (lutetium) has 175 particles in its nucleus (71 protons + 104 neutrons). Electrons are the only subatomic particles that aren't included in the nucleus, so lutetium has 71 particles outside of its nucleus.

(2) **16 neutrons, 15 electrons.** A nuclear mass of 31 amu means that the nucleus has 31 particles. Because 15 of them are protons, that leaves 16 amu for the neutrons. The numbers of protons and electrons are equal in a neutral atom (in this case, we're talking about phosphorus), so the atom has 15 electrons.

(3) $^{209}_{83}\text{Bi}$. Find bismuth on the periodic table or in Table 4-2 to get its chemical abbreviation and its atomic number. Because you already have its mass number (209), all you need to do is write all this information in $^A_Z X$ form.

(4) To fill out the chart, take the information presented in the isotope notation and break it into the individual pieces the chart asks for. Looking at hydrogen, you see that it has a mass number of 1 and an atomic number of 1. This means only one proton is present and no neutrons are present. The number of electrons is 1 because in a neutral atom, the number of electrons is always equal to the number of protons.

To solve chromium's line, notice that chromium has a mass number of 52 and an atomic number of 24. This tells you that the number of protons is 24, because the atomic number and the number of protons are the same. To determine the number of neutrons, you simply subtract the atomic number from the mass number: $52 - 24 = 28$ neutrons. Finally, the number of protons is equal to the number of electrons. Continue this process to fill in the info for iridium and molybdenum.

Element Name	Atomic Number	Mass Number	Number of Protons	Number of Electrons	Number of Neutrons
Hydrogen	1	1	1	1	0
Chromium	24	52	24	24	28
Iridium	77	192	77	77	115
Molybdenum	42	96	42	42	54

(5) $^{164}_{74}\text{W}$, $^{207}_{82}\text{Pb}$. Notice that you don't need most of the information in the table; all you really need to look up is the chemical symbol of each element.

(6) When you're given the atomic number, the number of protons, or the number of electrons, you automatically know the other two numbers because they're all equal. Each element in the periodic table is listed with its atomic number, so by locating the element, you can simply read off the atomic number and therefore know the number of protons and electrons. To calculate the atomic mass or the number of neutrons, you must be given one or the other. Calculate atomic mass by adding the number of protons to the number of neutrons. Alternatively, calculate the number of neutrons by subtracting the number of protons from the atomic mass.

Element Name	Atomic Number	Mass Number	Number of Protons	Number of Electrons	Number of Neutrons
Silver	47	108	47	47	61
Sulfur	16	32	16	16	16
Copper	29	64	29	29	35
Argon	18	40	18	18	22

(7) **Al^{3+}.** 13 protons corresponds to the element aluminum. To determine the charge on aluminum you would need to compare the 13 positive protons to the 10 negative electrons. You can even think of it as a basic math problem. $13 + (-10) = $ postive 3. This is why the aluminum has a 3+ charge.

(8) **16 protons and 18 electrons.** Sulfur has an atomic number of 16 so there are 16 protons. Since sulfur has a 2– charge that means there are 2 more electrons than protons, this means there are 18 electrons.

(9) **24.3 amu.** First, multiply the three mass numbers by their relative abundances in decimal form. Then add the resulting products to get the average atomic mass.

$$(24 \text{ amu})(0.789) = 18.94 \text{ amu}$$
$$(25 \text{ amu})(0.100) = 2.50 \text{ amu}$$
$$(26 \text{ amu})(0.111) = 2.89 \text{ amu}$$
$$18.94 \text{ amu} + 2.50 \text{ amu} + 2.89 \text{ amu} = 24.3 \text{ amu}$$

If you're ready to test your skills a bit more, take the following chapter quiz that incorporates all the chapter topics.

Whaddya Know? Chapter 4 Quiz

Quiz time! Complete each problem to test your knowledge on the various topics covered in this chapter. You can then find the solutions and explanations in the next section.

1. An atom has a mass number of 65 and 35 neutrons. How many protons are there and what is the identity of the element?

2. How many protons, neutrons, and electrons are there in a neutral atom of potassium with a mass number of 39?

3. How many protons and electrons are there in a fluorine ion, F^{1-}?

4. What is the mass number of an ion that has 56 protons, 81 neutrons, and 54 electrons?

5. What is the isotope notation for an atom of nitrogen–14?

6. How many protons, neutrons, and electrons does the following ion have: $^{28}_{13}Al^{3+}$

7. What is the isotope notation for an ion of sodium–23 with a charge of positive 1?

8. What is the number of protons, neutrons and electrons found in the following isotope: $^{33}_{16}S^{2-}$

9. What is the number of protons, neutrons and electrons found in the following isotope: $^{202}_{80}Hg^{2+}$

10. An atom has a mass number of 80 and 35 electrons. How many protons and neutrons does it have?

11. You have an imaginary element represented by the symbol X. It has 3 isotopes. Given the following relative abundances of the isotopes what is the average atomic mass, in amu, of element X?

Isotope	Mass Number	Relative Abundance
1	86	23.7%
2	87	45.2%
3	88	31.1%

Answers to Chapter 4 Quiz

1. **30 protons, zinc.** To determine the number of protons in the unknown element you need to subtract the number of neutrons from the mass number. Remember, mass number is the sum of the protons and the neutrons. $65 - 35 = 30$ protons. You then find that zinc has an atomic number of 30 on the periodic table.

2. **19 protons, 20 neutrons, 19 electrons.** The atomic number of potassium is 19 which means it has 19 protons. To calculate neutrons you can subtract 19 from the mass number of 39. $39 - 19 = 20$ neutrons. Since the atom is neutral that means the number of electrons is equal to the number of protons.

3. **9 protons, 10 electrons.** Fluorine has an atomic number of 9 so it contains 9 protons. This fluorine ion is shown to have a charge of 1−. This means that there is one more electron than proton. You can simply add 1 to the number of protons to determine the electron number. $9 + 1 = 10$ electrons.

4. **137.** The mass number of an atom is the sum of the protons and the neutrons. 56 protons + 81 neutrons = 137.

5. $_{7}^{14}\text{N}$. Nitrogen has an atomic number of 7. Its chemical symbol is N. The mass number of 19 is given to you in the problem. The top left number is the mass number, the bottom number is the atomic number.

6. **13 protons, 15 neutrons, 10 electrons.** The bottom number that you see when something is written in isotope notation represent the atomic number. To calculate the neutrons you simply subtract the protons, bottom number, from the top number, the mass number. $28 - 13 = 15$ neutrons. To calculate he electrons you see that this ion has a charge of 3+. This means the ion has 3 less electrons than protons. To calculate this you can do $13 - 3$ to determine there are 10 electrons.

7. $_{11}^{23}\text{Na}^{1+}$. Sodium has an atomic number of 11 so you put that on the bottom left. The mass number of 23 is given to you in the name sodium–23. You write that as the top number. The symbol of sodium is Na, which can be found on the periodic table. The 1+ charge is written because you are told it has a charge of positive one.

8. **16 protons, 17 neutrons, 18 electrons.** The bottom number of the isotope notation is 16, this is the atomic number. 33 is the mass number which is the sum of the protons and the neutrons. To calculate neutrons you do $33 - 16 = 17$ neutrons. The number of electrons can be calculated by adding two electrons to the number of protons since the negative 2 charge means there are 2 more electrons than protons. $16 + 2 = 18$ electrons.

9. **80 protons, 122 neutrons, 78 electrons.** The bottom number of the isotope notation is 80, this is the atomic number. 202 is the mass number which is the sum of the protons and the neutrons. To calculate neutrons you do $202 - 80 = 122$ neutrons. The number of electrons can be calculated by subtracting two electrons from the number of protons since the positive 2 charge means there are 2 less electrons than protons. $80 - 2 = 78$ electrons.

10. **35 protons, 45 neutrons.** Since this atom is described as an atom and not an ion, you can assume this is a neutral atom, which makes the number of protons equal to the number of electrons. The mass number given is the sum of the protons and the neutrons. Once you've determined that there are 35 protons, subtract that from the 80 given as the mass number to determine the number of neutrons, $80 - 35 = 45$.

(11) **87.07 amu.** To solve this problem you must calculate the weighted average of all of the given isotopes. You must first convert the percentages of the isotopes to their decimal equivalent by moving the decimal point 2 places to the left or simply by dividing each of them by 100. Once this is done you multiply each decimal percentage by their respective mass number and add all of them up.

$$(86)(.237)+(87)(.452)+(88)(.311)=87.07 \text{ amu}$$

2

The Periodic Table

In This Unit . . .

Chapter **5**

Surveying the Periodic Table of the Elements

There it hangs, looming ominously over the chemistry classroom, a formidable wall built of bricks with names like "C," "Ag," and "Tc." Behold the *periodic table of the elements!* But don't be fooled by its stern appearance or intimidated by its teeming details. The table is your friend, your guide, your key to making sense of chemistry. To begin to make friends with the table, concentrate on its trends. Start simply: Notice that the table has rows and columns. Keep your eye on the columns and rows, and soon you'll be making sense of things like atomic radii and valence electrons. Really.

Organizing the Periodic Table

In nature as well as in manmade systems, you may notice some repeating patterns. The seasons repeat their pattern of fall, winter, spring, and summer. The tides repeat their pattern of rising and falling. Tuesday follows Monday, December follows November, and so on. This pattern of repeating order is called *periodicity*.

In the mid-1800s, Dmitri Mendeleev, a Russian chemist, noticed a repeating pattern of chemical properties in the elements that were known at the time. Mendeleev arranged the elements in order of increasing atomic mass (see Chapter 3 for a description of atomic mass) to form something that fairly closely resembles the modern periodic table. He was even able to predict the properties of some of then-unknown elements. Later, the elements were rearranged in order of increasing *atomic number*, the number of protons in the nucleus of the atom (again, see Chapter 3). Figure 5-1 shows the modern periodic table.

PERIODIC TABLE OF THE ELEMENTS

	1 IA	2 IIA	3 IIIB	4 IVB	5 VB	6 VIB	7 VIIB	8 VIIIB	9 VIIIB	10 VIIIB	11 IB	12 IIB	13 IIIA	14 IVA	15 VA	16 VIA	17 VIIA	18 VIIIA
1	1 H Hydrogen 1.00797																	2 He Helium 4.0026
2	3 Li Lithium 6.939	4 Be Beryllium 9.0122											5 B Boron 10.811	6 C Carbon 12.01115	7 N Nitrogen 14.0067	8 O Oxygen 15.9994	9 F Fluorine 18.9984	10 Ne Neon 20.183
3	11 Na Sodium 22.9898	12 Mg Magnesium 24.312											13 Al Aluminum 26.9815	14 Si Silicon 28.086	15 P Phosphorus 30.9738	16 S Sulfur 32.064	17 Cl Chlorine 35.453	18 Ar Argon 39.948
4	19 K Potassium 39.102	20 Ca Calcium 40.08	21 Sc Scandium 44.956	22 Ti Titanium 47.90	23 V Vanadium 50.942	24 Cr Chromium 51.996	25 Mn Manganese 54.9380	26 Fe Iron 55.847	27 Co Cobalt 58.9332	28 Ni Nickel 58.71	29 Cu Copper 63.546	30 Zn Zinc 65.37	31 Ga Gallium 69.72	32 Ge Germanium 72.59	33 As Arsenic 74.9216	34 Se Selenium 78.96	35 Br Bromine 79.904	36 Kr Krypton 83.80
5	37 Rb Rubidium 85.47	38 Sr Strontium 87.62	39 Y Yttrium 88.905	40 Zr Zirconium 91.22	41 Nb Niobium 92.906	42 Mo Molybdenum 95.94	43 Tc Technetium (99)	44 Ru Ruthenium 101.07	45 Rh Rhodium 102.905	46 Pd Palladium 106.4	47 Ag Silver 107.868	48 Cd Cadmium 112.40	49 In Indium 114.82	50 Sn Tin 118.69	51 Sb Antimony 121.75	52 Te Tellurium 127.60	53 I Iodine 126.9044	54 Xe Xenon 131.30
6	55 Cs Cesium 132.905	56 Ba Barium 137.34	57 La Lanthanum 138.91	72 Hf Hafnium 179.49	73 Ta Tantalum 180.948	74 W Tungsten 183.85	75 Re Rhenium 186.2	76 Os Osmium 190.2	77 Ir Iridium 192.2	78 Pt Platinum 195.09	79 Au Gold 196.967	80 Hg Mercury 200.59	81 Tl Thallium 204.37	82 Pb Lead 207.19	83 Bi Bismuth 208.980	84 Po Polonium (210)	85 At Astatine (210)	86 Rn Radon (222)
7	87 Fr Francium (223)	88 Ra Radium (226)	89 Ac Actinium (227)	104 Rf Rutherfordium (261)	105 Db Dubnium (262)	106 Sg Seaborgium (266)	107 Bh Bohrium (264)	108 Hs Hassium (269)	109 Mt Meitnerium (268)	110 Ds Darmstadtium (269)	111 Rg Roentgenium (272)	112 Uub Ununbium (277)	113 Uut §	114 Uuq Ununquadium (285)	115 Uup §	116 Uuh Ununhexium (289)	117 Uus §	118 Uuo Ununoctium (293)

Lanthanide Series	58 Ce Cerium 140.12	59 Pr Praseodymium 140.907	60 Nd Neodymium 144.24	61 Pm Promethium (145)	62 Sm Samarium 150.35	63 Eu Europium 151.96	64 Gd Gadolinium 157.25	65 Tb Terbium 158.924	66 Dy Dysprosium 162.50	67 Ho Holmium 164.930	68 Er Erbium 167.26	69 Tm Thulium 168.934	70 Yb Ytterbium 173.04	71 Lu Lutetium 174.97
Actinide Series	90 Th Thorium 232.038	91 Pa Protactinium (231)	92 U Uranium 238.03	93 Np Neptunium (237)	94 Pu Plutonium (242)	95 Am Americium (243)	96 Cm Curium (247)	97 Bk Berkelium (247)	98 Cf Californium (251)	99 Es Einsteinium (254)	100 Fm Fermium (257)	101 Md Mendelevium (258)	102 No Nobelium (259)	103 Lr Lawrencium (260)

§ Note: Elements 113, 115, and 117 are not known at this time but are included in the table to show their expected positions.

FIGURE 5-1: The periodic table of the elements.

Examining the organization of the periodic table

Chemists can't imagine doing much of anything without having access to the periodic table. Instead of mastering the properties of 118+ elements (more are created almost every year), chemists — and chemistry students — can simply get a firm grasp of the properties of families of elements, thus saving a lot of time and effort. You can find the relationships among elements and figure out the formulas of many different compounds by referring to the periodic table. The table readily provides atomic numbers, mass numbers, and information about the number of valence electrons (the outermost s and p electrons; see Chapter 6 "The Electron").

Take a look at your new friend, the periodic table, in Figure 5-1. Notice the horizontal rows and the vertical columns of elements:

>> **Periods:** The periodic table is composed of horizontal rows called periods. The periods are numbered 1 through 7 on the left-hand side of the table. The atomic numbers increase from left to right in each period. Even though they're in the same period, these elements have chemical properties that are not all that similar. Consider the first two members of period 3: sodium (Na) and magnesium (Mg). In reactions, they both tend to lose electrons (after all, they are metals), but sodium loses one electron, whereas magnesium loses two. Chlorine (Cl), down near the end of the period, tends to gain an electron (it's a nonmetal).

>> **Groups:** The vertical columns are called *groups*, or *families*. The families may be labeled at the top of the columns in one of two ways. The older method uses Roman numerals and letters. Many chemists (especially old ones like me) prefer and still use this method. The newer method simply uses the numbers 1 through 18 (Figure 5-1 shows both systems of numbering). The older method is usually used in describing the features of the table because it helps relate the position of an element on the periodic table with its number of valence electrons more than the 1–18 grouping system does.

The members of a family do have similar properties. Consider the IA family, starting with lithium (Li) — don't worry about hydrogen, because it's unique and doesn't really fit anywhere — and going through francium (Fr). All these elements tend to lose only one electron in reactions. And all the members of the VIIA family tend to gain one electron. The elements within any group have very similar properties. The properties of the elements emerge mostly from their different numbers of protons and electrons (see Chapter 4) and from the arrangement of their electrons.

Each group on the periodic table is usually unique in its own way. The groups have family names, allowing for quick and easy identification of their respective elements. Take a look at Figure 5-2 to see a list of the most common groups and their elements.

Pay attention to these families:

>> The IA family is made up of the *alkali metals*. In reactions, these elements all tend to lose a single electron. This family contains some important elements, such as sodium (Na) and potassium (K). Both of these elements play an important role in the chemistry of the body and are commonly found in salts.

>> The IIA family is made up of the *alkaline earth metals*. All these elements tend to lose two electrons. Calcium (Ca) is an important member of the IIA family.

IA (1)	IIA (2)		VIIIA (18)
3 **Li** Lithium 6.939	4 **Be** Beryllium 9.0122	VIIA (17)	2 **He** Helium 4.0026
11 **Na** Sodium 22.9898	12 **Mg** Magnesium 24.312	9 **F** Flourine 18.9984	10 **Ne** Neon 20.183
19 **K** Potassium 39.102	20 **Ca** Calcium 40.08	17 **Cl** Chlorine 35.453	18 **Ar** Argon 39.948
37 **Rb** Rubidium 85.47	38 **Sr** Strontium 87.62	35 **Br** Bromine 79.904	36 **Kr** Krypton 83.80
55 **Cs** Cesium 132.905	56 **Ba** Barium 137.34	53 **I** Iodine 126.9044	54 **Xe** Xenon 131.30
87 **Fr** Francium (223)	88 **Ra** Radium (226)	85 **At** Astatine (210)	86 **Rn** Radon (222)
Alkali Metals	Alkaline Earth Metals	Halogens	Noble Gases

FIGURE 5-2: Groups on the periodic table.

» The VIIA family is made up of the *halogens*. They all tend to gain a single electron in reactions. Important members in the family include chlorine (Cl), used in making table salt and bleach, and iodine (I). Tincture of iodine is sometimes used as a disinfectant.

» The VIIIA family is made up of the *noble gases*. These elements are *very* unreactive. For a long time, the noble gases were called the *inert* gases, because people thought that these elements wouldn't react at all. Later, a scientist named Neil Bartlett showed that at least some of the inert gases could be reacted, but they required very special conditions. Since Bartlett's discovery, the gases have been referred to as noble gases.

EXAMPLE

Q. What group, using the Roman numeral numbering system and the numerical system, is calcium located within?

A. Calcium is located in group IIA. This group is also called group 2. The groups are the vertical columns found across the periodic table. They are numbered with a Roman numeral system IA through VIIIA, or a numerical system, 1–18.

Q. What period is bromine located in?

A. Bromine is located in period 4. Remember, periods are the rows down the periodic table. They are numbered 1–7 as you go down the table.

YOUR TURN

 1 What group, using the Roman numeral numbering system, are the following elements in?

(a) sodium

(b) gallium

(c) strontium

(d) oxygen

(e) xenon

2 What period are the following elements in?

(a) magnesium

(b) hydrogen

(c) francium

(d) titanium

(e) selenium

Meeting the metals, nonmetals, and metalloids

In addition to the groups (families) you see in Figure 5-2, there are other ways to classify elements on the periodic table. The most common way is to arrange elements based on their metallic properties. Most of the elements on the periodic table are considered *metals*. The metals have properties that you normally associate with the metals you encounter in everyday life. They are solid (with the exception of mercury, Hg, a liquid), shiny, good conductors of electricity and heat, *ductile* (they can be drawn into thin wires), and *malleable* (they can be easily hammered into very thin sheets). The large block of metals that is not part of the alkali or alkaline earth metals is called the *transition metal* block. Transition metals have properties that vary from extremely metallic, on the left side, to far less metallic, on the right side. Figure 5-3 shows the metals.

All the elements on the left-hand side and in the middle of the periodic table are metals, with one notable exception: hydrogen. It is the first element and is unique in its properties. Scientists stick it above lithium, but it doesn't react like a metal.

Meanwhile, except for a few elements that border the metals on the right (more on those in a moment), the elements on the right of the periodic table are classified as *nonmetals* (along with hydrogen). These elements are shown in Figure 5-4.

REMEMBER

Nonmetals have properties opposite those of the metals. The nonmetals are brittle, aren't malleable or ductile, are poor conductors of both heat and electricity, and tend to gain electrons in chemical reactions. Some nonmetals are liquids, and some are gases.

The elements that border the metals and the nonmetals are classified as *metalloids*, and they're shown in Figure 5-5.

FIGURE 5-3: The metals.

IA (1)	IIA (2)	IIIB (3)	IVB (4)	VB (5)	VIB (6)	VIIB (7)	VIIIB (8)	(9)	(10)	IB (11)	IIB (12)	13	14	15	16
3 Li Lithium 6.939	4 Be Beryllium 9.0122														
11 Na Sodium 22.9898	12 Mg Magnesium 24.312											13 Al Aluminum 26.9815			
19 K Potassium 39.102	20 Ca Calcium 40.08	21 Sc Scandium 44.956	22 Ti Titanium 47.90	23 V Vanadium 50.942	24 Cr Chromium 51.996	25 Mn Manganese 54.9380	26 Fe Iron 55.847	27 Co Cobalt 58.9332	28 Ni Nickel 58.71	29 Cu Copper 63.546	30 Zn Zinc 65.37	31 Ga Gallium 69.72			
37 Rb Rubidium 85.47	38 Sr Strontium 87.62	39 Y Yttrium 88.905	40 Zr Zirconium 91.22	41 Nb Niobium 92.906	42 Mo Molybdenum 95.94	43 Tc Technetium (99)	44 Ru Ruthenium 101.07	45 Rh Rhodium 102.905	46 Pd Palladium 106.4	47 Ag Silver 107.868	48 Cd Cadmium 112.40	49 In Indium 114.82	50 Sn Tin 118.69		
55 Cs Cesium 132.905	56 Ba Barium 137.34	57 La Lanthanum 138.91	72 Hf Hafnium 179.49	73 Ta Tantalum 180.948	74 W Tungsten 183.85	75 Re Rhenium 186.2	76 Os Osmium 190.2	77 Ir Iridium 192.2	78 Pt Platinum 195.09	79 Au Gold 196.967	80 Hg Mercury 200.59	81 Tl Thallium 204.37	82 Pb Lead 207.19	83 Bi Bismuth 208.980	84 Po Polonium (210)
87 Fr Francium (223)	88 Ra Radium (226)	89 Ac Actinium (227)	104 Rf Rutherfordium (261)	105 Db Dubnium (262)	106 Sg Seaborgium (266)	107 Bh Bohrium (264)	108 Hs Hassium (269)	109 Mt Meitnerium (268)	110 Ds Darmstadtium (281)	111 Rg Roentgenium (286)	112 Cn Copernicium (285)				

58 Ce Cerium 140.12	59 Pr Praseodymium 140.907	60 Nd Neodymium 144.24	61 Pm Promethium (145)	62 Sm Samarium 150.35	63 Eu Europium 151.96	64 Gd Gadolinium 157.25	65 Tb Terbium 158.924	66 Dy Dysprosium 162.50	67 Ho Holmium 164.930	68 Er Erbium 167.26	69 Tm Thulium 168.934	70 Yb Ytterbium 173.04	71 Lu Lutetium 174.97
90 Th Thorium 232.038	91 Pa Protactinium 231	92 U Uranium 238.03	93 Np Neptunium (237)	94 Pu Plutonium (242)	95 Am Americium (243)	96 Cm Curium (247)	97 Bk Berkelium (247)	98 Cf Californium (251)	99 Es Einsteinium (254)	100 Fm Fermium (257)	101 Md Mendelevium (258)	102 No Nobelium (259)	103 Lr Lawrencium (260)

REMEMBER

The metalloids, or *semimetals*, have properties that are somewhat of a cross between metals and nonmetals. They tend to be economically important because of their unique conductivity properties: They only partially conduct electricity, making them valuable in the semiconductor and computer chip industry. (Did you think the term *Silicon Valley* referred to a valley covered in sand? Nope. Silicon, one of the metalloids, is used in making computer chips.) Metalloids touch the black stairstep line that separates metals from nonmetals on the periodic table (Figure 5-1).

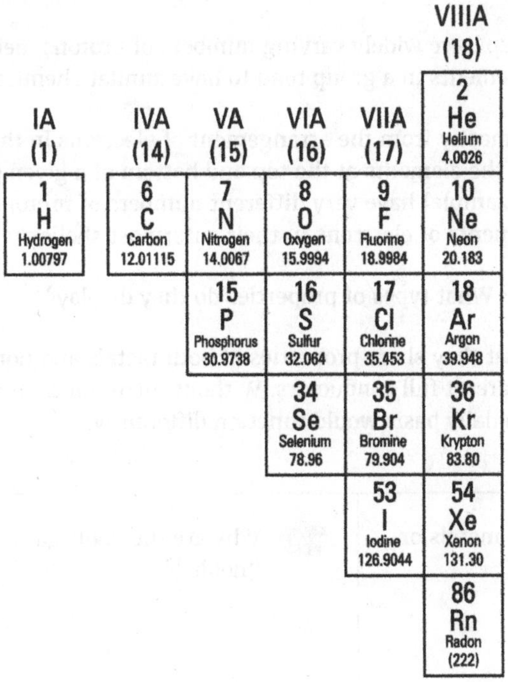

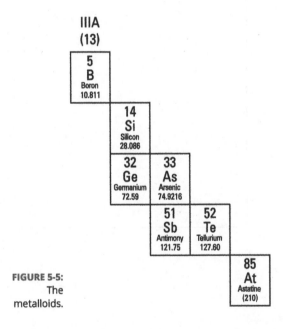

VIIIA (18)

IA (1)		**IVA (14)**	**VA (15)**	**VIA (16)**	**VIIA (17)**	**2** **He** Helium 4.0026
1 **H** Hydrogen 1.00797		**6** **C** Carbon 12.01115	**7** **N** Nitrogen 14.0067	**8** **O** Oxygen 15.9994	**9** **F** Fluorine 18.9984	**10** **Ne** Neon 20.183
			15 **P** Phosphorus 30.9738	**16** **S** Sulfur 32.064	**17** **Cl** Chlorine 35.453	**18** **Ar** Argon 39.948
				34 **Se** Selenium 78.96	**35** **Br** Bromine 79.904	**36** **Kr** Krypton 83.80
					53 **I** Iodine 126.9044	**54** **Xe** Xenon 131.30
						86 **Rn** Radon (222)

FIGURE 5-4:
The nonmetals.

IIIA (13)

5 **B** Boron 10.811			
14 **Si** Silicon 28.086			
32 **Ge** Germanium 72.59	**33** **As** Arsenic 74.9216		
	51 **Sb** Antimony 121.75	**52** **Te** Tellurium 127.60	
		85 **At** Astatine (210)	

FIGURE 5-5:
The metalloids.

Periods (rows) 6 and 7 have an added wrinkle. Elements with atomic numbers 58 through 71 and 90 through 103 are lifted out of the regular order and placed below the rest of the table. These two series are the *lanthanides* and the *actinides*, respectively. These elements are separated from the table for two main reasons:

>> Doing so prevents the table from being inconveniently wide.

>> The lanthanides all have pretty similar properties, as do the actinides.

Q. The elements within a group have widely varying numbers of protons, neutrons, and electrons. Why, then, do elements in a group tend to have similar chemical properties?

A. Chemical properties come mostly from the arrangement of electrons in the outermost shell of an atom. Although the elements at the top and bottom of a given group (like fluorine and astatine, for example) have very different numbers of protons, neutrons, and electrons, the arrangements of electrons in their outermost shells are very similar.

Q. Why are metalloids unique? What types of properties do they display?

A. Metalloids are unique in that they share properties of both metals and nonmetals. They can conduct electricity yet aren't full conductors. Without them, much of what you've come to know and use on a daily basis would function differently.

YOUR
TURN

3 Are the following elements metals or nonmetals?

a. selenium

b. fluorine

c. strontium

d. chromium

e. bismuth

4 Why are the noble gases referred to as "noble"?

Practice Questions Answers and Explanations

1 You identify the group location by referring to the number at the top of the periodic table for each column.

   a. sodium (Na): **IA**

   b. gallium (Ga): **IIIA**

   c. strontium (Sr): **IIA**

   d. oxygen (O): **VIA**

   e. xenon (Xe): **VIIIA**

2 Periods are determined by the numbered row that each element is found in. Periods are numbered 1–7.

   a. magnesium (Mg): **period 3**

   b. hydrogen (H): **period 1**

   c. francium (Fr): **period 7**

   d. titanium (Ti): **period 4**

   e. selenium (Se): **period 4**

3 Metals are found to the left of the staircase, with the exception of hydrogen, and nonmetals are found to the right of the staircase. Several elements touching the staircase are classified as metalloids.

   a. selenium (Se): **nonmetal**

   b. fluorine (F): **nonmetal**

   c. strontium (Sr): **metal**

   d. chromium (Cr): **metal**

   e. bismuth (Bi): **metal**

4 The noble gases are described as "noble" because they seem to consider it beneath themselves to react with other elements. Because these elements have completely filled valence shells, they have no energetic reason to react. They're already as stable as they can be.

Whaddya Know? Chapter 5 Quiz

Ready for a quiz? The 10 questions in this section will test the skills you learned in this chapter. When you're done, check out the section that follows for answers and explanations.

1. Is xenon a metal, nonmetal, or metalloid?

2. What is the name of the family that contains the least reactive elements found on the periodic table?

3. What is the group number of the halogens?

4. Name 3 elements classified as metalloids.

5. What period contains calcium and bromine?

6. What is the group number of the alkali metals?

7. Do elements share more properties with the group/family they are in or the period they are located within?

8. What is the sum of the group numbers (using the 1–18 group numbering system) of hydrogen, chlorine, and aluminum?

9. Where are metals generally located on the periodic table?

10. Is chlorine a metal, nonmetal, or metalloid?

Answers to Chapter 5 Quiz

(1) **Nonmetal.** Xenon is located on the far right of the periodic table in the noble gas family of elements, group 18. Nonmetals are generally located to the right of the metalloid elements on the periodic table. The major exception to this is hydrogen, which is located in group 1 but it is still a nonmetal.

(2) **Noble Gases.** The noble gasses are the least reactive family of elements on the periodic table. The noble gases have a filled outer shell of electrons and do not generally like to react with other compounds.

(3) **Group 17 or Group VIIA.** The halogens are column of elements that start with fluorine on the right side of the periodic table.

(4) **Boron, Silicon, Germanium, Arsenic, Antimony, Tellurium, Astatine.** These elements are classified as metalloids.

(5) **Period 4.** Calcium and bromine are both located in the fourth row of the periodic table. Rows are horizontal on the table. These rows are called periods.

(6) **Group 1 or group IA.** The alkali metals are the first column on the periodic table.

(7) Elements more closely share properties with the other elements in their family/group on the periodic table. Families/groups on the periodic table are the vertical columns that elements are found within. This is due to elements in the same family/group having a similar outer-most electron configuration.

(8) **31.** Hydrogen is in group 1. Aluminum is in group 13. Chlorine is in group 17. If you add $1+13+17$ you get 31.

(9) **Metals are located on the left side of the periodic table.** The left side is usually considered anything that is to the left of the metalloids on the table. The major exception to this is hydrogen, which is located in group 1 on the periodic table but hydrogen is not a metal.

(10) **Nonmetal.** Chlorine is located in group 17 and is found to the right of the metalloids making it a nonmetal.

IN THIS CHAPTER

» **Electrons and their orbitals**

» **Grasping the value of valence electrons**

» **Taking stock of electron configurations**

» **Dealing with quantum numbers**

» **Equating an electron's energy to light**

Chapter **6**

The Electron

The basics of an electron are pretty simple. Electrons are really small — basically the smallest thing that exists in the universe along with quarks and a few elementary particles. You already know from Chapter 4 that electrons are negatively charged. You might also be wondering how there can be an entire chapter dedicated to something so small and so negative, but as you read on you're going to discover just how much of chemistry is impacted by where electrons are found and how they interact with one another. Read on and pretty soon you'll be making sense of things like electron configurations, quantum numbers, and valence electrons.

Putting Electrons in Their Places: Electron Configurations

A neutral atom has the same number of electrons as protons (see Chapter 4), which equals the element's atomic number. After you know how many electrons an atom has, the next step is to figure out where those electrons live. Several schemes exist for depicting all this important information, but the *electron configuration* is a type of shorthand that captures much of the pertinent information.

Each numbered period (row) of the periodic table corresponds to a different *principal energy level*, with higher numbers indicating higher energy. Within each energy level, electrons can occupy different sublevels. Each sublevel is made up of different types of *orbitals*. Different types of orbitals have slightly different energy. Each orbital can hold up to two electrons, but electrons won't double up within an orbital unless no other open orbitals exist at the same energy level. Electrons fill up orbitals from the lowest energies to the highest.

There are four types of orbitals: s, p, d, and f:

» **s:** Period 1 consists of a single 1s orbital. A single electron in this orbital corresponds to the electron configuration of hydrogen, written as $1s^1$. The superscript written after the symbol for the orbital indicates how many electrons occupy that orbital. Filling the orbital with two electrons, $1s^2$, corresponds to the electron configuration of helium. Each higher principal energy level contains its own s orbital (2s, 3s, and so on), and these orbitals are the first to fill within those levels.

» **p:** In addition to s orbitals, principal energy levels 2 and higher contain p orbitals. There are three p orbitals at each level, accommodating a maximum of six electrons. Because the three p orbitals (also known as p_x, p_y, and p_z) have equal energy, they're each filled with a single electron before any receives a second electron. The elements in Periods 2 and 3 on the periodic table contain only s and p orbitals. The p orbitals of each energy level are filled only after the s orbital is filled.

» **d:** Period 4 and higher on the periodic table include d orbitals, of which there are five at each principal energy level, accommodating a maximum of ten electrons. The d orbital electrons are a major feature of the transition metals.

» **f:** Period 5 and higher include f orbitals, numbering seven at each level, accommodating a maximum of 14 electrons. The f orbital electrons are a hallmark of the lanthanides and the actinides (see the section "Examining the organization of the periodic table" in Chapter 5 for more on these rows).

TIP

Trying to visualize how electrons fill orbitals can get very confusing, so Figure 6-1 is a periodic table with the different orbitals put in place of the elements. It will help. A lot. To use the diagram, start at the upper left and read from left to right. When you reach the end of a row, go to the beginning of the next row down. Keep going until you reach your element.

As you can see in Figure 6-1, after you get to Period 4, the exact order in which you fill the energy levels can get a bit confusing. To keep things straight, the Aufbau filling diagram in Figure 6-2 is useful. To use the diagram, start at the bottom and work your way up, from the lowest arrows to the highest. For example, always start by filling 1s, then fill 2s, then 2p, then 3s, then 3p, then 4s, then 3d, and so on.

WARNING

Sadly, there are a few exceptions to the tidy picture presented by the Aufbau filling diagram. Copper, chromium, and palladium are notable examples. Without going into teeth-grinding detail, these exceptional electron configurations arise from situations in which electrons get transferred from their proper, Aufbau-filled orbitals to create half-filled or entirely filled sets of d orbitals; these half- and entirely filled states are more stable than the states produced by pure Aufbau-based filling.

To come up with a written electron configuration, you first determine how many electrons the atom in question actually has. Then you assign those electrons to orbitals, one electron at a time, from the lowest-energy orbitals to the highest. In a given type of orbital (like a 2p or 3d orbital, for example), you place two electrons within the same orbital only when there's no other choice. For example, suppose you want to find the electron configuration of carbon.

Carbon has six electrons (the same as the number of its protons, as its atomic number indicates), and it's in Period 2 of the periodic table. First, the s orbital of level 1 is filled. Then, the s orbital of level 2 is filled. These orbitals each accept two electrons, leaving two more with which to fill the p orbitals of level 2. Each of the remaining electrons would occupy a separate p orbital. You wind up with $1s^22s^22p^2$. (Only at oxygen, $1s^22s^22p^4$, would electrons begin to double up in the 2p orbitals, indicated by the superscript 4 on the p.)

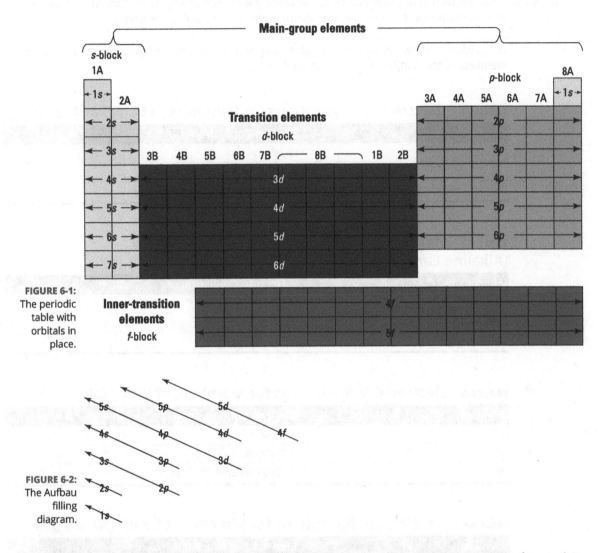

FIGURE 6-1:
The periodic table with orbitals in place.

FIGURE 6-2:
The Aufbau filling diagram.

TIP

Electron configurations can get a bit long to write. For this reason, you may see them written in a condensed form, like $[Ne]3s^23p^3$. This condensed configuration is the one for phosphorus. The expanded electron configuration for phosphorus is $1s^22s^22p^63s^23p^3$. To abbreviate the configuration, simply go backward in atomic numbers until you hit the nearest noble gas (which, in the case of phosphorus, is neon). The symbol for that noble gas, placed within brackets, becomes the new starting point for the configuration. Next, simply include the standard configuration for those electrons beyond the noble gas (which, in the case of phosphorus, includes two electrons in 3s and three electrons in 3p). The condensed form simply means that the atom's electron configuration is just like that of neon, with additional electrons filled into orbitals 3s and 3p as annotated.

REMEMBER

Ions have different electron configurations from their parent atoms because ions are created by gaining or losing electrons. Atoms tend to gain or lose electrons so they can achieve full valence octets, like those of the noble gases. Guess what! The resulting ions have precisely the same electron configurations as those noble gases. For example, by forming the Br⁻ anion, bromine achieves the same electron configuration as the noble gas krypton.

For quick reference, Tables 6-1 through 6-4 show the electron configurations of the first three members of the families IA, IIA, VIIA, and VIIIA.

TABLE 6-1 **Electron Configurations for Members of IA (Alkali Metals)**

Element	Electron Configuration
Li	$1s^22s^1$
Na	$1s^22s^22p^63s^1$
K	$1s^22s^22p^63s^23p^64s^1$

TABLE 6-2 **Electron Configurations for Members of IIA (Alkaline Earth Metals)**

Element	Electron Configuration
Be	$1s^22s^2$
Mg	$1s^22s^22p^63s^2$
Ca	$1s^22s^22p^63s^23p^64s^2$

TABLE 6-3 **Electron Configurations for Members of VIIA (Halogens)**

Element	Electron Configuration
F	$1s^22s^22p^5$
Cl	$1s^22s^22p^63s^23p^5$
Br	$1s^22s^22p^63s^23p^64s^23d^{10}4p^5$

TABLE 6-4 **Electron Configurations for Members of VIIIA (Noble Gases)**

Element	Electron Configuration
Ne	$1s^22s^22p^6$
Ar	$1s^22s^22p^63s^23p^6$
Kr	$1s^22s^22p^63s^23p^64s^23d^{10}4p^6$

Q. What is the electron configuration of titanium?

A. The electron configuration is $1s^22s^22p^63s^23p^64s^23d^2$. Titanium (Ti) is atomic number 22 and therefore has 22 electrons to match its 22 protons. These electrons fill orbitals from lowest to highest energy in the order shown by the Aufbau filling diagram in Figure 6-2. Note that the 3d orbitals fill only after the 4s orbitals have filled, so titanium has two valence electrons.

YOUR TURN

 1 What is the electron configuration of chlorine?

 2 What is the electron configuration of technetium?

 3 What is the electron configuration of iron?

4 What is the electron configuration of calcium?

5 What is the condensed electron configuration of bromine?

Quantifying Quantum Numbers

The arrangement and location of electrons is very important in determining the behavior of atoms. The more precisely you can determine where each electron is in an atom, the better you can predict how that atom will behave in any number of situations. The simple Bohr model (described in Chapter 4) was unable to explain observations made on complex atoms, so a more complex, highly mathematical model of atomic structure was developed to help account for the location of all electrons: the quantum mechanical model.

This model is based on *quantum theory*, which says matter also has properties associated with waves. According to quantum theory, knowing both the exact position and *velocity* (speed and direction) of an electron at the same time is impossible. This fact is known as the *uncertainty principle*. So scientists had to replace Bohr's orbits with *orbitals* (sometimes called *electron clouds*), volumes of space in which an electron is *likely* to be. In other words, certainty was replaced with probability.

The quantum mechanical model of the atom uses complex shapes of orbitals rather than Bohr's simple circular orbits. Without resorting to a lot of math (you're welcome), this section shows you some aspects of this newest model of the atom.

Four numbers, called *quantum numbers*, were introduced to describe the characteristics of electrons and their orbitals. You'll notice that they were named by totally top-rate techno-geeks:

>> Principal quantum number n

>> Angular momentum quantum number l

>> Magnetic quantum number m_l

>> Spin quantum number m_s

Table 6-5 summarizes the four quantum numbers. When the numbers are all put together, theoretical chemists have a pretty good description of the characteristics of a particular electron.

TABLE 6-5 **Summary of the Quantum Numbers**

Name	Symbol	Description	Allowed Values
Principal	n	Orbital energy	Positive integers (1, 2, 3, and so on)
Angular momentum	l	Orbital shape	Integers from 0 to $n - 1$
Magnetic	m_l	Orientation	Integers from $-l$ to 0 to $+l$
Spin	m_s	Electron spin	$-1/2$ or $+1/2$

The principal quantum number n

The principal quantum number n describes the average distance of the orbital from the nucleus — and the energy of the electron in an atom. It's really about the same as Bohr's energy-level numbers. It can have positive-integer (whole number) values: 1, 2, 3, 4, and so on. The larger the value of *n*, the higher the energy and the larger the orbital. Chemists sometimes call the orbitals *electron shells*.

The angular momentum quantum number *l*

The angular momentum quantum number *l* describes the shape of the orbital, and the shape is limited by the principal quantum number *n*: The angular momentum quantum number *l* can have integer values from 0 to *n* − 1. For example, if the *n* value is 3, three values are allowed for *l*: 0, 1, and 2.

REMEMBER

The value of *l* defines the shape of the orbital, and the value of *n* defines the size. Orbitals that have the same value of *n* but different values of *l* are called *subshells*. These subshells are given letters to help chemists distinguish them from each other. Table 6-6 shows the letters corresponding to the different values of *l*.

TABLE 6-6 **Letter Designation of the Subshells**

Value of l (Subshell)	Letter
0	s
1	p
2	d
3	f
4	g

When chemists describe one particular subshell in an atom, they can use both the *n* value and the subshell letter — 2p, 3d, and so on. Normally, a subshell value of 4 is the largest needed to describe a particular subshell. If chemists ever need a larger value, they can create subshell numbers and letters. Figure 6-3 shows the shapes of the s, p, and d orbitals.

Figure 6-3a has two s orbitals: one for energy level 1 (1s) and the other for energy level 2 (2s). S orbitals are spherical with the nucleus at the center. Notice that the 2s orbital is larger in diameter than the 1s orbital. In large atoms, the 1s orbital is nestled inside the 2s, just like the 2p is nestled inside the 3p.

Figure 6-3b shows the shapes of the p orbitals, and Figure 6-3c shows the shapes of the d orbitals. Notice that the shapes get progressively more complex.

The magnetic quantum number *m*ₗ

The magnetic quantum number m_l describes how the various orbitals are oriented in space. The value of m_l depends on the value of *l*. The values allowed are integers from −1 to 0 to +1. For example, if the value of *l* = 1 (p orbital; refer to Table 6-6), you can write three values for m_l: −1, 0, and +1. This means that there are three different p subshells for a particular orbital. The subshells have the same energy but different orientations in space.

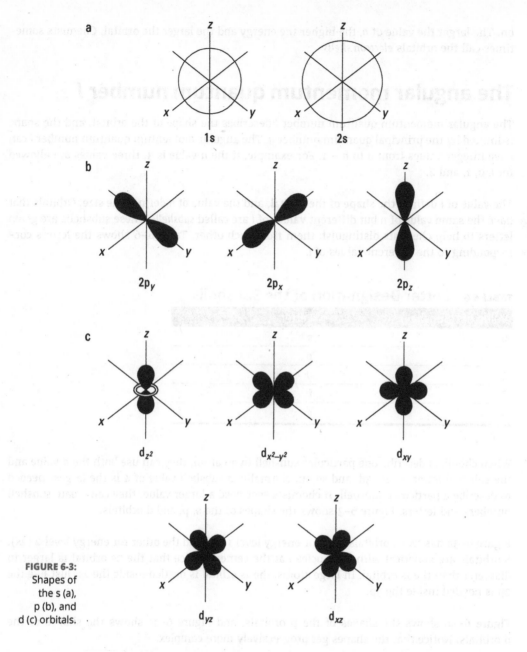

FIGURE 6-3:
Shapes of
the s (a),
p (b), and
d (c) orbitals.

Figure 6-3b shows how the p orbitals are oriented in space. Notice that the three p orbitals correspond to m_l values of -1, 0, and +1, oriented along the x, y, and z axes.

The spin quantum number m_s

The fourth and final quantum number is the spin quantum number m_s. This one you can think of as describing the direction the electron is spinning in a magnetic field — either clockwise or counterclockwise. Only two values are allowed for m_s: +1/2 or -1/2. Each subshell can have only two electrons, one with a spin of +1/2 and another with a spin of -1/2.

Put all the numbers together and whaddya get? (A pretty table)

We know. Quantum number stuff makes science nerds drool and normal people yawn. But, hey, sometime if the TV's on the blink and you've got some time to kill, take a peek at Table 6-7. You can check out the quantum numbers for each electron in the first two energy levels.

TABLE 6-7 **Quantum Numbers for the First Two Energy Levels**

n	l	Subshell Notation	m_l	m_s
1	0	1s	0	+1/2, −1/2
2	0	2s	0	+1/2, −1/2
	1	2p	−1	+1/2, −1/2
			0	+1/2, −1/2
			+1	+1/2, −1/2

Table 6-7 shows that energy level 1 ($n = 1$) has only an s orbital. It has no p orbital because an l value of 1 (p orbital) is not allowed. And notice that the 1s orbital can have only two electrons (m_s of +1/2 and −1/2). In fact, a maximum of only two electrons can be in any s orbital, whether it's 1s or 5s.

When you move from energy level 1 to energy level 2 ($n = 2$), both s and p orbitals can be present. If you write out the quantum numbers for energy level 3, you see s, p, and d orbitals. Each time you move higher in a major energy level, you add another orbital type.

Notice also that the 2p orbital has three subshells (m_l) (refer to Figure 6-3b) and that each holds a maximum of two electrons. The three 2p subshells can hold a maximum of six electrons.

The major energy levels have an energy difference (energy level 2 is higher in energy than energy level 1), but the energies of the different orbitals within an energy level also have differences. At energy level 2, both s and p orbitals are present. But the 2s is slightly lower in energy than the 2p. The three subshells of the 2p orbital have the same energy. Likewise, the five subshells of the d orbitals (refer to Figure 6-3c) have the same energy.

EXAMPLE

Q. If you have an electron in the 3p sublevel, what are all the possible values of each quantum number for that electron (n, l, ml, and ms)?

A. $n = 3$, $l = 1$, $m_l = -1, 0, +1$, and $m_s = +1/2, -1/2$.

Because your electron is in the third energy level, its principal quantum number, n, must be 3. Because it's in the p sublevel, the value of l can only be 1. In the p sublevel, ml has potential values from −1 to +1. This makes sense, because p sublevels have three orbitals. Finally, the electron could have a spin of +1/2 or −1/2, depending on its location in the orbital.

YOUR TURN

 6   How many m_l values are possible for an f sublevel?

7   If n is equal to 3, what are the possible values of l?

 8   What are the possible quantum numbers for an electron found in the 1s energy level?

Valence Electrons

The number of electrons found in an atom changes the reactivity of elements in predictable ways, based on how those electrons fill successive energy levels. Electrons in the highest energy level occupy the outermost shell of the atom and are called *valence electrons.* Valence electrons, which are involved in bonding (see Chapter 9), determine whether an element is reactive or unreactive. Because chemistry is really about the making and breaking of bonds, valence electrons are the most important particles for chemistry.

Chemists really only consider the electrons in the s and p orbitals in the energy level that is currently being filled as valence electrons. In the electron configuration for oxygen, $1s^2 2s^2 2p^4$, energy level 1 is filled, and in energy level 2, two electrons are in the 2s orbital and four electrons are in the 2p orbital for a total of 6 valence electrons. Those valence electrons are the ones lost, gained, or shared.

Look at the electron configurations for the alkali metals shown in the earlier section "Putting Electrons in Their Places: Electron Configurations" (Table 6-1). In lithium, energy level 1 is filled, and a single electron is in the 2s orbital. In sodium, energy levels 1 and 2 are filled, and a single electron is in energy level 3. All these elements have one valence electron in an s orbital. The alkaline earth elements (Table 6-2) each have two valence electrons. The halogens (Table 6-3) each have seven valence electrons (in s and p orbitals — d orbitals don't count), and the noble gases (Table 6-4) each have eight valence electrons, which fill their valence orbitals.

TIP

The Roman numeral numbering system for the groups in the periodic table can help you predict the valence number for any element. The Roman numeral indicates the valence number of each element family. In addition, most Group B metals, the transition metals, are said to have two valence electrons (with minor exceptions — which you can completely ignore — due to unpredictable electron filling due to stability reasons). Figure 6-4 shows this broken down by column in the periodic table. It is a good general reference to use; however, there are a few exceptions. Outside of the transition metals and the rare exception as mentioned above, the other major exception to this is the element helium. Helium is listed on the periodic table in group VIIIA, indicating it should have 8 valence electrons like all of the other noble gases but in reality it only has 2 valence electrons. This does make sense when you consider that helium only has 2 electrons and both of those electrons are found in the 1s orbital.

FIGURE 6-4:
Number of valence electrons for each family on the periodic table.

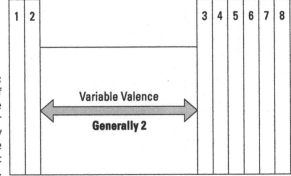

Atoms are most stable when their valence shells are completely filled with electrons. Chemistry happens because atoms attempt to fill the valence shells. Alkali metals in Group IA of the periodic table are so reactive because they need only to give up one electron to have a completely filled valence shell. Halogens in Group VIIA are so reactive because they need only to acquire one electron to have a completely filled valence shell. The elements within a group tend to have the same number of valence electrons and for that reason tend to have similar chemical properties. Elements in Groups IA and IIA tend to react strongly with elements in Group VIIA. Group B elements tend to react less strongly.

EXAMPLE

Q. How many valence electrons do the following elements have?

(a) Carbon

(b) Hydrogen

(c) Neon

(d) Iron

A. Use the elements location on the periodic table to determine the number of valence electrons. If the elements are in the A group then the Roman numeral abbreviation is equal to the number of valence electrons. If the element is in the transition metal B group you can assume the number of valence electrons are 2.

  a. Carbon has 4 valence electrons because it is in the IVA group.

  b. Hydrogen has 1 valence electron because it is in the IA group.

  c. Neon has 8 valence electrons VIIIA group.

  d. Iron has 2 valence electrons because it is a transition metal in the B group.

---

 9 How many valence electrons do the following elements have?

YOUR TURN

(a) Potassium

(b) Vanadium

(c) Oxygen

(d) Argon

(e) I

(f) He

(g) Ca

(h) Si

(i) Ge

 10 Why are valence electrons important?

Measuring the Amount of Energy (or Light) an Excited Electron Emits

To begin to grasp the nature of electrons, examining the nature of light is necessary. Visible light, X-rays, microwaves, radio waves, and so on are all various forms of electromagnetic radiation. *Electromagnetic radiation*, sometimes referred to as *radiant energy*, carries energy through space. If that space is a vacuum, all types of electromagnetic radiation travel at the speed of light, denoted by c, 3.00×10^8 m/s. These waves of electromagnetic radiation also have the following three properties (see Figure 6-5):

>> **Amplitude:** The *amplitude* of a wave is the height of the wave from the midpoint or baseline of the wave to its peak (highpoint). Think of it as a measure of the strength of the

electromagnetic radiation. For the visible part of the spectrum, you interpret the amplitude as the intensity or brightness of the light.

>> **Wavelength:** *Wavelength,* λ (lambda), is the distance between two identical adjacent points of a wave, such as peaks (high points) or troughs (low points). You may express the wavelength in any unit of length, although many times chemists choose a specific unit for a specific type of electromagnetic radiation (meters for radio and TV waves or angstroms [10^{-10} m] for X-rays).

>> **Frequency:** The *frequency,* ν (nu), is the number of waves that pass a given point during a specified time interval. Frequency has units of cycles per time, but chemists generally accept the cycles part as understood and omit it, leaving frequency expressed as reciprocal time (1/time). The SI base unit for time is the second, so frequency is measured in cycles per second and has the units of 1/s or s^{-1}.

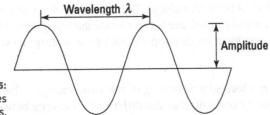

FIGURE 6-5:
Properties
of waves.

Because these electromagnetic waves are all traveling at the speed of light, the wavelength and the frequency of the light have an inverse relationship. If the wavelength is short, many waves pass a reference point per given amount of time and the frequency is high. If the wavelength is long, fewer waves pass the reference point during the same amount of time and the frequency is lower. You can see the relationship between the wavelength and frequency with this equation:

$$\lambda \nu = c$$

where λ is the wavelength (in meters), ν is the frequency (in s^{-1}), and c is the speed of light (in m/s).

In working problems with this direct relationship between wavelength and frequency, expressing the wavelength, frequency, and speed of light in the units of meters, cycles per second, and meters per second is easiest.

TIP

Figure 6-6 shows the electromagnetic spectrum in order of increasing wavelength. Notice that gamma rays have the shortest wavelength (highest frequency) and radio waves have the longest wavelength (lowest frequency). Also notice that the visible spectrum (the part of the electromagnetic spectrum that you can detect with your eyes) is a relatively small part of the entire electromagnetic spectrum. Scientists generally express the wavelength of the visible part of the spectrum in nanometers (10^{-9} meters), and the wavelength ranges from violet at about 400 nm to red at 750 nm.

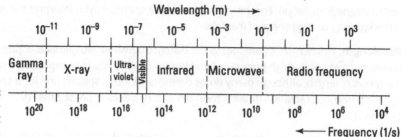

FIGURE 6-6: The electro-magnetic spectrum.

As mentioned previously, frequency is usually in cycles per second (or simply s⁻¹). This unit of cycles per second is a *hertz* (Hz), and thus frequency can be in hertz or any combination of hertz with SI prefixes (such as kilohertz). Chemists commonly express the frequency of radio stations in this fashion. Thus, 900 on the AM dial would be 900 kHz (kilohertz), or 900,000 s⁻¹.

Now where do electrons fit into all this? Interestingly, light has properties of both a particle and a wave. This concept is sometimes called the *wave–particle duality of light*. The upshot is that it's possible to measure the energy of an electron simply by measuring the wavelength of light it emits.

REMEMBER Electrons can jump between energy levels by absorbing or releasing energy. When an electron absorbs an amount of energy exactly equivalent to the difference in energy between levels (a *quantum* of energy), the electron jumps to the higher energy level. This jump in energy creates an *excited state*. Excited states don't last forever; the lower-energy *ground state* is more stable. Excited electrons tend to release amounts of energy equivalent to the difference between energy levels and thereby return to ground state. The discrete particles of light that correspond to quanta of energy are called *photons*.

REMEMBER To perform these calculations, you need the $\lambda v = c$ equation from earlier and the equation here:

$$\text{Energy}\,(E)=\text{Planck's constant}\,(h)\times\text{Frequency}\,(v)$$

where $h = 6.626 \times 10^{-34}$ J·s. Here, J stands for *joules*, the SI unit of energy. Frequency is expressed in hertz (Hz), where 1 hertz is 1 inverse second (s⁻¹). So multiplying a frequency by Planck's constant, h, yields *joules*, the units of energy.

Q. A hydrogen lamp emits blue light at a wavelength of 487 nanometers (nm). What is the frequency in hertz?

EXAMPLE

A. The frequency is 6.16×10^{14} s⁻¹. Wavelength and the speed of light are known quantities here; you must solve for frequency, $v = c/\lambda$. Don't forget to convert nanometers to meters (1 nm = 10⁻⁹ m), because the speed of light is given in meters per second:

$$v = \frac{c}{\lambda} = \frac{\left(3.00\times10^{8}\,\text{m/s}\right)}{\left(487\times10^{-9}\,\text{m}\right)} = 6.16\times10^{14}\,1/\text{s}$$

Q. What is the energy of emitted light with a frequency of 6.88×10^{14} Hz?

A. The energy is 4.56×10^{-19} J. You must recall here that Hz = s^{-1}.

$$E = h\nu = \left(6.626 \times 10^{-34} \text{ J} \cdot \text{s}\right)\left(6.88 \times 10^{14} \text{ 1/s}\right)$$
$$= 4.56 \times 10^{-19} \frac{\text{J} \cdot \text{s}}{\text{s}}$$
$$= 4.56 \times 10^{-19} \text{ J}$$

YOUR TURN

11 What is the frequency of a beam of light with wavelength 2.57×10^2 m?

12 One useful, high-energy emission from excited electrons is the X-ray. What is the wavelength (in meters) of an X-ray with a frequency of $\nu = 8.72 \times 10^{17}$ Hz?

---

13 What is the energy of a wave if it has a frequency of 3.91×10^8 Hz?

Practice Questions Answers and Explanations

(1) **$1s^2 2s^2 2p^6 3s^2 3p^5$.** Chlorine (Cl) has 17 electrons.

(2) **$1s^2 2s^2 2p^6 3s^2 3p^6 4s^2 3d^{10} 4p^6 5s^2 4d^5$.** Technetium (Tc) has 43 electrons, which fill according to the Aufbau diagram.

(3) **$1s^2 2s^2 2p^6 3s^2 3p^6 4s^2 3d^6$.** Iron (Fe) has 26 electrons, which fill according to the Aufbau diagram.

(4) **$1s^2 2s^2 2p^6 3s^2 3p^6 4s^2$.** Calcium (Ca) has 20 electrons, which fill according to the Aufbau diagram.

(5) **$[Ar]4s^2 3d^{10} 4p^5$.** Condense the configuration for bromine (Br) by summarizing all the previous entirely filled rows with a single noble gas in brackets. The expanded electron configuration of bromine is $1s^2 2s^2 2p^6 3s^2 3p^6 4s^2 3d^{10} 4p^5$. Bromine is in Row 4, so consolidate the electron configurations of Periods 1 through 3 as [Ar], because argon is the last element in Row 3. Then add to [Ar] the configuration of the remaining electrons, which fill $4s^2 3d^{10} 4p^5$.

(6) **Seven.** An f sublevel corresponds to an l value of 3. The possible values of m_l for an l of 3 are −3 to + 3 with 0 included in the middle. This equates to −3, −2, −1, 0, 1, 2, and 3 as potential values.

(7) **0, 1, and 2.** The possible values of l are equal to 0 through $n − 1$. So if $n = 3$, then the possible values are 0, 1, 2.

(8) **$n = 1$, $l = 0$, $m_l = 0$, and $m_s = +1/2$, $−1/2$.** The electron is in the first energy level, so its principal quantum number, n, must be 1. Because the electron is in the s sublevel, the value of l can only be 0. The potential values for m_l can range from −1 to +1. In this case, the electron is in the s sublevel, so m_l has only one orbital, with a value of 0. Finally, the electron could have a spin of +1/2 or −1/2, depending on its location in the s orbital.

(9) Use the elements location on the periodic table to determine the number of valence electrons. If the elements are in the A group then the Roman numeral abbreviation is equal to the number of valence electrons. If the element is in the transition metal group you can assume the number of valence electrons are 2.

   a. 1

   b. 2

   c. 6

   d. 8

   e. 7

   f. 2 (Helium is the one exception. It only has 2 electrons total, not 8 like the noble gas family.)

   g. 2

   h. 4

   i. 4

(10) Valence electrons are important because they occupy the highest energy level in the outermost shell of the atom. As a result, valence electrons are the electrons that see the most action, as in forming bonds or being gained or lost to form ions. The number of valence electrons in an atom largely determines the chemical reactivity of that atom and many other properties.

(11) $1.17 \times 10^6 \text{ s}^{-1}$.

$$v = \frac{c}{\lambda} = \frac{\left(3.0 \times 10^8 \text{ m/s}\right)}{\left(2.57 \times 10^2 \text{ m}\right)} = 1.17 \times 10^6 \text{ 1/s}$$

(12) $3.44 \times 10^{-10} \text{ m}$.

$$\lambda = \frac{c}{v} = \frac{\left(3.0 \times 10^8 \text{ m/s}\right)}{\left(8.72 \times 10^{17} \text{ 1/s}\right)} = 3.44 \times 10^{-10} \text{ m}$$

(13) $2.59 \times 10^{-25} \text{ J}$.

$$E = hv = \left(6.626 \times 10^{-34} \text{ J} \cdot \text{s}\right)\left(3.91 \times 10^8 \text{ 1/s}\right) = 2.59 \times 10^{-25} \text{ J}$$

If you're ready to test your skills a bit more, take the following chapter quiz that incorporates all the chapter topics.

Whaddya Know? Chapter 6 Quiz

Ready for a quiz? The 15 questions in this section will test the skills you learned in this chapter. When you're done, check out the section that follows for answers and explanations.

1. What is the electron configuration of sodium?

2. What is the electron configuration of titanium?

3. What is the shortened electron configuration of silicon?

4. What is the shortened electron configuration of barium?

5. What are the possible values of l if the value of n is equal to 2?

6. What are the possible values of m_l are there if you have an l value of 2?

7. What are the possible n, l, and m_l values for an electron in the 4d sublevel?

8. How many electrons are present in an atom that has a filled 3rd energy level ($n = 3$)?

9. What are the possible n, l, and m_l values for the highest energy electron in chlorine?

10. What are the possible n, l, and m_l values for the highest energy electron in calcium?

11. How many valence electrons do the following elements have:

    a. lithium

    b. Ba

    c. Cobalt

    d. I

12. What is the wavelength of a beam of light that has a frequency of 3.2×10^{15} Hz?

13. What type of electromagnetic radiation has a wavelength of 4.5×10^{-9} m? Hint: Refer to Figure 6–6 to answer this.

14. What is the wavelength of a photon of light that has an energy of 8.23×10^{-24} J?

15. What is the relationship between frequency and wavelength? If you increase the wavelength of a wave, what would happen to the frequency of the wave?

Answers to Chapter 6 Quiz

(1) **$1s^2 2s^2 2p^6 3s^1$.** Sodium has 11 electrons. That means you will need to have an electron configuration that adds up to a total of 11 electrons. Sodium fills according to the Aufbau diagram. You can also reference Figure 6-1 and locate sodium on it. You will see that sodium is located in the 3s sublevel. That means the highest energy electrons that sodium has are in 3s. You can then start from the beginning of the table and trace along from left to right until you get to sodium. You will pass through 1s, 2s, 2p and finally stop at 3s. This is your order of electron filling for sodium. Once you know this add electrons to each of the sublevels. Each s can have 2, and the p can have a total of 6. The 3s sublevel only has 1 electron though because sodium only has 11 electrons. If you add up the total electrons in 1s, 2s, and 2p you should get 10 electrons. This only leaves 1 to go into 3s.

(2) **$1s^2 2s^2 2p^6 3s^2 3p^6 4s^2 3d^2$.** Titanium has 22 electrons and fills according to the Aufbau diagram. If you prefer to use Figure 6-1 you'll see that titanium is located in the 3d sublevel.

(3) **[Ne]$3s^2 3p^2$.** Condense the configuration for silicon (Si) by summarizing all the previous entirely filled rows with a single noble gas in brackets. The expanded electron configuration of bromine is $1s^2 2s^2 2p^6 3s^2 3p^2$. Silicon is in Row 3, so consolidate the electron configurations of Periods 1 through 2 as [Ne], because neon is the last element in Row 2. Then add to [Ne] the configuration of the remaining electrons, which fill $3s^2 3p^2$.

(4) **[Xe]$6s^2$.** Condense the configuration for barium (Ba) by summarizing all the previous entirely filled rows with a single noble gas in brackets. The expanded electron configuration of barium is $1s^2 2s^2 2p^6 3s^2 3p^6 4s^2 3d^{10} 4p^6 5s^2 4d^{10} 5p^6 6s^2$. Silicon is in Row 6, so consolidate the electron configurations of Periods 1 through 5 as [Xe], because neon is the last element in Row 5. Then add to [Xe] the configuration of the remaining electrons, which only fill $6s^2$.

(5) **l could be equal to 0 or 1.** The possible values for l are always 0 through $n-1$. In this case if n is equal to 2 you would plug that into $n-1$ to determine the maximum value of l. As $2-1=1$ that makes the values possible as only 0 or 1.

(6) **−2, −1, 0, 1, 2.** The possible values of m_l are equal to $-l$ to $+l$. So in this case if l is equal to 2 that makes the possible values of m_l equal to −2, −1, 0, 1, 2. The most common error when dealing with m_l is to forget that 0 must always be accounted for when going from $-l$ to $+l$ so make sure you put it in there.

(7) **$n = 4$, $l = 2$, $m_l = -2, -1, 0, 1, 2$.** An electron in the 4d sublevel is in the 4th principal energy level. This means $n = 4$. The possible values of l based on the n value of 4 are 0, 1, 2, 3. In this case a d sublevel corresponds to an l value of 2. Remember, for the l quantum number you can determine its value based on the sublevel of where the electron is located. For an s sublevel $l = 0$, for a p sublevel $l = 1$, for a d sublevel $l = 2$, for an f sublevel $l = 3$. For an l value of 2 the possible m_l values are −2, −1, 0, 1, 2. Without knowing which orbital the electron is in you cannot localize which of the possible m_l values are exactly where the electron is.

(8) **18 electrons.** You can determine this because you know the 3rd energy level has s, p, and d sublevels present. An s sublevel holds 2 electrons, a p sublevel holds 6 electrons, and a d sublevel holds 10 electrons. To find the total number just add $2 + 6 + 10 = 18$ total electrons.

(9) **$n = 3$, $l = 1$, $m_l = -1, 0, 1$.** The highest energy electron in chlorine is in the 3p sublevel. This means the principal energy level $n = 3$. The l value is equal to 1. The possible values of m_l are −1, 0, 1.

(10) $n = 4, l = 0, m_l = 0$. The highest energy electron in calcium is in the 4s sublevel. This means the principal energy level $n = 4$. The l value is equal to 0. The possible values of m_l are 0.

(11) a. **1.** Lithium is in group IA, giving it 1 valence electron.

b. **2.** Barium is in group IIA, giving it 2 valence electrons.

c. **2.** Cobalt is in the group VIIIB. Since it is a B group metal, a transition metal, we can assume it has 2 valence electrons.

d. **7.** Iodine is in group VIIA giving it 2 valence electrons.

(12) 9.4×10^{-8} **m.** The units given to you for frequency are Hz. Remember that Hz can also be expressed as 1/s. When solving problems involving wavelength, frequency, energy, and the speed of light it is usually helpful to make that change. $\lambda = \dfrac{c}{v} = \dfrac{\left(3.0 \times 10^8 \text{ m/s}\right)}{\left(3.2 \times 10^{15} \text{ 1/s}\right)} = 9.4 \times 10^{-8}$ m

(13) **X-ray.** Looking at Figure 6-6 you can see that any electromagnetic radiation that has a wavelength of 10^{-9} is classified as X-ray radiation. Make sure you don't confuse frequency and wavelength on the chart. Pay attention to the units to help you on this.

(14) 2.4×10^{-2} **m.** To solve this problem you first must use the energy value given to you and convert that to frequency. Once you have determined the frequency you can plug that into the equation relating frequency and wavelength to the speed of light solving for wavelength.

$$E = 8.23 \times 10^{-24} \text{ J}$$
$$E = hv$$
$$v = \frac{E}{h}$$
$$v = \frac{\left(8.23 \times 10^{-24} \text{ J}\right)}{\left(6.626 \times 10^{-34} \text{ J} \cdot \text{s}\right)}$$
$$v = 1.24 \times 10^{10} \text{ s}^{-1}$$
$$c = \lambda v$$
$$\frac{c}{v} = \lambda$$
$$\frac{\left(3.0 \times 10^8 \text{ m/s}\right)}{\left(1.24 \times 10^{10} \text{ s}^{-1}\right)} = 2.4 \times 10^{-2} \text{ m}$$

(15) **Frequency and wavelength have an inverse relationship.** As the frequency of a wave increases the wavelength of that wave decreases. If you were to increase the wavelength of a wave, then the frequency of that wave would decrease.

IN THIS CHAPTER

» **Nuclear charge and how it impacts periodic trends**

» **Studying the size of atoms and their atomic radius**

» **Removing electrons based on ionization energy**

» **Investigating electronegativity**

Chapter 7

Periodic Trends

T he whole point of the periodic table, aside from providing interior decoration for chemistry classrooms, is to help predict and explain the properties of the elements. These properties change as the number of protons and electrons in an element change. Thankfully, as you may have noticed, the periodic table is arranged by the number of protons an element has. This makes it easy to see trends in how the properties of elements change as the number of protons or electrons change. These trends can be used to predict many different properties of an element based solely on its location on the periodic table.

In this chapter you will see how these trends impact an atom's size, desire for electrons, and other characteristics.

Nuclear Charge and Atomic Radius

When working with atoms and their parts, understanding the different sizes and their relationship to each other is essential. The simplest way of measuring an atom's size is called the atomic radius. The nucleus is tiny; the electrons in their energy levels around the nucleus really determine the boundaries of the atom. The occupied energy levels of the atom basically determine the atom's size and thus its radius. Effective nuclear charge has an effect on the electron and thus on the size of the atom, so it is something you should read on to discover before getting too far into the idea of atomic radius.

Comprehending effective nuclear charge

The nucleus of an atom contains protons and neutrons. As you saw in Chapter 4, the number of protons in a nucleus is the atomic number. Neutrons have no charge, whereas protons have a positive charge. If you add together the positive charges of all those protons, you have what's called the nuclear charge. This positive charge in the nucleus attracts electrons, which are negatively charged. (Opposite charges attract, and things that are similarly charged repel.) If the atom contains only a single electron, the attraction force is straightforward. However, when more than one electron is present (which is the case for almost every element), the situation becomes more complicated.

The negative charge of one electron interferes with the attractive force of the nucleus to other electrons. Electrons occupy "energy levels" that form a sort of layered shell around the nucleus. When electrons closer to the nucleus come between another electron and the nucleus, they cause a greater interference.

Therefore, electrons that are contained in the same energy level produce a minimum interference on each other, but the core electrons (inner-shell electrons) partially shield the outer electrons from the attraction of the positive charge in the nucleus, the nuclear charge. This shielding affects all electrons in an atom, but for trends in size, the shielding of the valence electrons (outermost electrons) is the key. Those outermost electrons are the electrons that determine the size of the atomic radius. If shielding lessens the attractive force of the nucleus to the valence electrons, they're free to move a little bit farther from the nucleus, and the atomic diameter increases.

The core electrons screen the nuclear charge from the valence electrons. The actual charge attracting the valence electrons is called the effective nuclear charge. The effective nuclear charge is the nuclear charge minus the shielding effect of the core electrons. The shielding effect is essentially the same for any period on the periodic table, but when going down a family (group), the effect increases as you move toward the bottom of the column. The greater the shielding, the less the effective nuclear charge, allowing the valence electrons to move a little farther away from the nucleus. This increases the size of the atom.

Atomic radius

The *atomic radius* is the distance from the center of the nucleus to the valence energy level. It's how chemists describe the size of the atom or ion. It's important in determining the type of crystal lattice formed, solubility, and so on.

On the periodic table, atomic radii (the plural of radius) increase from top to bottom within a family and increase from right to left within a period, as shown in Figure 7-1.

When you move down a family, the atomic radii of the atoms increase, making each atom significantly larger than the one above it. This increase in size is the result of electrons occupying energy levels of increasing distance from the nucleus, increasing n values. The nuclear charge increases as you go down a family (increased numbers of protons in the nucleus), but an increasing number of core electrons shield the valence electrons from the charge in the nucleus. The effective nuclear charge on the valence electrons is thus decreased, allowing the

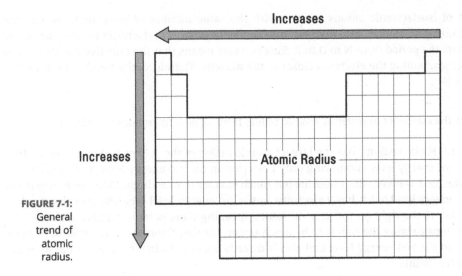

FIGURE 7-1: General trend of atomic radius.

Increases

Increases

Atomic Radius

electrons to move slightly away from the nucleus and countering the nuclear charge. Therefore, simply adding an energy level predominates in determining the size of the atomic radii.

Understanding the size changes as you move across a period isn't so simple. Unlike atoms in families, which add energy levels, atoms in periods add electrons to the same energy level. Because the added electrons don't come between other electrons and the nucleus, causing interference, you may think that the size of the atoms would remain basically constant. However, as you go from one element to the next, more protons are added to the nucleus, increasing the nuclear charge. The shielding effect of the core electrons is essentially constant as you move from left to right within a period, so the effective nuclear charge increases slightly with each move. This increased effective nuclear charge pulls the atom's electrons, especially the valence electrons, slightly closer to the nucleus, causing the atomic radii (size) of the atoms to decrease slightly when moving from left to right in a period.

Tracing tendencies of ionic radii

Atoms are neutral, but they may gain or lose electrons to form ions (atoms or groups of atoms that have an overall charge). To keep things simple, this chapter discusses only monatomic ions — ions that have only one atom. If an atom loses electrons, the chemical species that is left has more protons (more positive charges) than electrons (negative charges) and therefore has an overall positive charge. An ion with a positive charge is a cation. A cation has one positive charge for each electron lost.

A cation's radius size (its ionic radii) is smaller than an un-ionized atom. The loss of electrons leads to a decrease in the size when compared to the neutral atom because, at least for a representative element, the ion loses an entire energy level. Thus, a potassium ion, $K+$, is smaller than a potassium atom. The more electrons removed, the greater the decrease in radius becomes. Therefore, you observe the trend in radii of $Fe > Fe^{2+} > Fe^{3+}$. When an atom gains electrons, it produces an anion, which has a larger ionic radius than the original atom. You observe this trend because the effective nuclear charge (determined by the number of protons) is the same in both the atom and the anion but is spread over an increasing number of electrons. The more electrons added, the greater the increase in size. For oxygen and a couple of its anions, you see the following trend in radii: $O < O^- < O^{2-}$.

For a set of isoelectronic anions (anions with the same number of electrons), such as the anions of nitrogen, oxygen, and fluorine, the number of protons in the nucleus increases as you move across the period from N to O to F. This increase means that the effective nuclear charge also increases, pulling the electrons closer to the nucleus. Therefore, the trend in ionic radii is $F^- < O^{2-} < N^{3-}$.

Q. List the elements in order of atomic radius from largest to smallest: F, Na, Cl.

EXAMPLE **A.** Na, Cl, F. Na, sodium, has the largest atomic radius of the 3. This is due to it being in the 3rd energy level and having only 1 electron in the 3rd energy level. Chlorine has a radius that is bigger than fluorine but much smaller than sodium. Chlorine is also in the 3rd energy level but it has 7 electrons in the 3rd energy level (visually it is also much further to the right on the periodic table), meaning there is more repulsion among the electrons causing the radius to be much larger. Finally, fluorine is the smallest of the 3. It is in the 2nd energy level and also is directly above chlorine, meaning it has an even smaller radius.

YOUR
TURN

 List the elements in order of atomic radius from largest to smallest: Rb, Te, O.

 What would have the larger radius: S^{2-} or S?

Eyeing Trends in Ionization Energies

Ionization energy is the energy required to remove an electron from a gaseous atom in its ground state. Removing an electron from an atom always takes energy, so the ionization energy is always endothermic (requiring energy). The removal of a second electron requires even more energy (the second ionization energy), the removal of a third electron requires the third ionization energy, and so forth.

The following sections describe some of the trends in ionization energies that you may encounter.

Noting an increase in sequential energy

The amount of energy required for sequential ionization energies (removing more than one electron from an atom) increases with the number of electrons removed. You encounter this in the formation of cations of a +2 or +3 charge, corresponding to the loss of 2 and 3 electrons, respectively. This increase is due to the attractive force between the increasing positive charge of the cation and the electron that is in the process of being removed. Unfortunately, this increase isn't simply linear.

The general relationship between the value of an element's ionization energy and the position of the element on the periodic table is essentially the reverse of the atomic radii. In general, it can be said that ionization energy increases as you move from left to right on the periodic table and increases as you move up a family/group on the periodic table, as shown in Figure 7-2.

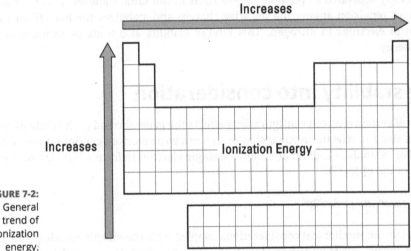

FIGURE 7-2:
General trend of ionization energy.

Small atoms have high ionization energies, and large atoms have low ionization energies. The size increase of an atom near the bottom of a column on the periodic table means that its outer electrons are further from the nucleus. The greater the distance between the positively charged protons and the negatively charged electrons, the weaker the attraction of the nucleus for the electrons (something physicists call the inverse square law). A weaker attraction holding electrons in an energy level means that you can use less energy to overcome the attractive force, which results in a lower ionization energy for the electron. For example, much more energy is required to remove an electron from lithium than from cesium (both alkali metals), because the electron being removed from lithium is much closer to the nucleus and therefore is held more strongly.

Within a period (a row on the periodic table), the ionization energy increases from left to right because the effective nuclear charge is increasing and therefore a greater attractive force must be overcome. However, this trend in ionization energy is more complicated than a simple increase in the effective nuclear charge.

The trends in ionization energy suggest that elements on the left side of the periodic table, especially the lower left, form cations (lose electrons) more readily than the elements on the right. The elements with lower ionization energies are the metals, and the elements with higher ionization energies are the nonmetals.

For example, if you examine the elements from lithium, Li, to neon, Ne, you can see the expected increase in ionization energies from left to right on the periodic table. However, the trend isn't linear; peaks occur at beryllium (Be) and nitrogen (N). Why are the ionization energy values for beryllium and nitrogen higher than expected? If you examine the electron configuration of beryllium, you can see that it is $1s^2 2s^2$ (see Chapter 6 for a discussion of electron configurations). The valence shell of beryllium has a filled 2s subshell. The filling of a subshell leads to additional stability and therefore requires more energy to pull the electron away.

In the case of nitrogen, the 2p sublevel is half-filled. The electrons are in different orbitals and are as widely separated as possible for electrons in the same sublevel. This wide separation minimizes repulsion among the negative charges and stabilizes the half-filled sublevel arrangement of electrons in nitrogen. This kind of stability also leads to an increase in the ionization energy.

Taking stability into consideration

The stability of certain electron configurations affects the general trend in the ionization energies based on size and effective nuclear charge. You can write electron configurations of cations to see this. For example, a sodium atom loses a single electron to form a Na^+ cation. The electron configuration of Na^+ is

Sodium cation (Na^+) $1s^2 2s^2 2p^6$

Because of quantum mechanical considerations coupled with the effective nuclear charge, the electrons are always lost from the level with the highest principal quantum number. For the representative elements, the last electron to enter is the first electron lost. In the case of the transition metals, the first electron lost is not the last electron to enter. If you use iron, Fe, for example, you can examine the electron configuration of the element and the configurations of the two common iron cations, Fe^{2+} and Fe^{3+}.

Fe $1s^2 2s^2 2p^6 3s^2 3p^6 4s^2 3d^6$

Fe^{2+} $1s^2 2s^2 2p^6 3s^2 3p^6 3d^6$

Fe^{3+} $1s^2 2s^2 2p^6 3s^2 3p^6 3d^5$

The formation of the iron(II) cation, Fe^{2+}, involves the loss of the two 4s electrons because these two electrons have a higher principal quantum number than the last electrons to enter (the 3d electrons). This factor explains why many of the transition metals have a stable 2+ ion. The iron 3d electrons are not affected until after the removal of the 4s electrons. And this removal deletes energy level 4, making the ions more stable because they have a lower overall energy.

Considering a few exceptions to the rule

For any element, ionization energy increases as an increasing number of electrons are removed. However, some elements do react differently in certain levels. The second ionization energy of sodium, the third ionization energy of magnesium, and the fourth ionization energy of aluminum are significantly higher than you would expect based on their proceeding ionization energies. Apparently, something else is going on. If you examine the electron configurations of these three elements, you find

Na	$1s^2 2s^2 2p^6 3s^1$
Mg	$1s^2 2s^2 2p^6 3s^2$
Al	$1s^2 2s^2 2p^6 3s^2 3p^1$

The first ionization energy of sodium, the first two ionization energies of magnesium, and the first three ionization energies of aluminum leave you with the following electron configurations:

Na^+	$1s^2 2s^2 2p^6$
Mg^{2+}	$1s^2 2s^2 2p^6$
Al^{3+}	$1s^2 2s^2 2p^6$

For these three elements, removal of an additional electron involves removal from the electron configuration $1s^2 2s^2 2p^6$, which contains only core electrons (the valence electrons having been already removed in the formation of the cation). This factor is at the root of all three high ionization energy values for the preceding stable ions: The removal of a core electron requires significantly more energy than the removal of valence electrons. In normal chemical reactions, sufficient energy is available to remove valence electrons but not core electrons. Thus, the simple removal of the valence electrons forms common cations of the representative elements.

Q. List the elements in order from highest ionization energy to lowest energy: Se, K, Br.

EXAMPLE

A. Br, Se, K. All of these elements are in the 4th period so the thing to focus on is how far they are to the right on the periodic table. Bromine has the highest ionization energy because it is furthest to the right on the periodic table. Bromine is very close to having a filled outer energy level. This means it very much does not want to give up an electron; instead, it very much wants to gain an electron, making its ionization energy quite high. Selenium is somewhat close to having a filled outer energy level, needing only 2 electrons, so it has a relatively high ionization energy as well. Potassium, on the other hand, needs to simply lose one electron and then it would move down to the 3rd energy level, in regard to electron configuration, and have a filled outer energy level. Sodium wants to get rid of that electron so its ionization energy is very low.

3 List the elements in order from lowest ionization energy to highest ionization energy: O, Na, Li.

4 Why do noble gases have the highest ionization energy?

Attracting Electrons: Electronegativities

Electronegativity is the strength an atom has to attract a bonding pair of electrons to itself. The larger the value of the electronegativity, the greater the atom's strength to attract a bonding pair of electrons. Figure 7–3 shows the electronegativity values of the various elements below each element symbol on the periodic table. Notice that, with a few exceptions, the electronegativities increase from left to right, in a period, and decrease from top to bottom, in a family.

This can be explained much the same as ionization energy. Smaller elements tend have a greater electronegativity than bigger elements. This can be explained by the fact that smaller elements have less shielding between their outermost electrons and the effective nuclear charge from nucleus. As you move down a group and increase the size of an atom, the nuclear charge increases due to an increase in atomic number, but this is overshadowed by there being more energy levels with electrons shielding the nuclear charge from the outer electrons.

When moving across a period from left to right, the electronegativity increases. This increase is seen because all of the elements in that period have the same number of energy levels present. In staying in the same period, there is no increase in shielding for any elements within the same period. As you move from left to right on the periodic table, though, atomic number increases, meaning the effective positive nuclear charge of the elements increases, which increases the ability of the atom to attract a negative electron. This increase is countered somewhat by there also being more electrons present in the atom, leading to some repulsive forces due to more electrons being present in the outer energy level. However, the impact of this is overshadowed by the increase in nuclear charge, leading to the trend you see.

You will of course notice that the noble gases are not listed on this chart. There is a good reason for that. Noble gases already have a filled outer energy level. They do not need any more electrons and have no desire to add them. They are stable and thus have effectively no electronegativity.

FIGURE 7-3: Electro-negativities of the elements.

Electronegativities of the Elements

Increasing →

Decreasing →

1 H 2.1																	
3 Li 1.0	4 Be 1.5											5 B 2.0	6 C 2.5	7 N 3.0	8 O 3.5	9 F 4.0	
11 Na 0.9	12 Mg 1.2											13 Al 1.5	14 Si 1.8	15 P 2.1	16 S 2.5	17 Cl 3.0	
19 K 0.8	20 Ca 1.0	21 Sc 1.3	22 Ti 1.5	23 V 1.6	24 Cr 1.6	25 Mn 1.5	26 Fe 1.8	27 Co 1.9	28 Ni 1.9	29 Cu 1.9	30 Zn 1.6	31 Ga 1.6	32 Ge 1.8	33 As 2.0	34 Se 2.4	35 Br 2.8
37 Rb 0.8	38 Sr 1.0	39 Y 1.2	40 Zr 1.4	41 Nb 1.6	42 Mo 1.8	43 Tc 1.9	44 Ru 2.2	45 Rh 2.2	46 Pd 2.2	47 Ag 1.9	48 Cd 1.7	49 In 1.7	50 Sn 1.8	51 Sb 1.9	52 Te 2.1	53 I 2.5
55 Cs 0.7	56 Ba 0.9	57 La 1.1	72 Hf 1.3	73 Ta 1.5	74 W 1.7	75 Re 1.9	76 Os 2.2	77 Ir 2.2	78 Pt 2.2	79 Au 2.4	80 Hg 1.9	81 Tl 1.8	82 Pb 1.9	83 Bi 1.9	84 Po 2.0	85 At 2.2
87 Fr 0.7	88 Ra 0.9	89 Ac 1.1														

Q. List the elements in order from highest electronegativity to lowest: B, N, Ne.

A. N, B, Ne. Looking at Figure 7-3, you can see that nitrogen has an electronegativity of 3.0, boron has an electronegativity of 2.0, and neon is not listed. In this case neon is not listed because it is a noble gas and noble gases have effectively zero electronegativity. This makes neon the least electronegative element of the 3.

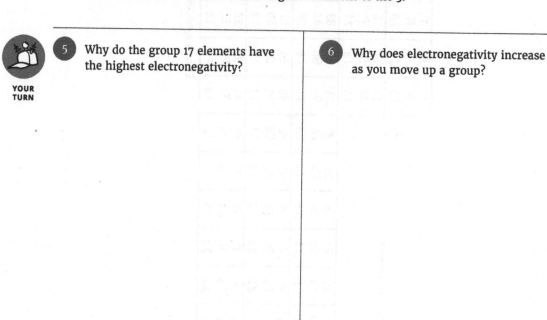

YOUR TURN

5   Why do the group 17 elements have the highest electronegativity?

6   Why does electronegativity increase as you move up a group?

Practice Questions Answers and Explanations

(1) Rb, Te, O. Rb, rubidium, has the largest atomic radius of the 3. This is due to it being in the 5th energy level and having only 1 electron in the 3rd energy level. Tellurium is also in the 5th energy level, but it has more protons than rubidium, meaning that the outer energy electrons have a greater effective nuclear charge reaching them, causing the radius to be smaller. Finally, oxygen is in the 2nd energy level, making it much smaller than either Rb or Te.

(2) S^{2-} has a much larger radius than S. Negative ions have a larger ionic radius than the corresponding neutral atomic radius of an atom. This is due to there being more electrons present in the outer energy level, without increasing proton number, leading to more repulsion happening between the electrons, causing the radius to increase.

(3) Li, Na, O. Oxygen is in the 2nd energy level along with lithium. However, oxygen is much further to the right, meaning it is close to having its outer energy level, its octet, filled. Oxygen does not want to lose an electron. This makes it have a very high ionization energy relative to lithium or sodium. Lithium has a higher ionization energy than sodium because it is in the 2nd energy level while sodium is in the 3rd energy level.

(4) Noble gases have the highest ionization energy because they have a filled outer energy level. They are totally stable. Due to this stable electron configuration they have no desire to gain or lose electrons. Relative to any other element on the periodic table this makes noble gases have the highest ionization energy.

(5) Group 17 elements have the highest electronegativity because they are 1 electron away from completing their octet. This means they need 1 electron to have a filled, stable valence shell electron configuration. Due to this they have the highest electronegativity, or ability to attract an electron.

(6) As you move up a group on the periodic table there are less energy levels between the outer electrons and the nucleus containing the positively charged protons that attract the negatively charged electrons. Since there is less shielding with all other things being equal, the electronegativity increases.

If you're ready to test your skills a bit more, take the following chapter quiz that incorporates all the chapter topics.

Whaddya Know? Chapter 7 Quiz

Quiz time! Complete each problem to test your knowledge on the various topics covered in this chapter. You can then find the solutions and explanations in the next section.

1. List the elements in order from smallest atomic radius to largest: O, N Sb

2. What would have a larger radius: Cl or Cl^{1-}?

3. What would have a larger radius: Na or Na^{1+}?

4. Why does atomic radius increase as you move from right to left across the periodic table?

5. Why do alkali metals have the lowest ionization energy?

6. Rank the following elements from highest to lowest ionization energy: Ba, S, Mg

7. The 1st ionization energy of magnesium has a value of 496 kj/mol. The second ionization energy of magnesium has a value of 1,450 kj/mol. The 3rd ionization energy of magnesium has a value of 7,730 kj/mol. Why is the 3rd ionization energy of magnesium dramatically higher than the 1st and 2nd values?

8. The ionic radius of a cation is _____ (larger than/smaller than/the same size as) the atomic radius of the neutral atom because the cation ion has _____ (more/fewer/an equal number of) electrons compared to the atom.

9. Rank the following elements in order of electronegativity from highest to lowest: F, Na, Cl, Cr

10. How does having more filled energy levels in an atom increase the shielding that occurs on the effective nuclear charge of the nucleus?

Answers to Chapter 7 Quiz

(1) **O, N, Sb.** Oxygen and nitrogen are right next to each other on the periodic table. Oxygen is further to the right, however. This means it has one more proton than nitrogen, making its atomic radius a slight bit smaller since they have the same number of energy levels. Antimony, Sb, is much further down on the periodic table. It has 5 energy levels, making it much larger than oxygen or nitrogen.

(2) **Cl^{1-}.** All negative ions have a larger radius than their respective neutral atoms. In this case Cl^{1-} gains an electron, leading to a greater repulsion of the electrons in the highest energy level. This makes the ionic radius increase relative to the neutral atomic radius.

(3) **Na.** All positive ions have a smaller radius than their respective neutral atoms. In this case Na^{1+} loses an electron. When it loses an electron, it goes from having its highest energy electrons in the 3rd energy level to having its most energetic electrons in the 2nd energy level. This makes it much smaller than its neutral atomic radius.

(4) Atomic radius increases as you move from right to left because the number of protons in an atom decreases as you move left on the periodic table. Elements that are to the right of a period have more protons than elements to the left of a period on the table. If staying within the same period, this reduction in effective nuclear charge leads to there being less pull on the outer electrons, which in turn leads to a larger atomic radius.

(5) Alkali metals have the lowest ionization energy because they only have 1 outer, valence, electron. If that outer valence electron were to be removed, the alkali metal would have a filled, lower, valence electron shell. It would be far more stable; thus, the alkali metals have a very low ionization energy.

(6) **S, Mg, Ba.** Sulfur has 6 valence electrons, meaning it needs to gain 2 electrons to fill its outer energy level. Due to this, sulfur is resistant to losing electrons, giving it a higher ionization energy. It is much further to right on the periodic table than either magnesium or barium. Magnesium has a higher ionization than barium because it is higher in the group. They both have the same number of valence electrons, but magnesium's are closer to the nucleus, meaning they are attracted more by the nuclear charge, making them harder to remove than barium's electrons.

(7) Magnesium's 3rd ionization energy is much higher than either of its first two. This occurs because magnesium has 2 valence electrons. When both of those outer electrons are removed, magnesium forms a 2+ ion. This ion has a filled octet and is stable. It becomes vastly more difficult to remove the electron due to this. The 2 outer electrons originally in magnesium, prior to being removed, could be found in the 3s sublevel. Once those are both removed, the next electron needing to be removed is in the 2p sublevel. This sublevel is also closer to the nucleus, making it much more difficult to remove the outer electron.

(8) The ionic radius of a cation is **smaller than** the atomic radius of the neutral atom because the cation has **fewer** electrons compared to the atom.

(9) **F, Cl, Cr, Na.** Fluorine has the highest electronegativity of any element with a value of 4.0. Chlorine has an electronegativity value of 3.0. Chromium has an electronegativity value of 1.6. Sodium has an electronegativity value of 0.9.

(10) Every filled energy level that exists between the outer valence electrons and the nucleus shields those outer electrons to some extent from the attractive pull of the nucleus. This occurs because the electrons found in those inner energy levels are all negatively charged and will block some of nuclear charge, the attractive force, from the nucleus and lessen the impact it has on the outermost electrons. Every energy level that exists between the nucleus and the outer electrons increases this shielding.

Chapter **8**

Doing Chemistry with Atomic Nuclei

Much of chemistry deals, in one way or the other, with chemical reactions. When you look at these reactions, you're always focused on how the valence electrons (the electrons in the outermost energy levels of atoms) are lost, gained, or shared. Very little is mentioned about the nucleus of the atom because, to a very large degree, it's not involved in chemical reactions. But in this chapter, we focus on the nucleus and the changes it can undergo.

Many elements on the periodic table exist in unstable versions called *radioisotopes*. These radioisotopes decay into other (usually more stable) elements in a process called *radioactive decay*. Because the stability of these radioisotopes depends on the composition of their nuclei, radioactivity is considered a form of nuclear chemistry. Unsurprisingly, *nuclear chemistry* deals with nuclei and nuclear processes. Nuclear fusion, which fuels the sun, and nuclear fission, which fuels a nuclear bomb, are examples of nuclear chemistry because they deal with the joining or splitting of atomic nuclei. In this chapter, you find out about nuclear decay, rates of decay called *half-lives*, and the processes of fusion and fission.

To understand nuclear chemistry, you need to know the basics of atomic structure. We cover atomic structure in Chapter 4 if you'd like to review.

Decaying Nuclei in Different Ways

For purposes of this book, we define *radioactivity* as the spontaneous decay of an unstable nucleus. An unstable nucleus may break apart into two or more other particles with the release of some energy. This breaking apart can occur in a number of ways, depending on the particular atom that's decaying. Sometimes the product of that nuclear decay is unstable itself and undergoes nuclear decay, too. For example, when U-238 (one of the radioactive isotopes of uranium) initially decays, it produces Th-234, which decays to Pa-234. The decay continues until, finally, after a total of 14 steps, Pb-206 is produced. Pb-206 is stable, and the decay sequence, or series, stops.

Before we show you how radioactive isotopes decay, we want to briefly explain why a particular isotope decays. The nucleus has all those positively charged protons shoved together in an extremely small space. All those protons are repelling each other. The forces that normally overpower the repelling protons and hold the nucleus together — that is, the *strong nuclear force* ("nuclear glue") — sometimes can't do the job, so the nucleus breaks apart, undergoing nuclear decay.

All elements with 84 or more protons are unstable; they eventually undergo decay. Other isotopes with fewer protons in their nucleus are also radioactive. The radioactivity corresponds to the neutron/proton ratio in the atom. If the neutron/proton ratio is too high (caused by too many neutrons or too few protons), the isotope is said to be *neutron rich* and is therefore unstable. Likewise, if the neutron/proton ratio is too low (it has too few neutrons or too many protons), the isotope is unstable. The neutron/proton ratio for a certain element must fall within a certain range for the element to be stable, which is why some isotopes of an element are stable and others are radioactive.

Naturally occurring radioactive isotopes decay in three primary ways: alpha, beta, or gamma decay.

In this section, we examine each of these types of decay.

Alpha decay

An *alpha particle* is defined as a positively charged particle of a helium nucleus. We hear ya: *Huh?* Try this: An alpha particle is composed of two protons and two neutrons, so it can be represented as a helium-4 atom. As an alpha particle breaks away from the nucleus of a radioactive atom, it has no electrons, so it has a +2 charge. Therefore and to wit, it's a *cation*, a positively charged ion (see Chapter 9).

But electrons are basically free — easy to lose and easy to gain. So normally, an alpha particle is shown with no charge because it very rapidly picks up two electrons and becomes a neutral helium atom instead of an ion.

Large, heavy elements, such as uranium and thorium, tend to undergo alpha emission. This decay mode relieves the nucleus of two units of positive charge (two protons) and four units of mass (two protons + two neutrons). What a process! Each time an alpha particle is emitted, four units of mass are lost. We wish we could find a diet that would allow us to lose 4 pounds at a time!

Radon-222 (Rn-222) is another alpha particle emitter, as shown in the following equation:

$$^{222}_{86}\text{Rn} \rightarrow {}^{218}_{84}\text{Po} + {}^{4}_{2}\text{He (an alpha particle)}$$

Here, radon-222 undergoes nuclear decay with the release of an alpha particle. The other remaining isotope must have a mass number of 218 (222 – 4) and an atomic number of 84 (86 – 2), which identifies the element as polonium (Po).

Beta decay

The second type of decay, called *beta decay* (β decay), comes in three forms, termed *betaplus*, *beta-minus*, and *electron capture*. All three involve emission or capture of an electron or a positron (a particle with the tiny mass of an electron but with a positive charge), and all three also change the atomic number of the daughter atom.

>> **Beta-plus:** In beta-plus decay, a proton in the nucleus decays into a neutron, a positron $\left({}^{0}_{+1}\text{e}\right)$, and a tiny, weakly interacting particle called a *neutrino*ν. This decay decreases the atomic number by 1. The mass number, however, does not change; both protons and neutrons are *nucleons* (particles in the nucleus), after all, each contributing 1 atomic mass unit. The general pattern of beta-plus decay is shown here:

$$^{A}_{Z}\text{X} \rightarrow {}^{A}_{Z-1}\text{Y} + {}^{0}_{+1}\text{e} + \nu$$

>> **Beta-minus:** Beta-minus decay essentially mirrors beta-plus decay. A neutron converts into a proton, emitting an electron and an *anti*neutrino (which has the same symbol as a neutrino except for the line on top). Particle and antiparticle pairs such as neutrinos and antineutrinos are a complicated physics topic, so we'll keep it basic here by saying that a neutrino and an antineutrino would annihilate one another if they ever touched, but they're otherwise very similar. Again, the mass number remains the same after decay because the number of nucleons remains the same. However, the atomic number increases by 1 because the number of protons increases by 1:

$$^{A}_{Z}\text{X} \rightarrow {}^{A}_{Z+1}\text{Y} + {}^{0}_{-1}\text{e} + \overline{\nu}$$

>> **Electron capture:** The final form of beta decay, electron capture, occurs when an inner electron — one in an orbital closest to the atomic nucleus — is "captured" by an atomic proton (see Chapter 6 for info on orbitals). By capturing the electron, the proton converts into a neutron and emits a neutrino. Here again, the atomic number decreases by 1:

$$^{A}_{Z}\text{X} + {}^{0}_{-1}\text{e} \rightarrow {}^{A}_{Z-1}\text{Y} + \nu$$

Gamma decay

Alpha and beta particles have the characteristics of matter: They have definite masses, occupy space, and so on. However, because there is no mass change associated with gamma emission, we refer to gamma emission as *gamma radiation emission*. Gamma radiation is similar to X-rays — high-energy, short-wavelength radiation. Gamma radiation commonly accompanies both alpha and beta emission, but it's usually not shown in a balanced nuclear reaction. Some isotopes, such as cobalt-60 (Co-60), give off large amounts of gamma radiation. Cobalt-60 is used in the radiation treatment of cancer. The medical personnel focus the gamma rays on the tumor, thus destroying it.

Gamma radiation allows the nucleus of a daughter atom to reach its lowest possible energy (most favorable) state. The general form of gamma decay is shown here, where $_Z^A X^*$ represents the excited state of the parent nucleus and the Greek letter gamma (γ) represents the gamma ray.

$$_Z^A X^* \rightarrow _Z^A Z + \gamma$$

When you think of dangerous radiation as portrayed in movies and on TV, you're thinking of gamma rays. They have an immense amount of energy and a very, very high frequency, which can kill livings cells easily. When you think of nuclear power, nuclear bombs, and other unsavory things you probably try to avoid on a daily basis, it's because you want to avoid their gamma rays. However, gamma rays also provide a wonderful benefit in cancer treatment because they can kill cancer cells.

Q. If a parent nucleus has decayed into $_{86}^{218}$Rn through alpha decay, what was the parent nucleus, and what other particles were produced from the decay?

EXAMPLE

A. The parent nucleus was radium, and an alpha particle was produced. Alpha decay lowers the mass number by 4 and the atomic number by 2, so the parent nucleus must have had a mass number of 222 and an atomic number of 88. Consulting your periodic table (Figure 5-1 in Chapter 5), you find that an atom with an atomic number of 88 is radium, so the parent nucleus must have been $_{88}^{222}$Ra.

YOUR TURN

 1 Write out the complete formula for the alpha decay of a uranium nucleus with 238 nucleons.

2 Sodium-22, a radioisotope of sodium with 22 nucleons, decays through electron capture. Write out the complete formula, including all emitted particles.

3 Classify the following reactions as alpha, beta, gamma, electron capture, or positron emission and supply the missing particles.

(a) $^{137}_{55}Cs \rightarrow$ ___ $+ \, ^{0}_{-1}e +$ ___   Type:

(b) $^{241}_{95}Am \rightarrow$ ___ $+ \, ^{4}_{2}He$ Type:

(c) $^{60}_{28}Ni^* \rightarrow$ ___ $+$ ___   Type:

(d) $^{11}_{6}C \rightarrow$ ___ $+$ ___ $+ \nu$ Type:

Measuring Rates of Decay: Half-Lives

The word *radioactive* sounds scary, but science and medicine are stuffed with useful, friendly applications for radioisotopes. Many of these applications are centered on the predictable decay rates of various radioisotopes.

If you could watch a single atom of a radioactive isotope — U–238, for example — you wouldn't be able to predict when that particular atom might decay. It might take a millisecond, or it might take a century. You simply have no way to tell.

But if you have a large enough sample — what mathematicians call a *statistically significant sample size* — a pattern begins to emerge. It takes a certain amount of time for half the atoms in a sample to decay. It then takes the same amount of time for half the remaining radioactive atoms to decay, and the same amount of time for half of those remaining radioactive atoms to decay, and so on. The amount of time it takes for one-half of a sample of an isotope to decay into daughter nuclei is called the *half-life* of the isotope, and it's given the symbol $t_{1/2}$. This process is shown in Table 8-1.

REMEMBER

The half-life decay of radioactive isotopes is not linear. For example, you can't find the remaining amount of an isotope as 7.5 half-lives by finding the midpoint between 7 and 8 half-lives. This decay is an example of an exponential decay, shown in Figure 8-1.

TABLE 8-1 Half-Life Decay of a Radioactive Isotope

Half-Life	Percent of Radioactive Isotope Remaining
0	100.00
1	50.00
2	25.00
3	12.50
4	6.25
5	3.12
6	1.56
7	0.78
8	0.39
9	0.19
10	0.09

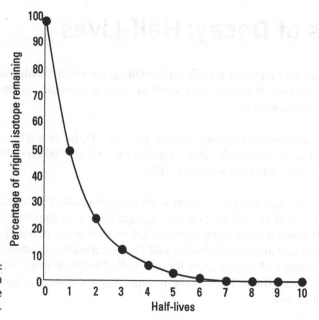

FIGURE 8-1: Decay of a radioactive isotope.

Table 8-2 lists some of the more useful radioisotopes, along with their half-lives and decay modes (we discuss these modes earlier in this chapter).

TIP

To calculate the remaining amount of an isotope after a given period of time or to determine the half-life of a substance, you can use the following formula:

$$A = A_0 (0.5)^{t/T}$$

TABLE 8-2 Common Radioisotopes, Half-Lives, and Decay Modes

Radioisotope	Half-Life	Decay Mode
Carbon-14	5.73×10^3 years	beta
Iodine-131	8.0 days	beta, gamma
Potassium-40	1.25×10^9 years	beta, gamma
Radon-222	3.8 days	alpha
Thorium-234	24.1 days	beta, gamma
Uranium-238	4.46×10^9 years	alpha

where A_0 is the amount of the isotope that existed originally, A is the amount after the decay time, t is the amount of time the sample has had to decay, and T is the half-life. Be sure to pay attention to order of operations when you use this formula to ensure you get the correct answer.

Q. You begin with a 500 g sample of radioactive material. After 4 half-lives have passed, how many grams of the sample remain?

EXAMPLE

A. There are two ways to solve this problem. Both get you to the same point; it's up to you which way you pick. Since a half-life is the amount of time required for a substance to decay by $\frac{1}{2}$ exactly, you can easily determine how much remains of the 500 g sample by simply dividing 500 g by 2, 4 times. This looks like the following:

$$\frac{500}{2} = 250 \text{ g}$$

$$\frac{250}{2} = 125 \text{ g}$$

$$\frac{125}{2} = 62.5 \text{ g}$$

$$\frac{62.5}{2} = 31.25 \text{ g remaining}$$

This is a perfectly fine way to solve the problem and it gets you to the correct result. However, there is another way to solve it involving a formula, which is the better approach because it doesn't rely on you performing a bunch of division problems in a row. (In general, as an overall rule for chemistry problems, the fewer steps you have to do to solve something, the better off you are. This is simply due to the fact that every time you add a step to a problem, you add a potential place where you can make an error when solving it. The fewer steps, the less chance for error.)

The formula for this problem can be written as follows:

$$\text{Remaining mass} = (\text{initial mass})(0.5)^{(n)}$$
$$n = \text{number of half lives}$$

To solve using this formula, you plug your values in as follows:

$$\text{Remaining mass} = (500 \text{ g})(0.5)^{(4)}$$
$$\text{Remaining mass} = 31.25 \text{ g}$$

Q. If a sample originally contained 1 g of thorium-234, how much of that isotope will the sample contain one year later?

A. The sample will contain 2.76×10^{-5} g. Table 8-2 tells you that thorium-234 has a half-life of 24.1 days, so that's T. The time elapsed (t) is 365 days, and the original sample was 1 g (A_0). Plugging these values into the half-life equation gives you

$$A = (1\,g)(0.5)^{365\ days/24.1\ days} = 2.76 \times 10^{-5}\,g$$

To follow the order of operations, you must first perform the division found in the exponent, then perform the exponent calculation, and then multiply the answer by 1.

Notice that the units in the exponent cancel out, so you're left with grams in the end. These units make sense because you're measuring the mass of the sample remaining.

YOUR TURN

4 After 3 half-lives, how many grams of a 100 g sample of radon-222 remain?

5 If you start with 0.5 g of potassium-40, how long will it take for the sample to decay to 0.1 g?

6 If a 50.0 g sample of a radioactive element has decayed into 44.3 g after 1,000 years, what is the element's half-life? Based on Table 8-2, which radioisotope do you think you're dealing with?

Making and Breaking Nuclei: Fusion and Fission

Fission and fusion differ from radioactive decay in that they generally require the nucleus of a parent atom to interact with an outside particle (some manmade isotopes have been known to undergo fission without bombardment — Fe-256, for example). Because the forces that hold atomic nuclei together are ridiculously powerful, the energy involved in splitting or joining two nuclei is tremendous.

Nuclear fission

In the 1930s, scientists discovered that some nuclear reactions can be initiated and controlled. Scientists usually accomplished this task by bombarding a large isotope with a second, smaller one — commonly a neutron. The collision caused the larger isotope to break apart into two or more elements, which is called *nuclear fission*. The nuclear fission of uranium-235 is shown in the following equation:

$$\ce{^{235}_{92}U} + \ce{^{1}_{0}n} \rightarrow \ce{^{142}_{56}Ba} + \ce{^{91}_{36}Kr} + 3\ce{^{1}_{0}n}$$

REMEMBER

Reactions of this type also release a lot of energy. Where does the energy come from? Well, if you make *very* accurate measurements of the masses of all the atoms and subatomic particles you start with and all the atoms and subatomic particles you end up with and then compare the two, you find that some mass is "missing." Matter disappears during the nuclear reaction. This loss of matter is called the *mass defect*. The missing matter is converted into energy.

Nuclear fission, the splitting of an atomic nucleus, doesn't occur in nature. Humans first harnessed the tremendous power of fission during the Manhattan Project, an intense, hush-hush effort by the United States that led to the development of the first atomic bomb in 1945. Fission has since been used for more-benign purposes in nuclear power plants. Nuclear power plants use a highly regulated process of fission to produce energy much more efficiently than is done in traditional, fossil fuel-burning power plants.

You can distinguish fission and fusion reactions from one another with a simple glance at products and reactants. If the reaction shows one large nucleus splitting into two smaller nuclei, then it's most certainly fission, whereas a reaction showing two small nuclei combining to make a single heavier nucleus is definitely fusion.

You can actually calculate the amount of energy produced during a nuclear reaction with a fairly simple equation developed by Einstein: $E = mc^2$. In this equation, E is the amount of energy produced, m is the "missing" mass, or the mass defect, and c is the speed of light, which is a rather large number. The speed of light is squared, making that part of the equation a very large number that, even when multiplied by a small amount of mass, yields a *large* amount of energy. For example, when a mole of U-235 decays to Th-234, the mass defect is 5×10^{-6} kg, which when converted to energy amounts to 5×10^{11} joules!

Notice that one neutron was used, but three were produced. These three neutrons, if they encounter other U–235 atoms, can initiate other fissions, producing even more neutrons. It's the old domino effect. In terms of nuclear chemistry, it's a continuing cascade of nuclear fissions called a *chain reaction*. The chain reaction of U–235 is shown in Figure 8-2.

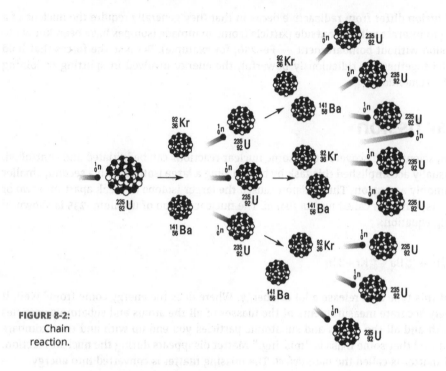

FIGURE 8-2:
Chain
reaction.

Nuclear Fusion

Soon after the fission process was discovered, another process, called *fusion*, was discovered. Fusion is essentially the opposite of fission. In fission, a heavy nucleus is split into smaller nuclei. With fusion, lighter nuclei are fused into a heavier nucleus.

The fusion process is the reaction that powers the sun. On the sun, in a series of nuclear reactions, four isotopes of hydrogen-1 are fused into a helium-4 with the release of a tremendous amount of energy. Here on Earth, two other isotopes of hydrogen are used: H–2, called *deuterium*, and H–3, called *tritium*. Deuterium is a minor isotope of hydrogen, but it's still relatively abundant. Tritium occurs naturally in minute amounts, but it can easily be produced by bombarding deuterium with a neutron. The fusion reaction is shown in the following equation:

$$_1^3H + _1^2H \rightarrow _2^4He + _0^1n$$

The first demonstration of nuclear fusion — the hydrogen bomb — was conducted by the military. A hydrogen bomb is approximately 1,000 times as powerful as an ordinary atomic bomb.

The isotopes of hydrogen needed for the hydrogen bomb fusion reaction were placed around an ordinary fission bomb. The explosion of the fission bomb released the energy needed to provide the *activation energy* (the energy necessary to initiate, or start, the reaction) for the fusion process.

Nuclear fusion as used in a hydrogen bomb may be fine for warfare, but to make the fusion process usable as an energy source for everyday life, the power has to be harnessed, much as it was in nuclear fission. However, developing a fusion power plant has proven to be much more difficult than a fission one.

The goal of scientists for the last 50 years has been the controlled release of energy from a fusion reaction. If the energy from a fusion reaction can be released slowly, it can be used to produce electricity. It will provide an unlimited supply of energy that has no wastes to deal with or contaminants to harm the atmosphere — just nonpolluting helium. But achieving this goal requires overcoming three problems:

>> Temperature

>> Time

>> Containment

Reaching the right temperature

The fusion process requires an extremely high activation energy. Heat is used to provide the energy, but it takes a *lot* of heat to start the reaction. Scientists estimate that the sample of hydrogen isotopes must be heated to approximately 40,000,000 K. (To get the Kelvin temperature, you add 273 to the Celsius temperature; see Chapter 2 for details.)

But 40,000,000 K is hotter than the sun! At this temperature, the electrons have long since left the building; all that's left is a positively charged *plasma*, bare nuclei heated to a tremendously high temperature. Presently, scientists are trying to heat samples to this high temperature through two ways: magnetic fields and lasers. Neither one has yet achieved the necessary temperature.

Maintaining for enough time

Time is the second problem scientists must overcome to achieve the controlled release of energy from fusion reactions. The charged nuclei must be held together close enough and long enough for the fusion reaction to start. Scientists estimate that the plasma needs to be held together at 40,000,000 K for about one second.

Containing the reaction

Containment is the major problem facing fusion research. At 40,000,000 K, everything is a gas. The best ceramics developed for the space program would vaporize when exposed to this temperature. Because the plasma has a charge, magnetic fields can be used to contain it — like a magnetic bottle. But if the bottle leaks, the reaction won't take place. And scientists have yet to create a magnetic field that won't allow the plasma to leak. Using lasers to zap the hydrogen isotope mixture and provide the necessary energy bypasses the containment problem, but scientists haven't figured out how to protect the lasers themselves from the fusion reaction.

Q. What type of nuclear reaction — fusion or fission — would you expect plutonium-239 to undergo, and why?

EXAMPLE

A. Fission is the expected reaction. Plutonium has an exceptionally large number of nucleons and is therefore likely to undergo fission. Fusion is impossible in all elements heavier than iron, and plutonium is much, much heavier than iron.

7   What type of nuclear reaction is shown in the following equation? How do you know? Where might such a reaction take place?

YOUR
TURN

$$^{235}_{92}U + ^{1}_{0}n \rightarrow ^{142}_{56}Ba + ^{91}_{36}Kr + 3^{1}_{0}n$$

8   What type of nuclear reaction is shown in the following equation? How do you know? Where might such a reaction take place? There's something atypical about the two hydrogen reactants. What is it?

$$^{3}_{1}H + ^{2}_{1}H \rightarrow ^{4}_{2}He + ^{1}_{0}n$$

9   What makes nuclear fusion such an appealing idea as a future power source for our society?

Practice Questions Answers and Explanations

(1) $^{238}_{92}\text{U} \rightarrow\ ^{234}_{90}\text{Th} + ^{4}_{2}\text{He}$. Alpha decay results in the emission of a helium nucleus from the parent atom, which leaves the daughter atom with two fewer protons and a total of four fewer nucleons. In other words, the atomic number of the daughter atom is two fewer than the parent atom, and the mass number is four fewer. Because the proton number defines the identity of an atom, the element with two fewer protons than uranium must be thorium.

(2) $^{22}_{11}\text{Na} + ^{0}_{-1}\text{e} \rightarrow\ ^{22}_{10}\text{Ne} + \nu$. Electron capture is a form of beta decay that results in the atomic number decreasing by 1 and the mass number remaining the same.

(3) Here are the missing particles and the types of radioactive decay:

   a. $^{137}_{55}\text{Cs} \rightarrow\ ^{137}_{56}\text{Ba} + ^{0}_{-1}\text{e} + \bar{\nu}$. **Type: beta.** You can identify this reaction as beta due to the emission of an electron $\left(^{0}_{-1}\text{e}\right)$. This means that a neutron converts to a proton, which requires increasing the atomic number by 1. You also have to change the chemical symbol to reflect the element that's now present due to the change in atomic number.

   b. $^{241}_{95}\text{Am} \rightarrow\ ^{237}_{93}\text{Np} + ^{4}_{2}\text{He}$. **Type: alpha.** This reaction is alpha decay due to the emission of an alpha particle, $^{4}_{2}\text{He}$. You simply need to adjust the atomic number and mass number to correspond to the loss of two neutrons and two protons. Thus, the mass number is reduced by 4, and the atomic number is reduced by 2. You then change the chemical symbol to reflect the element that's now present due to the change in atomic number.

   c. $^{60}_{28}\text{Ni}^{*} \rightarrow\ ^{60}_{28}\text{Ni} + \gamma$. **Type: gamma.** From the initial form of the equation, you can tell that an excited nucleus is present, so gamma decay will take place. To correctly write the equation, write the same isotope on the product side with a gamma ray being emitted.

   d. $^{11}_{6}\text{C} \rightarrow\ ^{11}_{5}\text{B} + ^{0}_{+1}\text{e} + \nu$. **Type: beta.** You see a neutrino in the products but no electron in the reactants, so this must be a beta reaction. A neutron changes to a proton, so the mass number remains the same, but one more proton is added to the atomic number. This causes the element to change from carbon to boron.

(4) **12.5 g.** Do not let the mention of radon-222 fool you. The type of isotope given here does not matter at all. All you need to focus on or care about is the starting mass of 100 g and the fact that 3 half-lives have passed. Simply plug into the equation and solve for the remaining mass:

Remaining mass $= 100(0.5)^{(3)}$

Remaining mass $= 12.5$ g

(5) **2.9×10^{9} years.** The problem tells you that $A = 0.1$ g and $A_0 = 0.5$ g. Table 20-2 tells you that the half-life of potassium-40 is 1.25×10^{9} years. Plugging these values into the equation, you get

$$A = A_0(0.5)^{t/T}$$

$$0.1 = 0.5(0.5)^{\left(t/1.25 \times 10^{9} \text{ years}\right)}$$

Divide both sides by 0.5:

$$0.2 = (0.5)^{\left(t/1.25 \times 10^{9} \text{ years}\right)}$$

Take the natural log of both sides:

$$\ln(0.2) = \ln(0.5)^{(t/1.25 \times 10^9 \text{ years})}$$

This step helps you isolate the exponent by pulling it out in front on the right-hand side:

$$\ln(0.2) = \left(\frac{t}{1.25 \times 10^9 \text{ years}}\right) \ln(0.5)$$

Now you can rearrange the equation to solve for t:

$$t = \frac{\ln 0.2}{\ln 0.5}\left(1.25 \times 10^9 \text{ years}\right)$$
$$= 2.9 \times 10^9 \text{ years}$$

(6) **5,710 years; carbon-14.** The problem tells you that $A_0 = 50.0$ g, $A = 44.3$ g, and $t = 1,000$ years. Use the same process as in Question 5 to isolate the exponent, this time solving for T instead of t.

$$A = A_0 (0.5)^{t/T}$$
$$44.3 \text{ g} = (50.0 \text{ g})(0.5)^{(1,000 \text{ years}/T)}$$
$$0.866 = (0.5)^{(1,000 \text{ years}/T)}$$
$$\ln(0.866) = \ln(0.5)^{(1,000 \text{ years}/T)}$$
$$\ln(0.866) = \left(\frac{1,000 \text{ years}}{T}\right) \ln(0.5)$$
$$T = 1,000 \text{ years} \left(\frac{\ln(0.5)}{\ln(0.866)}\right)$$
$$T = 5,710 \text{ years}$$

(7) This reaction is a fission reaction. It shows a heavy uranium nucleus being bombarded by a neutron and decaying into two lighter nuclei (barium and krypton). This is the very reaction that takes place in a nuclear reactor.

(8) This reaction is a fusion reaction. It shows two light nuclei combining to form one heavy nucleus. This reaction fuels the sun. The two hydrogen reactants are atypical because they're rare isotopes of hydrogen, called *tritium* and *deuterium*, respectively.

(9) Nuclear fusion can provide clean energy with nearly no waste products. Due to the efficient and clean nature of the reaction, nuclear fusion would be much safer and easier to run than any comparable source of power we currently have.

If you're ready to test your skills a bit more, take the following chapter quiz that incorporates all the chapter topics.

Whaddya Know? Chapter 8 Quiz

Ready for a quiz? The 15 questions in this section will test the skills you learned in this chapter. When you're done, check out the section that follows for answers and explanations.

1 What nuclear process involves two lighter elements combining to form a heavier nucleus? This is a process that results in a large release of energy.

2 Which type of nuclear process involves 1 heavy element splitting into two or more lighter elements along with a release of energy?

3 Write the correct notation for an alpha particle. What type of decay reaction are alpha particles found in?

4 Write the correct notation for a positron. What type of decay are positrons found in?

5 You have a 1000 g sample of iodine-131. After 24 days, what is the remaining mass of your sample? Reference Table 8-2 for the half-life of iodine-131.

6 After 4 half-lives a sample of thorium-234 has a mass of 20 g. What was the initial mass of the thorium-234? How much time has elapsed during the 4 half-lives? Reference Table 8-2 for the half-life of thorium-234.

7 If you start with a 1500 g sample of a radioactive element and find that after 20 years it has decayed to a mass of 980 g what is the element's half-life?

Fill in the missing information for the following reactions and identify the type of decay the reaction is showing:

8 $\underline{\quad} \rightarrow {}^{243}_{94}\text{Pu} + {}^{4}_{2}\text{He}$ Type:

9 $\underline{\quad} \rightarrow {}^{93}_{42}\text{Mo} + {}^{0}_{-1}\text{e}$ Type:

10 ${}^{224}_{88}\text{Ra} \rightarrow {}^{220}_{86}\text{Rn} + \underline{\quad}$ Type:

11 ${}^{192}_{77}\text{Ir} \rightarrow \underline{\quad} + {}^{0}_{0}\gamma$ Type:

12 ${}^{241}_{94}\text{Pu} \rightarrow {}^{0}_{-1}\text{e} + \underline{\quad}$ Type:

13 ${}^{218}_{84}\text{Po} \rightarrow \underline{\quad} + \underline{\quad}$ Type:

14 $\underline{\quad} \rightarrow {}^{214}_{84}\text{Po} + {}^{0}_{-1}\text{e}$ Type:

15 ${}^{99}_{43}\text{Tc} \rightarrow {}^{99}_{43}\text{Tc} + \underline{\quad}$ Type:

Answers to Chapter 8 Quiz

1. **Nuclear fusion** occurs when you two lighter elements combine to form a heavier element. This process is something you observe everyday if you happen to glance at the sun. When the two lighter elements combine there is a large amount of energy released.

2. **Nuclear fission** occurs when one heavier element splits into two or more lighter elements along with a release of neutrons. This process releases a substantial amount of energy. Nuclear fission is the process used in nuclear power plants to generate electricity.

3. ^4_2He. An alpha particle consists of two protons and four neutrons. This is effectively a helium atom, which you write as shown above. Alpha particles are found as a result of alpha decay.

4. $^0_{+1}\text{e}$. A positron is the antiparticle of an electron, effectively a positive electron. It has no neutrons or protons present within it. It is found in beta decay, specifically beta plus decay.

5. **125 g remaining.** From Table 8-2 you can see the half-life of iodine-131 is eight days exactly. If 24 days have passed, that means exactly three rounds of half-life decay of iodine-131 has occurred. You can determine this by simply dividing 24/8= 3 half-lives. Once you know this you simply need to plug it into the half-life formula and solve:

$$\text{Remaining mass} = 1000(0.5)^{(3)}$$
$$\text{Remaining mass} = 125 \text{ g}$$

6. **320 g. 96.4 days.** To solve this you need to rearrange your equation to solve for the initial mass:

$$\text{Remaining mass} = \text{initial mass}$$
$$\frac{\text{Remaining mass}}{(0.5)^{(n)}} = \text{initial mass}$$
$$\frac{20 \text{ g}}{(0.5)^{(4)}} = 320 \text{ g initially of thorium-234}$$

To determine the amount of time that elapsed during those 4 half-lives you need to determine the time it takes for one half-life of thorium-234 to elapse. Table 8-2 shows it takes 24.1 days for one half-life to occur. Knowing this you can multiply 24.1 days by 4 to get your answer.

$$24.1 \times 4 = 96.4 \text{ days}$$

(7) **32.5 years.** This problem tells you that $A_0 = 1500$ g and $A = 980$ g and $t = 20$ years. Plug into your equation to solve for T to determine the half-life.

$$A = A_0 (0.5)^{\frac{t}{T}}$$

$$980 = 1500 (0.5)^{\left(\frac{20 \text{ years}}{T}\right)}$$

$$0.653 = (0.5)^{\left(\frac{20 \text{ years}}{T}\right)}$$

$$\ln(.653) = \ln(0.5)^{\left(\frac{20 \text{ years}}{T}\right)}$$

$$\ln(.653) = \left(\frac{20 \text{ years}}{T}\right) \ln(0.5)$$

$$T = 20 \text{ years} \frac{\ln(0.5)}{\ln(.653)}$$

$$T = 32.5 \text{ years}$$

(8) $^{247}_{96}\text{Cm} \rightarrow \,^{243}_{94}\text{Pu} + \,^{4}_{2}\text{He}$. **Type: Alpha decay.** You can identify this reaction as alpha decay because an alpha particle (helium) is emitted from the decay. This means, in simpler terms, an alpha particle/helium is present as a product in the reaction. To determine the unknown reactant, you need to add 4 to the mass number and 2 to the atomic number of the $^{243}_{94}\text{Pu}$ in the products. When you change its atomic number from 94 to 96, that means you also need to adjust the chemical symbol to match the number of protons which is Cm.

(9) $^{93}_{41}\text{Nb} \rightarrow \,^{93}_{42}\text{Mo} + \,^{0}_{-1}\text{e}$. **Type: Beta decay.** This reaction shows a type of beta decay, specifically beta-minus. You can tell this because there is an emission of an electron present in the products of the reaction. To determine the unknown reactant, you need to remember that in beta decay a neutron converts into a proton and emits an electron. The electron emission plays no role in determining the mass number or atomic number of the proton, but the conversion of neutron to proton does matter. A neutron converting to a proton does not change the mass number of 93 as both proton and neutron have a mass of 1 amu. It does increase the atomic number by 1, though. In this case you need to work backwards and subtract 42 – 1 to realize the atomic number of the reactant should be 41. This corresponds to the element Nb. The mass number remains the same, 93.

(10) $^{224}_{88}\text{Ra} \rightarrow \,^{220}_{86}\text{Rn} + \,^{4}_{2}\text{He}$. **Type: Alpha Decay**

To determine the unknown product you can compare the mass number and atomic number of the reactant Ra and the product of Rn. You can see the mass number decreases by 4 and the atomic number decreases by 2. This corresponds to an emission of something with four neutrons and two protons, a helium, which is an alpha particle.

(11) $^{192}_{77}\text{Ir} \rightarrow \,^{192}_{77}\text{Ir} + \,^{0}_{0}\gamma$. **Type: Gamma decay**

This is a gamma decay reaction because a gamma ray is emitted during the reaction. Remember, a gamma ray has no mass. This means the mass number and atomic number of the reactant does not change at all, so you simply write the same symbol again.

(12) $^{241}_{94}\text{Pu} \rightarrow \,^{0}_{-1}\text{e} + \,^{241}_{95}\text{Am}$. **Type: Beta Minus decay**

This reaction shows an electron is emitted from the decay and a neutron converts into a proton. This indicates beta minus decay. This increases the atomic number from 94 to 95 and changes the element from Pu to Am. However, it does not change the mass number, so 241 remains.

(13) $^{218}_{84}Po \rightarrow {}^{214}_{82}Pb + {}^{4}_{2}He$. **Type: Alpha Decay**

This is identified for you as alpha decay. That means you will have two protons and two neutrons emitted as an alpha particle/helium. The mass number decreases by 4 and the atomic number reduces by 2. To write the correct isotope for your answer, you need to subtract $218 - 4 = 214$ as the new mass number and $84 - 2$ as the new atomic number. The new element is Pb. You also need to write the alpha particle, a helium atom with a mass number of 4 and an atomic number of 2.

(14) $^{214}_{83}Bi \rightarrow {}^{214}_{84}Po + {}^{0}_{-1}e$. **Type: Beta minus decay**

This is beta minus decay because an electron is emitted and a neutron converts to a proton. To determine the unknown, work backwards from the Pollonium-214. You know the mass number does not change in beta decay so you keep the 214. Next, you know that in beta minus you have one more proton in your product than in your reactants, so you need to subtract $84 - 1$ to find an atomic number of 83, which corresponds to the element Bi.

(15) $^{99}_{43}Tc \rightarrow {}^{99}_{43}Tc + {}^{0}_{0}\gamma$. **Type: Gamma decay**

This is gamma decay because the mass number and atomic number of the element do not change. They remain the same. To complete the equation you write a gamma ray being emitted, as shown here. Remember that a gamma ray has no mass, so the mass number and atomic number are both zero.

3
Making and Breaking Bonds

In This Unit . . .

Chapter **9**

Building Bonds

Many atoms are prone to public displays of affection, pressing themselves against other atoms in an intimate electronic embrace called *bonding.* Atoms bond with one another by playing various games with their valence electrons. In this chapter, we describe the basic rules of those games and how those rules lead to the many different chemical compounds that exist.

Because valence electrons are so important to bonding, chemistry problems involving bonding sometimes make use of *electron dot structures,* symbols that represent valence electrons as dots surrounding an atom's chemical symbol. You should be able to draw and interpret electron dot structures for atoms, as in Figure 9-1. This figure shows the electron dot structures for elements in the periodic table's first two rows. Notice that the valence shells progressively fill moving from left to right.

TIP

To determine the electron dot structure of any element, count the number of electrons in that element's valence shell. Then draw that number of dots around the chemical symbol for the element. To do so, imagine the chemical symbol as a square. Start from the top of the symbol and, going clockwise, put one dot on each side until you run out of valence electrons. Don't double up on any side until you've gone around the square once.

The next thing to understand before jumping too far into chemical bonding are the factors that determine whether atoms gain or lose electrons to form ions. Electron dot structures can help you do this, but there are easier ways to determine whether an atom gains or loses electrons to fill its valence shell. That's what you'll cover in this chapter.

FIGURE 9-1:
Electron dot
structures
for elements
in the first
two rows of
the periodic
table.

IA	IIA	IIIA	IVA	VA	VIA	VIIA	VIIIA
H •							He :
Li •	• Be •	• $\dot{B}$ •	• $\dot{C}$ •	: $\dot{N}$:	: $\ddot{O}$:	: $\ddot{F}$:	: $\ddot{Ne}$:

Forming Ions

You can predict what kind of ion many elements will form simply by looking at their position on the periodic table. With the exception of Row 1 (hydrogen and helium), all main group elements are most stable (think "happiest") with a full shell of eight valence electrons, known as an *octet*. Atoms tend to take the shortest path to a complete octet, whether that means ditching a few electrons to achieve a full octet at a lower energy level or grabbing extra electrons to complete the octet at their current energy level. In general, metals on the left side of the periodic table (and in the middle) tend to lose electrons, and nonmetals on the right tend to gain electrons.

As a reminder, when atoms become ions, they lose the one-to-one balance between their protons and electrons and therefore acquire an overall charge:

>> **Cations:** Atoms that lose electrons (like metals) acquire positive charge, becoming *cations*, such as Na^+ or Mg^{2+}.

>> **Anions:** Atoms that gain electrons (like nonmetals) acquire negative charge, becoming *anions*, such as Cl^- or O^{2-}.

The superscripted numbers and signs in the atoms' symbols indicate the ion's overall charge. Cations have superscripts with + signs, and anions have superscripts with – signs. When the element sodium, Na, loses an electron, it loses one negative charge and is left with one overall positive charge because it now has one more proton than electron. So Na becomes Na^+.

TIP You can often determine the charge an ion normally has by the element's position on the periodic table. For example, all the alkali metals (the IA elements) lose a single electron to form a cation with a +1 charge. In the same way, the alkaline earth metals (IIA elements) lose two electrons to form a +2 cation. Aluminum, a member of the IIIA family, loses three electrons to form a +3 cation.

By the same reasoning, the halogens (VIIA elements) all have seven valence electrons. All the halogens gain a single electron to fill their valence energy level. And all of them form an anion with a single negative charge. The VIA elements gain two electrons to form anions with a –2 charge, and the VA elements gain three electrons to form anions with a –3 charge.

Table 9-1 shows the family, element, ion name, and ion symbol for some common monatomic (one-atom) cations, and Table 9-2 gives the same information for some common monatomic anions.

TABLE 9-1 **Some Common Monatomic Cations**

Family	Element	Ion Name	Ion Symbol
IA	Lithium	Lithium cation	Li^+
	Sodium	Sodium cation	Na^+
	Potassium	Potassium cation	K^+
IIA	Beryllium	Beryllium cation	Be^{2+}
	Magnesium	Magnesium cation	Mg^{2+}
	Calcium	Calcium cation	Ca^{2+}
	Strontium	Strontium cation	Sr^{2+}
	Barium	Barium cation	Ba^{2+}
IB	Silver	Silver cation	Ag^+
IIB	Zinc	Zinc cation	Zn^{2+}
IIIA	Aluminum	Aluminum cation	Al^{3+}

TABLE 9-2 **Some Common Monoatomic Anions**

Family	Element	Ion Name	Ion Symbol
VA	Nitrogen	Nitride anion	N^{3-}
	Phosphorus	Phosphide anion	P^{3-}
VIA	Oxygen	Oxide anion	O^{2-}
	Sulfur	Sulfide anion	S^{2-}
VIIA	Fluorine	Fluoride anion	F^-
	Chlorine	Chloride anion	Cl^-
	Bromine	Bromide anion	Br^-
	Iodine	Iodide anion	I^-

REMEMBER

Determining the number of electrons that members of the transition metals (the B families) lose is more difficult. In fact, many of these elements lose a varying number of electrons, so they form two or more cations with different charges.

The electrical charge that an atom achieves is sometimes called its *oxidation state*. Many of the transition metal ions have varying oxidation states. Table 9-3 shows some common transition metals that have more than one oxidation state.

Ions aren't always *monoatomic* (composed of just one atom). Ions can also be *polyatomic*, composed of a group of atoms. We cover polyatomic ions and how you can write formulas with them in Chapter 11.

TABLE 9-3 Some Common Metals with More Than One Oxidation State

Family	Element	Ion Name	Ion Symbol
VIB	Chromium	Chromium(II) or chromous	Cr^{2+}
		Chromium(III) or chromic	Cr^{3+}
VIIB	Manganese	Manganese(II) or manganous	Mn^{2+}
		Manganese(III) or manganic	Mn^{3+}
VIIIB	Iron	Iron(II) or ferrous	Fe^{2+}
		Iron(III) or ferric	Fe^{3+}
	Cobalt	Cobalt(II) or cobaltous	Co^{2+}
		Cobalt(III) or cobaltic	Co^{3+}
IB	Copper	Copper(I) or cuprous	Cu^{+}
		Copper(II) or cupric	Cu^{2+}
IIB	Mercury	Mercury(I) or mercurous	Hg_2^{2+}
		Mercury(II) or mercuric	Hg^{2+}

Q. Fluorine and sodium are only two atomic numbers apart on the periodic table. Why, then, does fluorine form an anion, F^-, whereas sodium forms a cation, Na^+?

A. Fluorine (F) is in Group VIIA, just one group to the left of the noble gases, and therefore needs to gain only one electron to complete a valence octet. Sodium (Na) lies just one group to the right of the noble gases, having wrapped around into Group IA of the next energy level. Therefore, sodium needs to lose only one electron to achieve a full valence octet.

1 How many electrons will be gained or lost by the following elements when forming an ion?

(a) lithium

(b) selenium

(c) boron

(d) oxygen

(e) chlorine

2 What type of ion is nitrogen most
likely to form?

 3 What type of ion is beryllium most
likely to form?

Pairing Charges with Ionic Bonds

REMEMBER

Atoms of some elements, like metals, can easily lose valence electrons to form *cations* (atoms
with positive charge) that have stable electron configurations. Atoms of other elements, like
the halogens, can easily gain valence electrons to form *anions* (atoms with negative charge)
with stable electron configurations. Cations and anions experience *electrostatic attraction* to one
another due to the opposite charges present. So a cation will snuggle up to an anion, given
the chance. The attraction between cations and anions is called *ionic bonding,* and it happens
because the energy of the ionically bonded ions is lower than the energy of the ions when
they're separated.

Metals (like sodium) tend to give up their electrons to nonmetals (like chlorine) because nonmetals are much more *electronegative* (they more strongly attract electrons within a bond to
themselves). The greater the difference in electronegativity between the two ions, the more
ionic the bond that forms between them (where *ionic* means completely uneven in the sharing
of electrons). You can think of an ionic bond as resulting from the transfer of an electron from
one atom to another, as Figure 9-2 shows for sodium and chlorine.

FIGURE 9-2:
The transfer
of an
electron
from sodium
to chlorine
to form an
ionic bond
between the
Na⁺ cation
and the
Cl⁻ anion.

$$Na\cdot \rightsquigarrow \cdot\overset{\displaystyle\cdot}{\underset{\displaystyle\cdot}{Cl}}: \longrightarrow Na^+ \quad :\overset{\displaystyle\cdot}{\underset{\displaystyle\cdot}{Cl}}:^-$$

Forming Sodium Chloride

Sodium is a fairly typical metal. It's silvery, soft, and a good conductor. It's also highly reactive. Sodium is normally stored under oil to keep it from reacting with the water in the atmosphere. If you melt a freshly cut piece of sodium and put it into a beaker filled with greenish–yellow chlorine gas, something very impressive happens. The molten sodium begins to glow with a white light that gets brighter and brighter. The chlorine gas swirls, and soon the color of the gas begins to disappear. In a couple of minutes, the reaction is over, and the beaker can be safely uncovered. You find a white crystalline substance, table salt (NaCl), deposited on the inside of the beaker.

In the following sections, we show you what happens during the chemical reaction to create table salt and, more importantly, why it occurs. Understanding these concepts will go a long way in your investigation of ionic bonding.

Meeting the components

If you really stop and think about it, the process of creating table salt is pretty remarkable. You take the following two substances, which are both hazardous (the Germans used chlorine gas against the opposing troops during World War I), and from them you make a substance that's necessary for life.

>> **Sodium:** Sodium is an alkali metal, a member of the IA family on the periodic table. The Roman numerals at the top of the A families show the number of valence electrons (s and p electrons in the outermost energy level) in the particular element (see Chapter 6 for details). So sodium has 1 valence electron and 11 total electrons because its atomic number is 11.

   You can use an energy-level diagram to represent the distribution of electrons in an atom. Sodium's energy-level diagram is shown in Figure 9-3. (If energy-level diagrams are new to you, check out Chapter 4. A number of minor variations are commonly used in writing energy-level diagrams, so don't worry if the diagrams in Chapter 4 are slightly different from the ones we show you here.)

>> **Chlorine:** Chlorine is a member of the halogen family — the VIIA family on the periodic table. It has 7 valence electrons and a total of 17 electrons. The energy-level diagram for chlorine is also shown in Figure 9-3.

TIP

If you want, instead of using the bulky energy–level diagram to represent the distribution of electrons in an atom, you can use the electron configuration. (For a complete discussion of electron configurations, see Chapter 6.) Write, *in order,* the energy levels being used, the orbital types (s, p, d, and so on), and — in superscript — the number of electrons in each orbital. Here are the electronic configurations for sodium and chlorine:

Sodium (Na)	$1s^2 2s^2 2p^6 3s^1$
Chlorine (Cl)	$1s^2 2s^2 2p^6 3s^2 3p^5$

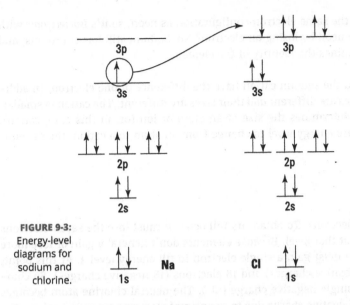

FIGURE 9-3:
Energy-level diagrams for sodium and chlorine.

Understanding the reaction

The noble gases are the VIIIA elements on the periodic table. They're extremely unreactive because each atom's *valence* energy level (outermost energy level) is filled. Achieving a filled (complete) valence energy level is a driving force in nature in terms of chemical reactions, because that's when elements become stable, or *unreactive.* They don't lose, gain, or share electrons.

The other elements in the A families on the periodic table do gain, lose, or share valence electrons in order to fill their valence energy levels and become stable. Because this process, in most cases, involves filling the outermost s and p orbitals, it's sometimes called the *octet rule* — elements gain, lose, or share electrons to reach a full octet (eight valence electrons: two in the s orbital and six in the p orbital).

Sodium's role

Sodium has one valence electron; by the octet rule, sodium becomes stable when it has eight valence electrons. Two possibilities exist for sodium to become stable: It can gain seven more electrons to fill energy level 3, or it can lose the one 3s electron so that energy level 2 (which is already filled with eight electrons) becomes the valence energy level. In general, the loss or gain of one, two, or sometimes even three electrons can occur, but an element doesn't ordinarily lose or gain more than three electrons. So to gain stability, sodium loses its 3s electron. At this point, it has 11 protons (11 positive charges) and 10 electrons (10 negative charges). The once-neutral sodium atom now has a single positive charge (+11 plus –10 equals +1). It's now an *ion*, an atom that has a charge due to the loss or gain of electrons. And ions that have a positive charge (such as sodium) due to the loss of electrons are called *cations.* You can write an electron configuration for the sodium cation:

$$Na^+ \quad 1s^2 2s^2 2p^6$$

The sodium ion (cation) has the same electron configuration as neon, so it's *isoelectronic* with neon. So has sodium become neon by losing an electron? No. Sodium still has 11 protons, and the number of protons determines the identity of the element.

The neutral sodium atom and the sodium cation have the difference of one electron. In addition, their chemical reactivities are different *and* their sizes are different. The cation is smaller. The outermost energy level determines the size of an atom or ion (or, in this case, cation). Because sodium loses an entire energy level to change from an atom to a cation, the cation is smaller.

Chlorine's role

Chlorine has seven valence electrons. To obtain its full octet, it must lose the seven electrons in energy level 3 or gain one at that level. Because elements don't generally gain or lose more than three electrons, chlorine must gain a single electron to fill energy level 3. At this point, chlorine has 17 protons (17 positive charges) and 18 electrons (18 negative charges). So chlorine becomes an ion with a single negative charge (Cl^-). The neutral chlorine atom becomes the chloride ion. Ions with a negative charge due to the gain of electrons are called *anions*. The electronic configuration for the chloride anion is

$$Cl^- \qquad 1s^2 2s^2 2p^6 3s^2 3p^6$$

The chloride anion is isoelectronic with argon. The chloride anion is also slightly larger than the neutral chlorine atom. To complete the octet, the one electron gained went into energy level 3, but now 17 protons are attracting 18 electrons. The attractive force on each electron has been reduced slightly, and the electrons are free to move outward a little, making the anion a little larger.

In general, a cation is smaller than its corresponding atom, and an anion is slightly larger.

REMEMBER

Ending up with a bond

Sodium can achieve its full octet and stability by losing an electron. Chlorine can fill its octet by gaining an electron. If the two are in the same container, then the electron sodium loses can be the same electron chlorine gains. We show this process in Figure 9-2, indicating that the 3s electron in sodium is transferred to the 3p orbital of chlorine.

The transfer of an electron creates ions — cations (positive charge) and anions (negative charge) — and opposite charges attract each other. The Na^+ cation attracts the Cl^- anion and forms the compound NaCl, or table salt. This is an example of an *ionic bond*, which is a *chemical bond* (a strong attractive force that keeps two chemical elements together) that comes from the *electrostatic attraction* (attraction of opposite charges) between cations and anions.

Ionic Salts

Although ions are often individual charged atoms, there are many examples of *polyatomic ions*, which are charged particles made up of more than one atom. Common polyatomic ions include ammonium, $NH_4{}^+$, and sulfate, $SO_4{}^{2-}$.

When cations and anions associate in ionic bonds, they form *ionic compounds*. At room temperature, most ionic compounds are solid because of the strong electrostatic forces that hold together the ions within them. The ions in ionic solids pack together in a *lattice*, a highly organized, regular arrangement that allows for the maximum electrostatic interaction while reducing the repulsive forces between anions and between cations. The geometric details of the packing can differ among different ionic compounds, but you can see a simple lattice structure in Figure 9-4. Flip to Chapter 11 for full details on polyatomic ions.

FIGURE 9-4:
The lattice structure of an ionic solid, sodium chloride.

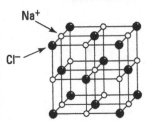

The strong electrostatic forces that hold together ionic lattices result in the high melting and boiling points that are common among ionic compounds. Although disrupting ionic bonds may take a great deal of thermal energy, many ionic compounds are easily dissolved in water or in other *polar solvents* (fluids made up of molecules that have unevenly distributed charge).

When the solvent molecules are polar, they can engage in favorable interactions with the ions, helping to compensate for disrupting the ionic bonds. For example, polar water molecules can interact well with both sodium cations (Na^+) and chlorine anions (Cl^-). Water molecules are polar because they have distinct and separate bits of positive and negative charge. Water molecules can orient their positive bits toward Cl^- and their negative bits toward Na^+. Positive charges attract negative charges and vice versa, so these kinds of interactions are favorable; that is, they require less energy. So water dissolves solid NaCl quite well because the water–ion interactions can compete with the Na^+–Cl^- interactions.

When ionic compounds are melted or dissolved so that the individual ions can move about, the resulting liquid is a very good conductor of electricity. Ionic solids, however, are often poor conductors of electricity.

Salts are a common variety of ionic compound. A salt can form from the reaction between a base and an acid. For example, hydrochloric acid reacts with sodium hydroxide to form sodium chloride (a salt) and water:

$$HCl(aq) + NaOH(aq) \rightarrow NaCl(aq) + H_2O(l)$$

Note that *aq* indicates that the substance is dissolved in water, in an *aqueous* solution.

Q. Why do metals tend to form ionic compounds with nonmetals?

EXAMPLE **A.** Metals are much less electronegative than nonmetals, meaning that they give up valence electrons much more easily. Nonmetals (especially Group VIIA and VIA nonmetals) very easily gain new valence electrons. So metals and nonmetals tend to form bonds in which the metal atoms entirely surrender valence electrons to the nonmetals. Bonds with extremely unequal electron–sharing are called *ionic bonds*.

YOUR TURN

④ Draw the electron dot structure of potassium fluoride and show how the electron is transferred between the elements potassium and fluorine to create the ionic compound.

⑤ The ionic compound lithium sulfide forms between the elements lithium and sulfur. In which direction are electrons transferred to form ionic bonds, and how many electrons are transferred?

⑥ Magnesium chloride is dissolved in a beaker of water and in a beaker of rubbing alcohol until no more compound will dissolve. An electrical circuit is set up for each beaker, with wires leading from a battery into the solution and a separate set of wires leading from the solution to a light bulb. The bulb connected to the aqueous solution circuit glows more brightly than the bulb connected to the alcohol solution circuit. Why?

Sharing Electrons with Covalent Bonds

REMEMBER

Sometimes the way for atoms to reach their most stable, lowest-energy states is to share valence electrons. When atoms share valence electrons, chemists say that they're engaged in *covalent bonding.* The very word *covalent* means "together in valence." Compared to ionic bonding, covalent bonding tends to occur between atoms of similar electronegativity, especially between nonmetals.

Just as ionic bonds tend to form in such a way that both atoms end up with completely filled valence shells, the atoms involved in covalent bonds tend to share electrons in such a way that each ends up with a completely filled valence shell. The shared electrons are attracted to the nuclei of both atoms, forming the bond.

Considering a hydrogen example

Hydrogen is number 1 on the periodic table — upper left corner. The hydrogen found in nature is often not composed of an individual atom. It's primarily found as H_2, a *diatomic* (two-atom) element. (Taken one step further, because a *molecule* is a combination of two or more atoms, H_2 is called a *diatomic molecule.*)

Hydrogen has one valence electron. It'd love to gain another electron to fill its 1s energy level, which would make it *isoelectronic* with helium (because the two would have the same electronic configuration), the nearest noble gas. Energy level 1 can hold only two electrons in the 1s orbital, so gaining another electron fills it. The driving force of hydrogen is filling the valence energy level and achieving the same electron arrangement as the nearest noble gas.

Imagine one hydrogen atom transferring its single electron to another hydrogen atom. The hydrogen atom receiving the electron fills its valence shell and reaches stability while becoming an anion (H^-). However, the other hydrogen atom now has no electrons (H^+) and moves further away from stability. This process of electron loss and gain simply won't happen, because the driving force of *both* atoms is to fill their valence energy level. So the H_2 compound can't result from the loss or gain of electrons. What *can* happen is that the two atoms share their electrons. At the atomic level, this sharing is represented by the overlap of the electron orbitals (sometimes called *electron clouds*). The two electrons (one from each hydrogen atom) "belong" to both atoms. Each hydrogen atom feels the effect of the two electrons; each has, in a way, filled its valence energy level. A *covalent bond* is formed — a chemical bond that comes from the sharing of one or more electron pairs between two atoms. The overlapping of the electron orbitals and the sharing of an electron pair are represented in Figure 9-5a.

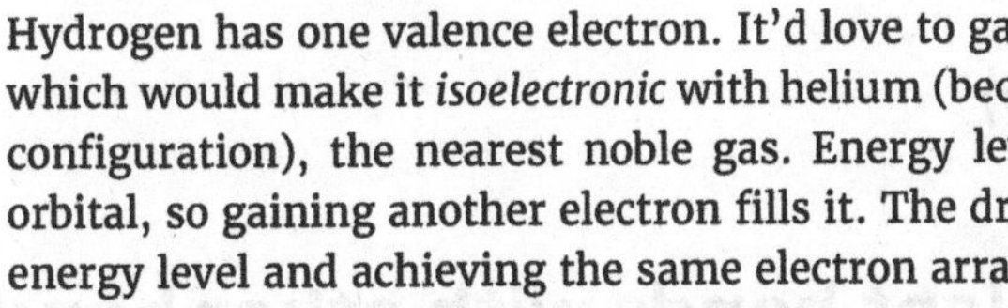

FIGURE 9-5:
The formation of a covalent bond in hydrogen.

Another way to represent this process is through the use of an *electron (Lewis) dot formula*. In this type of formula, valence electrons are represented as dots surrounding the atomic symbol, and the shared electrons are shown between the two atoms involved in the covalent bond. The electron dot formula representations of H_2 are shown in Figure 9-5b.

Most of the time, we use a slight modification of the electron dot formula called the *Lewis structural formula*; it's basically the same as the electron dot formula, but the shared pair of electrons (the covalent bond) is represented by a dash. The Lewis structural formula of diatomic hydrogen is shown in Figure 9-5c.

REMEMBER

In addition to hydrogen, six other elements are found in nature in the diatomic form: oxygen (O_2), nitrogen (N_2), fluorine (F_2), chlorine (Cl_2), bromine (Br_2), and iodine (I_2). So when we talk about oxygen gas or liquid bromine, we're talking about the diatomic element (diatomic molecule).

For another example of using the electron dot formula to represent the shared electron pair of a diatomic compound, look at bromine (Br_2), which is a member of the halogen family (see Figure 9-6). The two halogen atoms, each with seven valence electrons, share an electron pair and fill their octet.

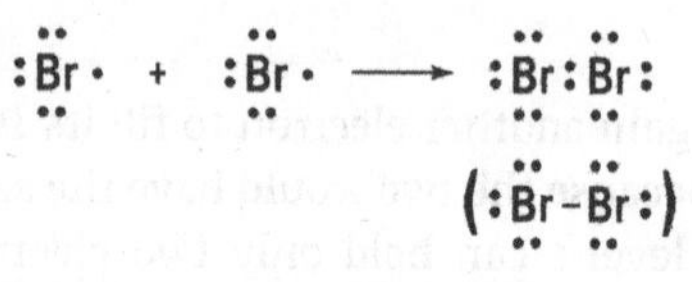

FIGURE 9-6:
The covalent bond formation of Br_2.

$$:\!\overset{\bullet\bullet}{Br}\!\cdot \quad + \quad :\!\overset{\bullet\bullet}{Br}\!\cdot \quad \longrightarrow \quad :\!\overset{\bullet\bullet}{Br}\!:\!\overset{\bullet\bullet}{Br}\!:$$

$$\left(:\!\overset{\bullet\bullet}{Br}\!-\!\overset{\bullet\bullet}{Br}\!:\right)$$

Comparing covalent bonds with other bonds

Ionic bonding occurs between a metal and a nonmetal. Covalent bonding, on the other hand, occurs between two nonmetals. The properties of these two types of compounds are different. Ionic compounds are usually solids at room temperature, whereas covalently bonded compounds can be solids, liquids, or gases. There's more. Ionic compounds (salts) usually have a much higher melting point than covalent compounds. In addition, ionic compounds tend to be electrolytes, and covalent compounds tend to be nonelectrolytes.

We know just what you're thinking: "If metals react with nonmetals to form ionic bonds, and nonmetals react with other nonmetals to form covalent bonds, do metals react with other metals?" The answer is yes and no.

Metals don't really react with other metals to form compounds. Instead, metals combine to form *alloys*, solutions of one metal in another. But in a situation called *metallic bonding*, which affects both alloys and pure metals, the valence electrons of each metal atom are donated to an electron pool, commonly called a *sea of electrons*, and are shared by all the atoms in the metal. These valence electrons are free to move throughout the sample instead of being tightly bound to an individual metal nucleus. The ability of the valence electrons to flow throughout the entire metal sample is why metals tend to be conductors of electricity and heat and is responsible for the luster of metals.

Understanding multiple bonds

Covalent bonding is the sharing of one *or more* electron pairs. In hydrogen and most other diatomic molecules, only one electron pair is shared. But in many covalent bonding situations, more than one electron pair is shared. This section shows you an example of a molecule in which more than one electron pair is shared.

Energy is needed to break a covalent bond. The resistance of the bond to breaking is called its *bond strength.* In general, more energy is needed to break a double bond than a single bond if the same elements are involved. For example, a carbon-to-carbon double bond (two shared pairs of electrons) has a higher bond strength (requires more energy to break the bond) than a carbon-to-carbon single bond. The double bond isn't twice as strong as a single bond, but its strength is considerably greater. And a triple bond is stronger yet. Chemists also observe that multiple bonds are shorter in bond length (the distance between the nuclei of the bonded atoms) than single bonds — double bonds are shorter than single bonds, and triple bonds are shorter than double bonds.

Nitrogen (N_2) is a diatomic molecule in the VA family on the periodic table, meaning that it has five valence electrons (see Chapter 5 for a discussion of families on the periodic table). So nitrogen needs three more valence electrons to complete its octet. A nitrogen atom can fill its octet by sharing three electrons with another nitrogen atom, forming three covalent bonds, a triple bond. The triple-bond formation of nitrogen is shown in Figure 9-7.

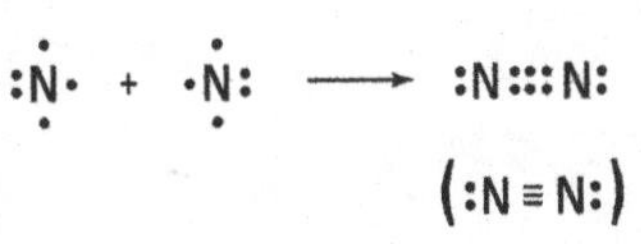

FIGURE 9-7:
Triple bond formation in N_2.

$$:\!N\!\cdot \;+\; \cdot N\!: \longrightarrow \;:\!N\!:::\!N\!:$$
$$(:\!N \equiv N\!:)$$

A triple bond isn't quite three times as strong as a single bond, but it's a very strong bond. In fact, the triple bond in nitrogen is one of the strongest bonds known. This strong bond is what makes nitrogen gas very stable and resistant to reaction with other chemicals. It's also why many explosive compounds (such as TNT and ammonium nitrate) contain nitrogen. When these compounds break apart in a chemical reaction, nitrogen gas (N_2) is formed and a large amount of energy is released.

Carbon dioxide (CO_2) is another example of a compound containing a multiple bond. Carbon can react with oxygen to form carbon dioxide. Carbon has four valence electrons, and oxygen has six. Carbon can share two of its valence electrons with each of the two oxygen atoms, forming two double bonds. These double bonds are shown in Figure 9-8.

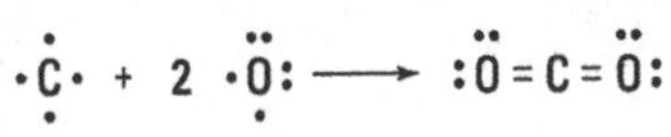

FIGURE 9-8:
Formation of carbon dioxide.

$$\cdot C \cdot \;+\; 2 \;\; \cdot O\!: \longrightarrow \;:\!O = C = O\!:$$

Here's something to remember when working with ionic and covalent compounds: A *molecule* is a compound that's covalently bonded. Referring to sodium chloride, which has ionic bonds, as a molecule is technically incorrect, but lots of chemists (and chemistry students) do it anyway. The mistake is kind of like using the wrong fork at a formal dinner. Some people may notice, but most don't notice or don't care. But just so you know, the correct term for ionic compounds is *formula unit*.

Q. What type of bond would you expect to form between a carbon atom and an oxygen atom? Why?

EXAMPLE

A. You would expect them to form a covalent bond. This is because carbon and oxygen are both nonmetals. When they bond they will share their electrons between one another rather than transferring an electron. This results in a covalent bond.

---

7 Identify the following compounds as having ionic or covalent:

YOUR TURN

(a) NO_2

(b) $CaCl_2$

(c) Li_3N

(d) H_2O

(e) NaCl

---

8 What is the difference between an ionic and covalent bond in regards to the electrons and how they form a bond?

Practice Questions Answers and Explanations

1 Elements seek stable valence shells by gaining or losing electrons. Whether an element gains or loses electrons has to do with where that element sits on the periodic table. For elements on the left side, losing just a few electrons is easier than gaining many more; for these elements, the Roman numeral of the A group to which they belong tells you how many electrons they lose — and therefore tells you their positive charge. For elements on the right side of the table, gaining a few electrons is easier than losing many more; these elements gain the number of electrons necessary to create valence shells like those in Group VIIIA (the noble gases). An element in Group VIA therefore gains two electrons.

   a. Lithium (Li) loses one electron, forming the lithium cation Li^+.

   b. Selenium (Se) gains two electrons to form the selenide anion Se^{2-}.

   c. Boron (B) loses three electrons to form the boron cation B^{3+}.

   d. Oxygen (O) gains two electrons to form the oxide anion O^{2-}.

   e. Chlorine (Cl) gains one electron to form the chloride anion Cl^-.

2 **N^{3-},** also known as *nitride* or the *nitrogen anion*. By gaining three electrons, nitrogen, which is in Group VA, assumes a full octet, like neon.

3 **Be^{2+},** also known as the *beryllium cation*. By losing two electrons, beryllium, which is in Group IIA, assumes a full octet valence shell, with two electrons like helium.

4 Potassium (K) transfers its single valence electron to fluorine (F), yielding an ionic bond between K^+ and F^-, as in the following figure:

$$K \cdot \quad \rightarrow \cdot \overset{\cdot\cdot}{\underset{\cdot\cdot}{F}} : \quad \longrightarrow \quad K^+ \; : \overset{\cdot\cdot}{\underset{\cdot\cdot}{F}} :^-$$

5 **Two lithium atoms each transfer a single electron to one sulfur atom to yield the ionic compound Li_2S.** As an alkali metal (Group IA), lithium easily gives up its single valence electron. As a Group VIA nonmetal, sulfur readily accepts two additional electrons into its valence shell.

6 As a salt, magnesium chloride ($MgCl_2$) is certainly an ionic compound. You can tell this easily because magnesium is a metal and chlorine is a nonmetal. Therefore, $MgCl_2$ dissolves to a greater extent in more polar solvents. Dissolved ions act as *electrolytes*, conducting electricity in solutions. The more brightly glowing bulb in the circuit containing the aqueous (water-based) solution suggests that more electrolytes are present in that solution. More salt dissolves in water than in rubbing alcohol because water is the more polar solvent.

7 Ionic compounds involve bonds between positive ions and negative ions. In practical terms this usually corresponds to a bond between a metal and a nonmetal. Covalent compounds involve bonds between nonmetal atoms that do not have an ionic charge.

   a. Covalent. Nitrogen is a nonmetal and oxygen is a nonmetal.

   b. Ionic. Calcium is a metal and chlorine is a nonmetal.

   c. Ionic. Lithium is a metal and nitrogen is a nonmetal.

   d. Covalent. Hydrogen is a nonmetal and oxygen is a nonmetal.

   e. Ionic. Sodium is a metal and chlorine is a nonmetal.

(8) Ionic bonds involve a transfer of electrons and covalent bonds involve a sharing of electrons between atoms. When ionic bonds are formed electrons are transferred from one atom to another. This transfer of electron(s) occurs so that both atoms will have a filled valence shell. When the electrons are transferred positive and negative ions are formed causing an electrostatic bond to form. Covalent compounds form bonds by sharing electrons. This sharing of electrons is done to fill the valence shell of the atoms as well. However, since the electrons are shared there are no ions formed.

Whaddya Know? Chapter 9 Quiz

Ready for a quiz? The 15 questions in this section will test the skills you learned in this chapter. When you're done, check out the section that follows for answers and explanations.

1 How many electrons would you expect the following atoms to gain or lose when they form an ion?

a. Nitrogen

b. Calcium

c. Helium

d. Chlorine

2 What is the expected ionic charge for the following elements?

a. Barium

b. Sodium

c. Bromine

d. Sulfur

e. Xenon

f. Aluminum

3 What are the 7 diatomic elements?

4 Does ionic bonding involve a sharing of electrons or a transfer of electrons?

5 Does covalent bonding involve a sharing of electrons or a transfer of electrons?

6 What family of elements has a filled octet without the need to bond or form an ion?

7 Are metals or nonmetals more likely to give up their electron in an ionic bond? Why?

8 Identify the bonds found in the following compounds as ionic or covalent:

a. CsBr

b. CO_2

c. CS_2

d. $CuCl_2$

e. CaI

f. NiO

g. SO_3

Answers to Chapter 9 Quiz

(1) a. Nitrogen would gain three electrons. Nitrogen has five valence electrons, so it is easier for nitrogen to gain three valence electrons to fill its octet.

b. Calcium would lose two electrons. Calcium has two valence electrons so it is easier for calcium to lose those two electrons to have a filled octet.

c. Helium will neither gain nor lose electrons. It is a noble gas and already has a filled outer energy level. It doesn't have a complete octet, however, as its highest energy level is only a 1s sublevel. The maximum number of electrons this sublevel can hold is two electrons. Helium has two electrons in its neutral state so it doesn't form an ion as it is already stable.

d. Chlorine gains one electron. Chlorine has seven valence electrons, so it is much easier for chlorine to gain one electron than to lose seven.

(2) a. Ba^{2+}

b. Na^{1+}

c. Br^{-1}

d. S^{2-}

e. Xe. Xenon does not have a charge as it is a noble gas. It does not gain or lose electrons to fill its octet. It already has a filled octet of eight valence electrons.

f. Al^{3+}

(3) H_2, N_2, O_2, F_2, Cl_2, Br, I_2. There are a few ways to help you remember these elements. The first, as a reminder, is the world BrINClHOF. It sounds like "brinklehof." This pneumonic is reasonably useful for helping you remember the seven elements. In addition, six of the seven elements form a seven in the top right of the periodic table. The only element that does not is hydrogen, which is by itself on the top left of the table.

(4) Ionic bonding involves a **transfer** of electrons. When an ionic bond forms, one or more electrons are transferred from one of atom to another. This ionic transfer creates a positive and negative ion, which generates an electrostatic attraction between the ions, forming a bond.

(5) Covalent bonding involves a **sharing** of electrons. When a covalent bond forms, one or more electrons are shared between two atoms. This sharing of the electrons allows both atoms to fill their octet.

(6) The noble gases are the family of elements on the periodic table that already have a filled valence/shell. This means they always have a filled octet. The noble gases do not need to gain, lose, or share any electrons to fill their octet. Due to this, the noble gases very rarely form bonds.

(7) Metals are more likely to give up their electrons when forming an ionic bond. Metals form positive ions, meaning they tend to lose their electrons when bonding ionically. This is due to most metals having two or fewer valence electrons, so it is far easier for them to lose an electron or two than it is for them to gain six or seven electrons.

(8) a. Ionic bond. Cesium is a metal and bromine is a nonmetal, making this an ionic bond.

b. Covalent bond. Carbon and oxygen are both nonmetals, so they form a covalent bond.

c. Covalent bond. Carbon and sulfur are both nonmetals, so they form a covalent bond.

d. Ionic bond. Copper is a metal and chlorine is a nonmetal, making this an ionic bond.

e. Ionic bond. Calcium is a metal and iodine is a nonmetal, making this an ionic bond.

f. Ionic bond. Nickel is a metal and oxygen is a nonmetal, making this an ionic bond.

g. Covalent bond. Sulfur is a nonmetal and oxygen is a nonmetal, making this a covalent bond.

Chapter **10**

The Shape of Molecules

O nce you understand the types of bonds that form and why those bonds form, the next step is to look at the shape that different molecules take. The shape of molecules plays a huge role in giving certain molecules some of their unique properties. Water, for example, has some very unique properties, like its strong surface tension and its ability to dissolve so many things. These properties are due to its asymmetrical molecular shape. As you read on you'll see how to determine the molecular structure of different molecules, and then use that information to predict their three-dimensional geometry and finally the properties that can arise from that geometry.

Drawing the Structural Formulas of Molecules

To write a formula that stands for the exact compound you have in mind, you often must write the structural formula instead of the molecular formula. The *structural formula* shows the elements in the compound, the exact number of each atom in the compound, and the bonding pattern for the compound. The electron dot formula and Lewis formula are examples of structural formulas.

Writing the electron dot formula for water

The following steps explain how to write the electron dot formula for a simple molecule — water — and provide some general guidelines and rules to follow:

1. **Write a skeletal structure showing a reasonable bonding pattern using just the element symbols.**

   The skeletal structure involves only the atomic symbols and not the valence electrons. Often, most atoms are bonded to a single atom. This atom is called the *central atom*. Hydrogen and the halogens are very rarely, if ever, central atoms. Carbon, silicon, nitrogen, phosphorus, oxygen, and sulfur are always good candidates, because they form more than one covalent bond to fill their valence energy level. In the case of water, H_2O, oxygen is the central element and the hydrogen atoms are both bonded to it. The bonding pattern looks like this:

   $$O\ H$$
   $$H$$

   The hydrogen atoms can go anywhere around the oxygen. We put the hydrogen atoms at a 90–degree angle to each other, but it really doesn't matter when writing electron dot (or Lewis) formulas.

2. **Take all the valence electrons from all the atoms and throw them into an electron pot.**

   Each hydrogen atom has 1 electron, and the oxygen atom has 6 valence electrons (VIA family), so you have 8 electrons in your electron pot. You use those electrons to make your bonds and complete each atom's octet.

   electron pot

3. **Use the $N - A = S$ equation to figure the number of bonds in this molecule.**

   In this equation,

   N equals the sum of the number of valence electrons needed by each atom. N has only two possible values — 2 or 8. If the atom is hydrogen, it's 2; if it's anything else, it's 8.

   A is the number of valence electrons in your electron pot — the sum of the number of valence electrons available for each atom. If you're doing the structure of an ion, you add one electron for every unit of negative charge if it's an anion or subtract one electron for every unit of positive charge if it's a cation.

   S equals the number of electrons shared in the molecule. And if you divide S by 2, you have the number of covalent bonds in the molecule.

   So in the case of water,

   $N = 8 + 2(2) = 12$ (8 valence electrons for the oxygen atom, plus 2 each for the two hydrogen atoms)

   $A = 6 + 2(1) = 8$ (6 valence electrons for the oxygen atom, plus 1 for each of the two hydrogen atoms)

$S = 12 - 8 = 4$ (four electrons shared in water), and $S/2 = 4/2 = 2$ bonds

You now know that water molecules have two bonds (two shared pairs of electrons).

4. **Distribute the electrons from your electron pot to account for the bonds.**

   You use 4 electrons from the 8 in the pot, which leaves you with 4 to distribute later. At least one bond must connect your central atom to the atoms surrounding it.

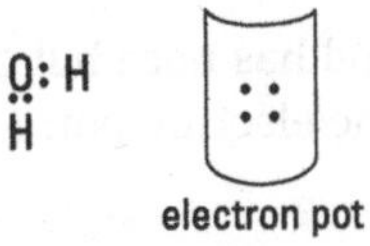

O : H
H

::

electron pot

5. **Distribute the rest of the electrons (normally in pairs) so that each atom achieves its full octet of electrons.**

   Remember that hydrogen needs only 2 electrons to fill its valence energy level. In this case, each hydrogen atom has 2 electrons, but the oxygen atom has only 4 electrons, so the remaining 4 electrons are placed around the oxygen, emptying your electron pot. The completed electron dot formula for water is shown in Figure 10-1.

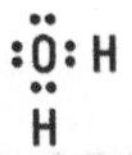

FIGURE 10-1:
Electron dot
formula of
H_2O.

:Ö: H
 H

REMEMBER Notice that this structural formula actually shows two types of electrons: *bonding electrons*, the electrons that are shared between two atoms, and *nonbonding electrons*, the electrons that are not being shared. The last 4 electrons (2 electron pairs) that you put around oxygen are not being shared, so they're nonbonding electrons.

Writing the Lewis formula for water

If you want the Lewis formula for water, all you have to do is substitute a dash for every bonding pair of electrons. This structural formula is shown in Figure 10-2.

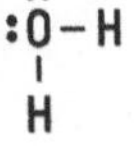

FIGURE 10-2:
The Lewis
formula for
H_2O.

:Ö – H
     |
     H

Writing the Lewis formula for C_2H_4O

Here's an example of a Lewis formula that's a little more complicated: C_2H_4O.

The compound has the following framework:

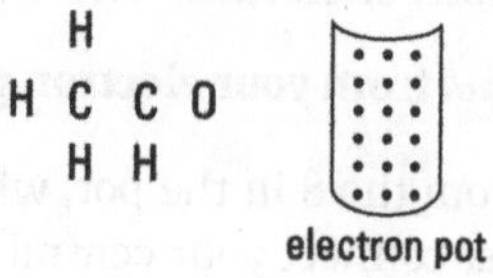

H
H C C O
H H

electron pot

Notice that the compound has not 1 but 2 central atoms — the 2 carbon atoms. You can put 18 valence electrons into the electron pot: 4 for each carbon atom, 1 for each hydrogen atom, and 6 for the oxygen atom.

Now apply the $N - A = S$ equation:

$N = 2(8) + 4(2) + 8 = 32$ (2 carbon atoms with 8 valence electrons each, plus 4 hydrogen atoms with 2 valence electrons each, plus an oxygen atom with 8 electrons)

$A = 2(4) + 4(1) + 6 = 18$ (4 electrons for each of the two carbon atoms, plus 1 electron for each of the 4 hydrogen atoms, plus 6 valence electrons for the oxygen atom)

$S = 32 - 18 = 14$, and $S/2 = 14/2 = 7$ bonds

Add single bonds between the carbon atoms and the hydrogen atom, between the 2 carbon atoms, and between the carbon atom and oxygen atom. That's 6 of your 7 bonds.

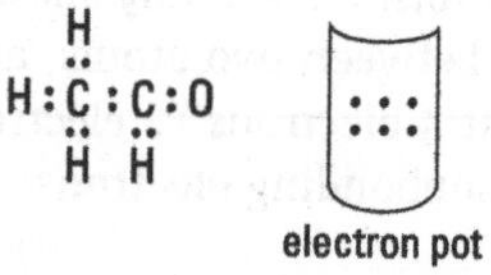

H
H:C:C:O   :·:
H H

electron pot

The seventh bond can go only one place, and that's between the carbon atom and the oxygen atom. It can't be between a carbon atom and a hydrogen atom, because that would overfill hydrogen's valence energy level. And it can't be between the two carbon atoms, because that would give the carbon on the left 10 electrons instead of 8. So there must be a double bond between the carbon atom and the oxygen atom. The 4 remaining electrons in the pot must be distributed around the oxygen atom, because all the other atoms have reached their octet. The electron dot formula is shown in Figure 10-3.

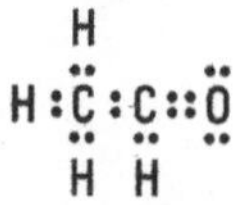

FIGURE 10-3:
Electron dot
formula of
C_2H_4O.

H
H:C:C::Ö
H H

If you convert the bonding pairs to dashes, you have the Lewis formula of C_2H_4O, as shown in Figure 10-4.

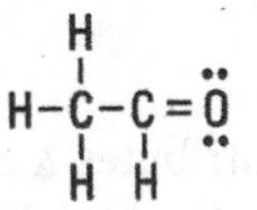

FIGURE 10-4:
The Lewis formula for C_2H_4O.

$$H-\overset{\overset{\displaystyle H}{|}}{\underset{\underset{\displaystyle H}{|}}{C}}-C=\overset{..}{\underset{..}{O}}$$

We like the Lewis formula because it enables you to show a lot of information without having to write all those little dots. But it, too, is rather bulky. Sometimes chemists (who are, in general, a lazy lot) use *condensed structural formulas* to show bonding patterns. They may condense the Lewis formula by omitting the nonbonding electrons and grouping atoms together and/or by omitting certain dashes (covalent bonds). A couple of condensed formulas for C_2H_4O are shown in Figure 10-5.

FIGURE 10-5:
Condensed structural formulas for C_2H_4O.

$CH_3 - CH = O$

CH_3CHO

Sometimes a covalent bond forms when one atom donates both electrons to the bond, with the other atom contributing no electrons. This kind of bond is called a *coordinate covalent bond*. Atoms with lone pairs are capable of donating both electrons to a coordinate covalent bond. A *lone pair* consists of two electrons paired within the same orbital that aren't used in bonding. Even though covalent bonding usually occurs between nonmetals, metals can engage in coordinate covalent bonding. Usually, the metal receives electrons from an electron donor called a *ligand*.

REMEMBER

Sometimes a given set of atoms can covalently bond with each other in multiple ways to form a compound. This situation leads to something called *resonance*. Each of the possible bonded structures is called a *resonance structure*. The actual structure of the compound is a *resonance hybrid*, a sort of weighted average of all the resonance structures. For example, if two atoms are connected by a single bond in one resonance structure and the same two atoms are connected by a double bond in a second resonance structure, then in the resonance hybrid, those atoms are connected by a bond that is worth 1.5 bonds. A common example of resonance is found in ozone, O_3, shown in Figure 10-6.

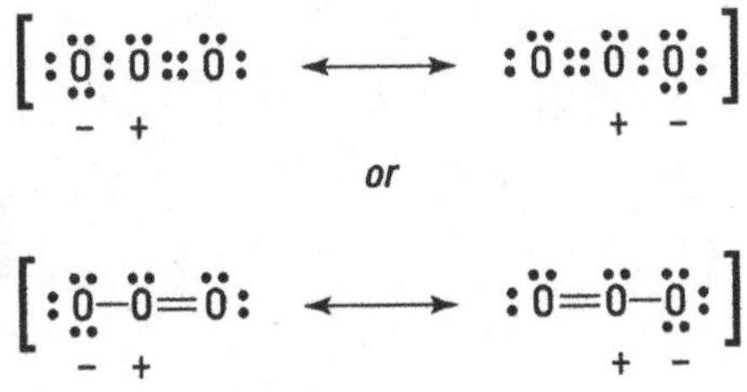

FIGURE 10-6:
Two representations of resonance structures of ozone.

$$\left[\;:\overset{..}{\underset{..}{O}}:\overset{..}{O}::\overset{..}{\underset{..}{O}}:\;\underset{-\quad+}{}\;\longleftrightarrow\;:\overset{..}{\underset{..}{O}}::\overset{..}{O}:\overset{..}{\underset{..}{O}}:\;\underset{+\quad-}{}\right]$$

or

$$\left[\;:\overset{..}{\underset{..}{O}}-\overset{..}{O}=\overset{..}{\underset{..}{O}}:\;\underset{-\quad+}{}\;\longleftrightarrow\;:\overset{..}{\underset{..}{O}}=\overset{..}{O}-\overset{..}{\underset{..}{O}}:\;\underset{+\quad-}{}\right]$$

Q. Draw a Lewis structure for propene, C_3H_6.

EXAMPLE **A.** First, add up the total valence electrons. Each carbon atom contributes 4 electrons, and each hydrogen contributes 1, for a total of 18 valence electrons. Next, pick a central atom. The only choice available is carbon, because hydrogen can only have one bond and thus can never be a central atom in a Lewis structure. You have three carbons, so just connect the three into a carbon chain. In formulas containing just carbon and hydrogen, you'll always find the carbons bonded together in a chain with the hydrogens bonded around the outside.

With the three carbons connected together, you've used up four of the total valence electrons, because each bond counts as two electrons. You then can place the hydrogens around the carbons as you see fit, as long as you remember that each carbon can only have four total bonds or lone electron pairs. After placing the hydrogens, you end up with eight total bonds in the structure for a total of 16 valence electrons. This leaves two electrons that you can place on a carbon atom that does not have four total bonds or electron pairs present.

You're now finished with the simple part. One carbon atom in the structure still requires two additional electrons to fill its valence shell. The only way to fill this shell is to take the lone pair of electrons you added to one of the carbons and instead use it to create a double bond between two of the carbons. You then need to move the hydrogens around to ensure that each carbon has a total of four bonds. Only one arrangement of hydrogen atoms to the three carbons allows you to fill all the carbon valence shells, as you can see in the following figure:

$$
\begin{array}{ccccc}
 & H & H & & H \\
 & | & | & & \diagup \\
H & - C & - C & = & C \\
 & | & & & \diagdown \\
 & H & & & H
\end{array}
$$

YOUR TURN

1 Bertholite is the common name for dichlorine (Cl_2), a toxic gas that has been used as a chemical weapon. Why is bertholite most certainly a covalently bonded compound? What is the most likely electron dot structure of this compound?

When aluminum chloride salt is dissolved in water, aluminum(III) cations become surrounded by clusters of six water molecules to form a "hexahydrated" aluminum cation,
$Al(H_2O)_6{}^{3+}$. Being a Group IIIA metal, aluminum easily gives up its valence electrons.
The oxygen atom in water possesses two lone pairs. What kind of bonding most likely
occurs between the aluminum and the hydrating water molecules?

 Benzene, C_6H_6, is a common industrial solvent. The benzene molecule is based on a
ring of covalently bonded carbon atoms. Draw two acceptable Lewis structures for benzene. Based on the structures, describe a likely resonance hybrid structure for benzene.

Occupying and Overlapping Molecular Orbitals

Another covalent bonding model is molecular orbital theory. In *molecular orbital (MO) theory*,
atomic orbitals on the individual atoms combine to form molecular orbitals (MOs), which
aren't hybrid orbitals. A molecular orbital covers the entire molecule and has a definite shape
and energy. The combination of two atomic orbitals produces two molecular orbitals. (Like in
hybridization, the total number of orbitals never changes.)

>> **Bonding molecular orbital:** The bonding MO has a lower energy than the original atomic
orbitals.

>> **Antibonding molecular orbital:** The antibonding molecular orbital has a higher energy
than the original atomic orbitals.

As shown in Figure 10-7, lower-energy orbitals are more stable than higher-energy orbitals. Notice that the bonding molecular orbital, because it's of lower energy than the original atomic orbitals, strengthens the bond.

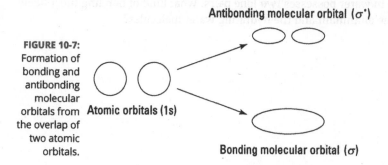

Antibonding molecular orbital (σ^*)

FIGURE 10-7:
Formation of bonding and antibonding molecular orbitals from the overlap of two atomic orbitals.

Atomic orbitals (1s)

Bonding molecular orbital (σ)

The end-to-end overlap of two p orbitals yields a σ bonding and a σ^* antibonding molecular orbital. The side-by-side overlapping of p orbitals yields a π bond composed of one π bonding molecular orbital and one π^* antibonding molecular orbital, as you can see in Figure 10-8.

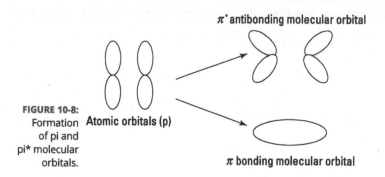

π^* antibonding molecular orbital

FIGURE 10-8:
Formation of pi and pi* molecular orbitals.

Atomic orbitals (p)

π bonding molecular orbital

Before going any further into molecular orbital theory, make sure you know exactly what sigma and pi bonds are:

>> **Sigma bond:** When atomic orbitals overlap head to head in such a way that the resulting molecular orbital is symmetric with respect to the *bond axis* (the line connecting the two bonded atoms), chemists say that a σ bond *(sigma bond)* is formed.

>> **Pi bond:** When atomic orbitals overlap side to side in such a way that the resulting molecular orbital is symmetric with the bond axis in only one plane, chemists say that a π bond *(pi bond)* is formed.

Sigma bonds are stronger than pi bonds because the electrons within sigma bonds lie directly between the two atomic nuclei. The negatively charged electrons in sigma bonds therefore experience favorable (as in low-energy) attraction to the positively charged nuclei. Electrons in pi bonds are farther away from the nuclei, so they experience weaker attraction.

Sigma bonds form when s or p orbitals overlap in a head-on manner. Single bonds are usually sigma bonds. Pi bonds are usually double or triple bonds. Figure 10-9 depicts these situations.

FIGURE 10-9:
Formation of a
sigma bond (σ)
from two s
orbitals and
formation of a
pi bond (π)
from two
adjacent
p orbitals.

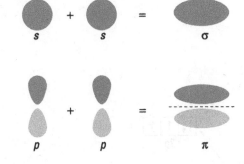

After the molecular orbitals form, you put electrons in. You add electrons using the same rules you use for electron configurations:

>> The lower-energy orbitals fill first.

>> Each orbital has a maximum of two electrons.

>> Orbitals of equal energy half-fill before pairing electrons.

>> When two s atomic orbitals combine, two sigma (σ) molecular orbitals form. One is sigma bonding (σ), and the other is sigma antibonding (σ*).

Figure 10-10 shows the molecular orbital diagram for H_2.

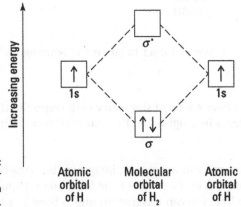

FIGURE 10-10:
Molecular
orbital diagram
for H_2.

Notice that the two electrons (one from each hydrogen atom) have both gone into the sigma bonding MO. You can determine the bonding situation in the molecular orbital theory by calculating the MO bond order. The *MO bond order* is the number of electrons in bonding MOs minus the number of electrons in antibonding MOs, divided by 2. A stable bonding situation exists between two atoms when the bond order is greater than zero. The larger the bond order, the stronger the bond. For H_2 in Figure 10-10, the bond order is (2 − 0)/2 = 1.

When two sets of p orbitals combine, one sigma bonding MO and one sigma antibonding MO are formed along with two pi bonding MOs and two pi antibonding MOs. Figure 10-11 shows the MO diagram for O_2. For the sake of simplicity, you don't see the 1s orbitals of each oxygen or MO here, just the valence electron orbitals.

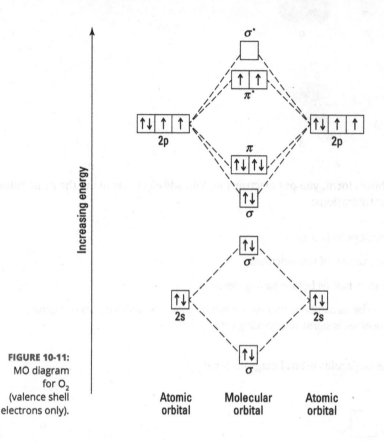

FIGURE 10-11:
MO diagram
for O_2
(valence shell
electrons only).

The bond order for O_2 is $(10 - 6)/2 = 2$. (Don't forget to count the bonding and antibonding electrons at energy level 1.)

EXAMPLE

Q. Both sigma bonds and pi bonds have a kind of symmetry with respect to the two atoms in the bond. What is the difference in a sigma bond's symmetry versus a pi bond's symmetry?

A. In any bond between two atoms, you can imagine an imaginary line (the bond axis) that connects the center of one atom to the center of the other atom. Sigma bonds are perfectly symmetric about this line; you can imagine the sigma bond as a kind of tube that wraps around the bond axis. If you were to rotate the bonded atoms around the imaginary line, the bond would look the same all the way around.

Pi bonds, on the other hand, are symmetric to the bond axis *in only one plane*. You can imagine the two atoms pressed onto a flat surface; the bond axis is the imaginary line on this surface. The pi bonds connect the two atoms above the line and below the line. If you were to rotate the bonded atoms around the imaginary line, the pi bonds would rise up off the surface and sink below the surface as you rotated them, like the planks on a paddlewheel rise above the surface of the water and then sink below the surface of the water.

YOUR TURN

4 Draw a molecular orbital diagram for the hypothetical molecule dihelium (He_2).

5 Based on the molecular orbital diagram of dihelium (He_2), explain why dihelium is far less likely to exist than hydrogen (H_2).

6 Double bonds involve one sigma bond and one pi bond. A simple molecule that contains a double bond is ethene, $H_2C=CH_2$. Ethene reacts with water to form ethanol:

$$H_2C = CH_2 + H_2O \rightarrow H_3C\text{-}COH$$

This reaction is favorable, meaning that it progresses on its own, without any input of energy. Why might this be the case?

Polarity: Sharing Electrons Unevenly

The *shape* (spatial orientation of the atoms) of a molecule may determine a great deal of its properties. For example, the shape of many organic molecules determines whether they're reactive in certain circumstances. The shape is especially indicative of reactivity for *enzymes*, biological catalysts that are found in the human body. If the shape of a particular enzyme is altered, that enzyme becomes useless in helping a certain biochemical reaction occur. Shape is also important in the complex molecules of drugs; it makes the drugs active but also leads to side effects. This section examines the role polarity plays in the shapes of molecules.

Polarity and electronegativity

One of our favorite lines from the book *Animal Farm* is "All animals are equal, but some animals are more equal than others." The same is true of covalent bonds — electron pairs may be shared, but not always equally.

When a chlorine atom covalently bonds to another chlorine atom, the shared electron pair is shared equally. The electron density that comprises the covalent bond is located halfway between the two atoms. Each atom attracts the two bonding electrons equally because each nucleus has the same number of protons.

But what happens when the two atoms involved in a bond aren't the same? The two positively charged nuclei have different attractive forces; they "pull" on the electron pair to different degrees. The end result is that the electron pair is shifted toward one atom. But the question is "Which atom does the electron pair shift toward?" To answer this question you can think back to electronegativities, which are covered in Chapter 7.

REMEMBER

Remember, larger electronegativity values result in an atom attracting bonding electron pairs to itself more than smaller electronegativities. You can look at Figure 7-3 in Chapter 7 for the entire electronegativity chart. As a reminder, they tend to increase from left to right on the periodic table and to increase from the bottom to the top of groups on the periodic table. Electronegativities are useful because they give information about what will happen to the bonding pair of electrons when two atoms bond. Basically, three types of bonds can form:

>> **Nonpolar covalent bond:** This bond has an electron pair that's equally shared. You have a nonpolar covalent bond anytime the two atoms involved in the bond are the same or anytime the difference in the electronegativities of the atoms involved in the bond is very small. For example, look at the Cl_2 molecule. Chlorine has an electronegativity value of 3.0, as shown in Figure 7-3. Each chlorine atom attracts the bonding electrons with a force of 3.0. Because there's an equal attraction, the bonding electron pair is shared equally between the two chlorine atoms and is located halfway between the two atoms.

>> **Polar covalent bond:** The electron pair in this bond is shifted toward one atom. The atom that more strongly attracts the bonding electron pair is slightly negative, and the other atom is slightly positive. The larger the difference in the electronegativities, the more negative and positive the atoms become.

Consider hydrogen chloride (HCl). Hydrogen has an electronegativity of 2.1, and chlorine has an electronegativity of 3.0. The electron pair that is bonding HCl together shifts toward the chlorine atom because it has a larger electronegativity value. Check out the next section for more information on these types of bonds.

>> **Ionic bond:** In this case, the bonding electrons are totally removed from one of the atoms and ions are formed. Now look at a case in which the two atoms have extremely different electronegativities — sodium chloride (NaCl). Sodium chloride is ionically bonded (see the earlier sections "Pairing Charges with Ionic Bonds" and "Forming Sodium Chloride"). An electron has transferred from sodium to chlorine. Sodium has an electronegativity of 0.9, and chlorine has an electronegativity of 3.0. That's an electronegativity difference of 2.1 (3.0 – 0.9), making the bond between the two atoms very, very polar. In fact, the electronegativity difference provides another way of predicting the kind of bond that will form between two elements.

The following table breaks down these three types of bonds that are formed and shows their electronegativity difference:

Electronegativity Difference	Type of Bond Formed
0.0 to 0.2	Nonpolar covalent
0.3 to 1.4	Polar covalent
>1.5	Ionic

The presence of a polar covalent bond in a molecule can have some pretty dramatic effects on the properties of a molecule.

Polar covalent bonding

If the two atoms involved in the covalent bond aren't the same, the bonding pair of electrons is pulled toward one atom, with that atom taking on a slight (partial) negative charge and the other atom taking on a partial positive charge. In many cases, the molecule then has a positive end and a negative end and can be referred to as a *dipole* (think of a magnet). The polar bonds within a molecule add to create polarity in the molecule as a whole. The precise way in which the individual bonds contribute to the overall polarity of the molecule depends on the shape of the molecule:

>> **Two very polar bonds pointing in opposite directions:** Their polarities cancel out.

>> **Two polar bonds pointing in the same direction:** Their polarities add.

>> **Two polar bonds pointing at each other so they're diagonal to one another:** Their polarities cancel in one direction but add in a perpendicular direction.

Figure 10-12 shows a couple of examples of molecules in which dipoles have formed.

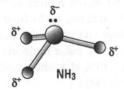

FIGURE 10-12: Polar covalent bonding in HF and NH₃.

In hydrogen fluoride (HF), the bonding electron pair is pulled much closer to the fluorine atom than to the hydrogen atom, so the fluorine end becomes partially negatively charged and the hydrogen end becomes partially positively charged. The same thing takes place in ammonia (NH_3); the nitrogen has a greater electronegativity than hydrogen, so the bonding pairs of electrons are more attracted to it than to the hydrogen atoms. The nitrogen atom takes on a partial negative charge, and the hydrogen atoms take on a partial positive charge.

The presence of a polar covalent bond explains why some substances act the way they do in a chemical reaction: Because this type of molecule has a positive end and a negative end, it can attract the part of another molecule with the opposite charge.

In addition, this type of molecule can act as a weak electrolyte because a polar covalent bond allows the substance to act as a conductor. So if a chemist wants a material to act as a good *insulator* (a device used to separate conductors), the chemist looks for a material with as weak a polar covalent bond as possible.

To explore an example of polarity, start by looking at Figure 10-13, which shows the Lewis structures of two diatomic molecules, the hydrogen molecule, H_2, and the chlorine fluoride molecule, ClF.

FIGURE 10-13:
Lewis structures of hydrogen and chlorine fluoride with electro-negativity arrows.

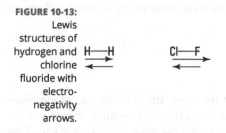

The arrows beneath the two structures illustrate the relative electronegativities of the elements. In the hydrogen molecule, the two arrows are the same length because both atoms have equal electronegativities. In the chlorine fluoride molecule, the arrows are of different lengths because the electronegativities are different. The electronegativity of fluorine is greater than that of chlorine, so the arrow pointing to the fluorine is longer than the arrow pointing to the chlorine.

If you combine the two equal arrows in the hydrogen molecule, they cancel. Hydrogen atoms, due to their identical electronegativities, pull equally on the shared electrons, resulting in an equal sharing of the electrons. Therefore, H_2 is nonpolar, because the charges on either end of the molecule are the same and equal.

However, if you combine the unequal arrows in chlorine fluoride, you're left with a net arrow pointing toward the fluorine end of the molecule. This remaining arrow indicates that the shared negative electrons are pulled closer to the fluorine. The unequal sharing of the electrons leads to a slight (partial) negative charge on the fluorine, leaving a slight (partial) positive charge on the chlorine. Thus, ClF is a polar molecule. (If the difference in electronegativities were greater, the more electronegative element would strip the electron from the less electronegative element to form an ionic bond.)

A lowercase Greek letter delta, δ, indicates the presence of a partial charge. If you add the partial charges to the picture of the chlorine fluoride molecule, you get the structure shown in Figure 10-14.

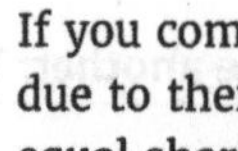

FIGURE 10-14:
The Lewis structure of chlorine fluoride shows partial charges.

$$\delta^+ \qquad \delta^-$$
$$\underset{\longrightarrow}{Cl \!-\!\!-\! F}$$

The arrow below the molecule is an alternative to the partial charges. The arrow has a cross at the positive end and points toward the more negative end.

Predicting polarity

Predicting the polarity of diatomic molecules is simple. A diatomic molecule is polar if the bond is a polar covalent bond, and it's nonpolar if the covalent bond is nonpolar. Thus, molecules such as HCl, NO, and ClF are polar, and molecules such as H_2, Cl_2, and N_2 are nonpolar. Refer to the preceding sections for a discussion of electronegativities and polar covalent bonding.

REMEMBER

A polar molecule is a dipole. *Dipole* refers to the two poles, the partially positive pole and the partially negative pole. The magnitude of the dipole is related to the difference in the electronegativities (and, to a lesser degree, to the distance between the two atoms). The *dipole moment* is a measure of the magnitude of the dipole. Nonpolar molecules have a zero dipole moment, and all polar molecules have a nonzero dipole moment.

Although the prediction of the polarity of a diatomic molecule is relatively simple, what happens when there are more than two atoms? We use water to illustrate this problem because water clearly demonstrates that, if you want to know about the polarity of a molecule, you must first know the *molecular geometry*, the arrangement of the atoms and electrons in three-dimensional space. Figure 10-15 shows the correct Lewis structures for the water molecule.

FIGURE 10-15:
Two possible Lewis structures of water with the partial positive and negative charges shown.

$$\delta^+ \qquad \delta^- \qquad \delta^+$$
$$H—\overset{\overset{\displaystyle \cdot\cdot}{}}{\underset{\cdot\cdot}{O}}—H$$

$$H—\overset{\cdot\cdot}{\underset{|}{O}}\overset{\delta^-}{:}$$
$$\underset{\delta^+}{H}$$

Due to the difference in electronegativities between the hydrogen and the oxygen, all the bonds are polar covalent. The oxygen atom, being more electronegative than the hydrogens, has a partial negative charge, leaving a partial positive charge on each hydrogen atom. The linear arrangement in the Lewis structure on the left has a partial positive charge at each end and a partial negative charge in the center. Because the molecule doesn't have a partial positive end and a partial negative end, this molecule is nonpolar. The bent molecule, however, is polar because it has a partial negative oxygen on one end of the molecule and two hydrogens with partial positives at the other end.

The properties of water indicate which structure is more likely. A nonpolar water molecule would have properties similar to other nonpolar molecules of approximately the same molecular mass, such as methane. For example, it should boil at about –100°C. Because water boils 200 degrees higher (+100°C), you can reasonably assume that water must be polar instead of nonpolar.

Q. Why doesn't it make sense to ask whether an element (like hydrogen or fluorine) engages in polar bonds versus nonpolar bonds?

EXAMPLE

A. You know whether an element engages in polar or nonpolar bonds only with respect to specific bonds with other elements. For example, hydrogen engages in a perfectly nonpolar covalent bond with another hydrogen atom in the molecule H_2. Likewise, fluorine engages in a nonpolar bond with another fluorine atom in F_2. On the other hand, the bond between hydrogen and fluorine in the compound HF is very polar because of the large electronegativity difference between the two atoms.

YOUR
TURN

7 Predict whether bonds between the following pairs of atoms are nonpolar covalent, polar covalent, or ionic:

(a) H and Cl

(b) Ga and Ge

(c) O and O

(d) Na and Cl

(e) C and O

8 Tetrafluoromethane (CF_4) contains four covalent bonds. Water (H_2O) contains two covalent bonds. Which molecule has bonds with more polar character?

Shaping Molecules: VSEPR Theory and Hybridization

One method to predict the shape of molecules is the *valence-shell electron-pair repulsion (VSEPR) theory*. The basis of this theory is that the valence-shell electron pairs around a central atom try to move as far away from each other as possible (like two supermodels showing up at a party wearing the same dress). Electrons spread out to minimize the repulsion between their like (negative) charges. This theory includes both electrons in bonds and nonbonding or lone-pair electrons.

With this VSEPR method, we show you how to actually determine two geometries. The two geometries are as follows:

>> **Electron-pair geometry:** Sometimes called the *electron-group geometry,* it considers all electron pairs surrounding a nucleus.

>> **Molecular geometry:** The nonbonding electrons become "invisible," and you consider only the arrangement of the atomic nuclei.

In determining geometry, double and triple bonds count the same as single bonds.

TIP

To determine the electron-group and molecular geometry, follow these steps:

1. **Write the Lewis electron dot formula of the compound.**

   Refer to the earlier section "Drawing the Structural Formulas of Molecules" for the rules of writing Lewis structures.

2. **Determine the number of electron pair groups surrounding the central atom(s).**

   Remember that double and triple bonds count the same as a single bond.

3. **Determine the geometric shape that maximizes the distance between the electron groups.**

   Check out Table 10-1 for electron-pair geometry, the shapes associated with the number of electron pairs.

4. **Mentally allow the nonbonding electrons to become invisible.**

   They are still present and are still repelling the other electron pairs; however, you just don't "see" them.

5. **Determine the molecular geometry from the arrangement of bonding pairs around the central atom by referring to Table 10-1.**

Even though you normally don't have to worry about more than four electron pairs around the central atom (because of the octet rule), you can find some of the less common exceptions to the octet rule in Table 10-1. Figure 10-16 shows some of the more common shapes mentioned in the table.

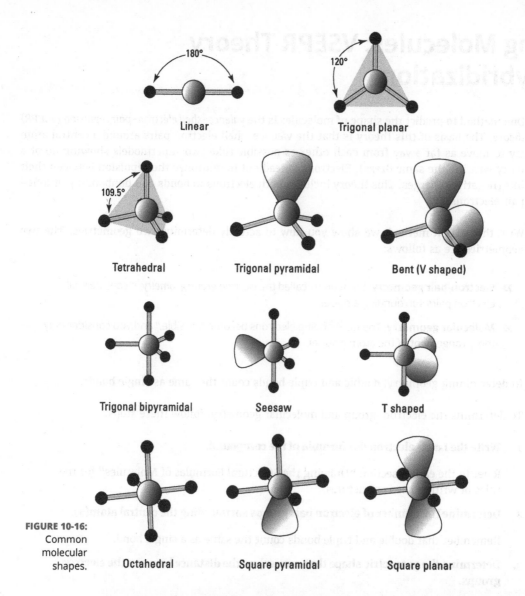

FIGURE 10-16: Common molecular shapes.

TABLE 10-1 **Predicting Molecular Shape with the VSEPR Theory**

Total Number of Electron Pairs	Number of Bonding Pairs	Electron-Pair Geometry	Molecular Geometry
2	2	Linear	Linear
3	3	Trigonal planar	Trigonal planar
3	2	Trigonal planar	Bent, V shaped
3	1	Trigonal planar	Linear
4	4	Tetrahedral	Tetrahedral
4	3	Tetrahedral	Trigonal pyramidal
4	2	Tetrahedral	Bent, V shaped

Total Number of Electron Pairs	Number of Bonding Pairs	Electron-Pair Geometry	Molecular Geometry
5	5	Trigonal bipyramidal	Trigonal bipyramidal
5	4	Trigonal bipyramidal	Seesaw
5	3	Trigonal bipyramidal	T shaped
5	2	Trigonal bipyramidal	Linear
6	6	Octahedral	Octahedral
6	5	Octahedral	Square pyramidal
6	4	Octahedral	Square planar

Now consider the geometry of water (H_2O) and ammonia (NH_3).

The first thing you have to do is determine the Lewis formula for each compound. Follow the rules outlined earlier in "Drawing the Structural Formulas of Molecules" and write the Lewis formulas, as shown in Figure 10-17.

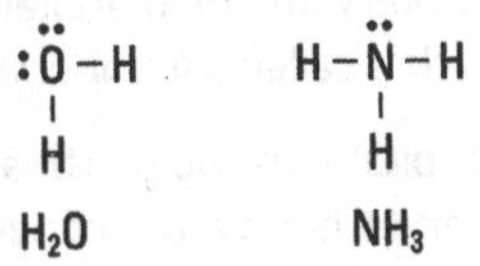

FIGURE 10-17: Lewis formulas for H_2O and NH_3.

$$:\overset{..}{O}-H$$
$$\;\;\;|$$
$$\;\;\;H$$
$$H_2O$$

$$H-\overset{..}{N}-H$$
$$\;\;\;\;|$$
$$\;\;\;\;H$$
$$NH_3$$

For water, four electron pairs are around the oxygen atom, so the electron-pair geometry is *tetrahedral* (refer to Figure 10-16). Only two of these four electron pairs are involved in bonding, so the molecular shape is bent or V shaped. Because the molecular shape for water is V shaped, we always show water with the hydrogen atoms at about a 90-degree angle to each other — it's a good approximation of the actual shape.

Ammonia also has four electron pairs around the nitrogen central atom, so its electron-pair geometry is tetrahedral as well. Only one of the four electron pairs is nonbonding, however, so its molecular shape is *trigonal pyramidal.* This shape is like a three-legged milk stool, with the nitrogen being the seat, the three bonding pairs of electrons being the legs, and the "invisible" lone pair of nonbonding electrons sticking straight up from the seat. You'd get a surprise if you sat on an ammonia stool!

Another method to determine molecular geometry involves using the valence bond theory. *Valence bond theory* explains covalent bonding in terms of the blending of atomic orbitals to form new types of orbitals: hybrid orbitals. *Hybrid orbitals* are orbitals formed when the atomic orbitals of the central atom combine when forming a compound. However, the total number of orbitals doesn't change; the number of hybrid orbitals equals the number of atomic orbitals used. The number and type of atomic orbitals used determine what type of hybrid orbitals form. Figure 10-18 shows the hybrid orbitals resulting from the mixing of s, p, and d orbitals. The atoms share electrons through the overlapping of their orbitals.

	Linear	Trigonal planar	Tetrahedral	Trigonal bipyramidal	Octahedral
Atomic orbitals mixed	one s one p	one s two p	one s three p	one s three p one d	one s three p two d
Hybrid orbitals formed	two sp	three sp²	four sp³	five sp²d	six sp³d²
Unhybridized orbitals remaining	two p	one p	none	four d	three d
Orientation					

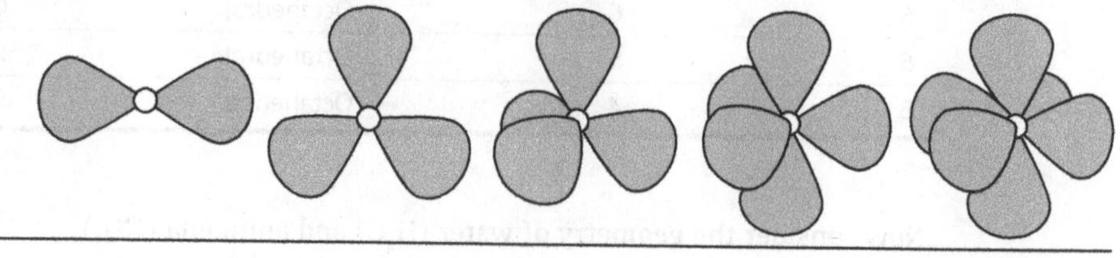

FIGURE 10-18: Hybridization involving the s, p, and d orbitals.

The type of hybridization formed depends on the type and number of atomic orbitals that are involved, and this in turn affects the shape of the resulting molecule:

» **sp hybridization:** This results from the overlap of one s orbital with one p orbital. Two sp-hybrid orbitals form with a bond angle of 180 degrees, which is called a *linear orientation*.

» **sp² hybridization:** This results from the overlap of one s orbital with two p orbitals. Three sp²-hybrid orbitals form with a trigonal planar orientation and a bond angle of 120 degrees. This type of bonding occurs in the formation of the C-to-C double bond, as in $CH_2=CH_2$.

» **sp³ hybridization:** This results from the combination of one s orbital and three p orbitals, resulting in four sp³-hybrid orbitals with a tetrahedral geometric orientation. You find this sp³ hybridization in carbon when it forms four single bonds.

» **sp³d hybridization:** This results from the blending of one s orbital, three p orbitals, and one d orbital. The result is five sp³d orbitals with a trigonal bipyramidal orientation. This type of bonding occurs in compounds like PCl_5, an exception to the octet rule.

» **sp³d² hybridization:** This occurs when one s orbital, three p orbitals, and two d orbitals come together to create an octahedral arrangement. SF_6 is an example. SF_6 is an exception to the octet rule. If one of the bonding pairs in an sp³d² hybridization becomes a lone pair, a square pyramidal shape results; two lone pairs results in a square planar shape.

Figure 10-19 shows the hybridization found in ethylene, $H_2C=CH_2$. Each carbon has sp² hybridization. On each carbon, two of the hybrid orbitals overlap with an s orbital on a hydrogen atom to form a carbon-to-hydrogen covalent bond. The third sp²-hybrid orbital overlaps with the sp² hybrid on the other carbon to form a carbon-to-carbon covalent bond. Note that each carbon has a remaining p orbital that has not undergone hybridization. These p orbitals also overlap above and below a line joining the carbons.

Ethylene has two types of bonds. *Sigma* (σ) *bonds* have the overlap of the orbitals on a line between the two atoms involved in the covalent bond. In ethylene, the C–H bonds and one of the C–C bonds are sigma bonds. *Pi* (π) *bonds* have the overlap of orbitals above and below a line through the two nuclei of the atoms involved in the bond. A double bond is always composed of one sigma bond and one pi bond. A carbon-to-carbon triple bond results from the overlap of one sp-hybrid orbital and two p orbitals on one carbon and the same overlap on the other carbon, resulting in one sigma bond (overlap of the sp-hybrid orbitals) and two pi bonds (overlap of two sets of p orbitals). More on sigma and pi bonds can be found earlier this chapter in the section "Occupying and Overlapping Molecular Orbitals."

REMEMBER

In a multiple bond, one of the bonds must always be a sigma bond and the others are pi.

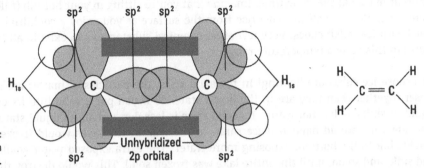

FIGURE 10-19:
Hybridization
in ethylene.

EXAMPLE

Q. Methane, CH_4, has four hydrogen atoms bonded to a central carbon atom. Ammonia, NH_3, has three hydrogen atoms bonded to a central nitrogen atom. Using VSEPR theory, predict the molecular geometry of each compound.

A. After drawing out the Lewis structure of each molecule, you see that CH_4 has four bonded pairs and zero lone pairs of electrons (though, hopefully, you can tell that just from looking at the formulas). This results in a tetrahedral geometry. NH_3 has three bonded pairs of electrons and one lone pair of electrons around the central nitrogen, resulting in a trigonal pyramidal geometry.

**YOUR
TURN**

9 What's the hybridization of carbon in the following molecules?

(a) Carbon dioxide (CO_2)

(b) Formaldehyde (CH_2O)

(c) Methyl bromide (H_3CBr)

10 Use Lewis structures and VSEPR theory to predict the geometry of the following molecules.

(a) Water (H_2O)

(b) Ethyne (C_2H_2)

(c) Carbon tetrachloride (CCl_4)

Wondering About Water and Intermolecular Forces

Water (H_2O) has some very strange chemical and physical properties. It can exist in all three states of matter at the same time. Imagine that you're sitting in your hot tub (filled with *liquid* water) watching the steam (*gas*) rise from the surface as you enjoy a cold drink from a glass filled with ice (*solid*) cubes. Very few other chemical substances can exist in all these physical states in this close a temperature range.

And those ice cubes are floating! In the solid state, the particles of matter are usually much closer together than they are in the liquid state. So if you put a solid into its corresponding liquid, it sinks — but not water. Its solid state is less dense than its liquid state, so it floats. Imagine what would happen if ice sank. In the winter, the lakes would freeze, and the ice would sink to the bottom, exposing more water. The extra exposed water would then freeze and sink, and so on, until the entire lake was frozen solid. This would destroy the aquatic life in the lake in no time. Fortunately, instead, the ice floats and insulates the water underneath it, protecting aquatic life. And water's boiling point is unusually high. Other compounds similar in weight to water have a *much* lower boiling point.

Another unique property of water is its ability to dissolve a large variety of chemical substances. It dissolves salts and other ionic compounds as well as polar covalent compounds such as alcohols and organic acids. In fact, water is sometimes called the universal solvent because it can dissolve so many things. It can also absorb a large amount of heat, which allows large bodies of water to help moderate the temperature on earth.

Water has many unusual properties because of its polar covalent bonds. Oxygen has a larger electronegativity than hydrogen, so the electron pairs are pulled in closer to the oxygen atom, giving it a partial negative charge. Subsequently, both of the hydrogen atoms take on a partial positive charge. The partial charges on the atoms created by the polar covalent bonds in water are shown in Figure 10-20.

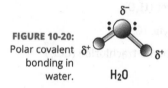

FIGURE 10-20:
Polar covalent bonding in water.

Water is a dipole and acts like a magnet, with the oxygen end having a negative charge and the hydrogen end having a positive charge. These charged ends can attract other water molecules. The partially negatively charged oxygen atom of one water molecule can attract the partially positively charged hydrogen atom of another water molecule. This attraction between the molecules occurs frequently and is a type of *intermolecular force* (force between different molecules).

REMEMBER

Intermolecular forces can be of three different types:

» **London force:** Also called the *dispersion force,* this force is a very weak type of attraction that generally occurs between nonpolar covalent molecules, such as nitrogen (N_2), hydrogen (H_2), or methane (CH_4). It results from the ebb and flow of the electron orbitals, giving a very weak and very brief charge separation around the bond.

» **Dipole-dipole interaction:** This intermolecular force occurs when the positive end of one dipole molecule is attracted to the negative end of another dipole molecule. It's much stronger than a London force, but it's still pretty weak.

» **Hydrogen bond:** The third type of interaction is really just an extremely strong dipole-dipole interaction that occurs when a hydrogen atom on one molecule is bonded to one of three extremely electronegative elements — O, N, or F — on another molecule. These three elements have a very strong attraction for the bonding pair of electrons, so the atoms involved in the bond take on a large amount of partial charge. This bond turns out to be highly polar — and the higher the polarity, the more effective the bond. It's only about 5 percent of the strength of an ordinary covalent bond, but that's still very strong for an intermolecular force. The hydrogen bond is the type of interaction that's present in water (see Figure 10-21).

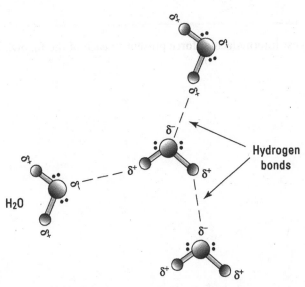

FIGURE 10-21: Hydrogen bonding in water.

Water molecules are stabilized by these hydrogen bonds, so breaking up (separating) the molecules is very hard. The hydrogen bonds account for water's high boiling point and ability to absorb heat. When water freezes, the hydrogen bonds lock water into an open framework that includes a lot of empty space. In liquid water, the molecules can get a little closer to each other, but when the solid forms, the hydrogen bonds result in a structure that contains large holes. The holes increase the volume and decrease the density. This process explains why the density of ice is less than that of liquid water (the reason ice floats). The structure of ice is shown in Figure 10-22, with the hydrogen bonds indicated by dotted lines.

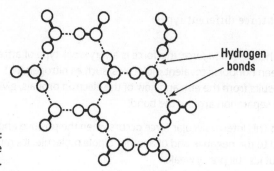

FIGURE 10-22:
The structure
of ice.

Hydrogen
bonds

Q. Which types of intermolecular forces are present in any diatomic molecule?

EXAMPLE **A.** Diatomic molecules have London dispersion forces. The first thing you need to consider when trying to identify intermolecular forces in any molecule is the polarity of that molecule. In this case, all diatomic molecules are nonpolar. You know this because diatomic molecules are two of the same element bonded together; therefore, they have an electronegativity difference of zero. Their pull for electrons is exactly the same, so the distribution is even. Because the molecules are nonpolar, only London dispersion forces are present, because all other intermolecular forces rely on polarity.

YOUR
TURN

 11 Identify the strongest intermolecular force present in each of the following molecules:

a. HF

b. H_2S

c. CH_4

Practice Questions Answers and Explanations

(1) Dichlorine is Cl_2, a compound formed when one chlorine atom bonds to another. Because each atom in the compound is of the same element, the two atoms have the same electronegativity. The difference in electronegativity between the two atoms is zero, so the bond between the two chlorine atoms must be covalent. Another easy way to tell that the bond is covalent is to recognize that chlorine is a nonmetal and is bonded to itself and that two nonmetals bonded together usually form a covalent bond. The electron dot structure of dichlorine is in the following figure:

$$: \overset{..}{\underset{..}{Cl}} : \overset{..}{\underset{..}{Cl}} :$$

(2) **A coordinate covalent bond forms between the aluminum and the hydrating water molecules.** Aluminum is a Group IIIA element, so the aluminum(III) cation (with a charge of +3) formally has no valence electrons. The oxygen of water has lone pairs. Therefore, water molecules most likely hydrate the cation by donating lone pairs to form coordinate covalent bonds. In this respect, you can call the water molecules *ligands* of the metal ion.

(3) You can see resonance structures for benzene in the following figure. Adjacent carbons in the ring are held together with either single or double covalent bonds, depending on the resonance structure. So in the resonance hybrid structure, each carbon–carbon bond is identical and is like a one–and–a–half bond rather than a single or double bond.

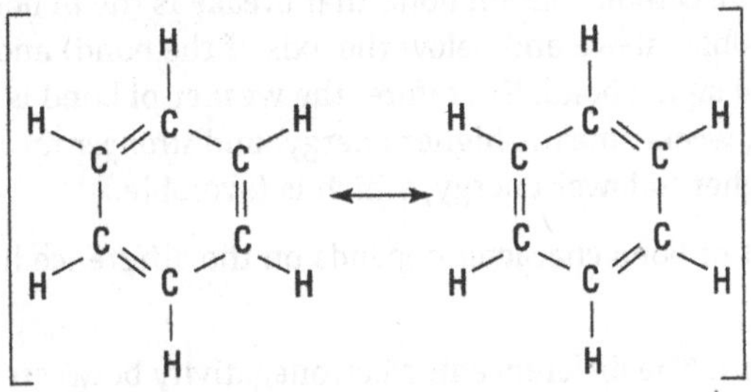

(4) Take a look at the molecular orbital diagram for dihelium in the following figure. Each helium atom contributes two electrons to the molecular orbitals for a total of four electrons. Each molecular orbital holds two electrons, so both the low-energy bonding orbital and the high-energy antibonding orbitals are filled.

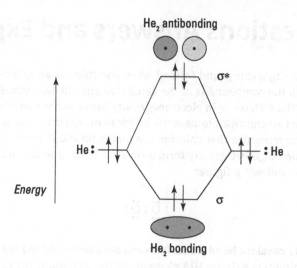

He₂ antibonding

σ*

He:

Energy

:He

σ

He₂ bonding

5 The total energy change to make dihelium from two separate helium atoms is the sum of the changes due to bonding and antibonding. Putting two electrons in the antibonding orbital costs more energy than is saved by putting two electrons in the bonding orbital. Therefore, going from two separate helium atoms to one molecule of dihelium requires an input of energy. Reactions spontaneously go to lower energy, not higher energy, so dihelium is unlikely to exist under normal conditions.

6 The reaction breaks one of the carbon–carbon bonds, replacing the double bond with two single bonds. The carbon–carbon bond that breaks is the pi bond, because the pi electrons are more accessible (above and below the axis of the bond) and because the pi bond is weaker than the sigma bond. Therefore, the weaker pi bond is replaced by a stronger sigma bond. In bonds, *weaker* means higher energy, and *stronger* means lower energy. The reaction moves from higher to lower energy, which is favorable.

7 The assignment of bond character depends on the difference in electronegativity between the atoms:

a. **Polar covalent.** The difference in electronegativity between H and Cl is 0.9, so the bond is a polar covalent bond.

b. **Nonpolar covalent.** The difference in electronegativity between Ga and Ge is 0.2, so the bond is a nonpolar covalent bond.

c. **Nonpolar covalent.** The difference in electronegativity between O and O is 0.0, so the bond is a nonpolar covalent bond.

d. **Ionic.** The difference in electronegativity between Na and Cl is 2.1, so the bond is an ionic bond.

e. **Polar covalent.** The difference in electronegativity between C and O is 1.0, so the bond is a polar covalent bond.

8 **Tetrafluoromethane (CF_4).** Both molecules contain polar covalent bonds based on the electronegativity difference between the atoms. The C–F bonds of CF_4 are slightly more polar than the H–O bonds of water due to the greater difference in electronegativity between carbon and fluorine (1.5 versus 1.4), but the bonds are very close to being the same.

9. Here's the hybridization of carbon in the given molecules:

   a. **In CO_2, carbon is sp hybridized.** CO_2 is linear, with the carbon double-bonded to each oxygen so that four electrons are constrained in each of two bond axes. Remember, even though two double bonds are present in carbon dioxide, each double bond counts as only one bonding domain.

   b. **In CH_2O, carbon is sp_2 hybridized,** so the molecule is trigonal planar. The four electrons of the double bond are constrained within one bond axis, and two electrons each are within bonds to hydrogen.

   c. **In H_3CBr, carbon is sp_3 hybridized,** so the molecule is shaped like a tetrahedron, similar to the shape of methane, CH_4. Each bond is separated from all the others by approximately 109°. Two electrons are constrained within each of four bonds.

10. The Lewis structures for the three molecules are in the following figure.

   a. **Bent.** The oxygen of water is sp^3 hybridized, with two single bonds and two lone pairs, resulting in a bent shape.

   b. **Linear.** Each carbon of ethyne is sp hybridized, with six electrons devoted to a triple bond and two devoted to a single bond with hydrogen, resulting in a linear shape.

   c. **Tetrahedral.** Carbon tetrachloride is sp^3 hybridized, with all electron pairs involved in single bonds, resulting in a tetrahedral shape.

11. In the case of each molecule, you need to determine the polarity so you can then determine which intermolecular forces are present. Remember also that if you have hydrogen bonded to an atom of F, O, or N (That spells FON — almost sounds like FUN!), then hydrogen bonding is present.

   a. **Hydrogen bonding.** HF is a polar molecule with a hydrogen bonded to a fluorine atom. This indicates hydrogen bonding is taking place.

   b. **Dipole–dipole bonding.** H_2S is a polar molecule. You don't have an F, O, or N present, so the strongest force acting here is dipole–dipole.

   c. **London dispersion forces.** CH_4 (methane) is a nonpolar molecule, so the only force acting here is London dispersion. The electrons shared between the carbon and the four hydrogen atoms are evenly distributed, preventing any partial positive or negative charge from developing to create the dipole.

If you're ready to test your skills a bit more, take the following chapter quiz that incorporates all the chapter topics.

Whaddya Know? Chapter 10 Quiz

Ready for a quiz? The 10 questions in this section will test the skills you learned in this chapter. When you're done, check out the section that follows for answers and explanations.

1   Draw the Lewis structure of methane (CH_4).

2   Draw the Lewis structure of carbon dioxide (CO_2).

3   Draw the Lewis structure of phosphorous trichloride (PCl_3).

4   Draw the Lewis structure of dihydrogen sulfide (H_2S).

5   Predict whether bonds between the following pairs of atoms are nonpolar covalent, polar covalent, or ionic:

  a. H and Cl

  b. Na and Br

  c. C and O

6   Which atom(s) would have the partial positive charge and which atom(s) would have the partial negative charge in the following molecules:

  a. HCl

  b. N_2

  c. H_2S

7   Compare the covalent bonds found in methane and the carbon tetrachloride. Which of the bonds have more polar character?

8   What is the molecular geometry of the following molecules?

  a. CCl_4

  b. PCl_3

  c. H_2S

9   What is the hybridization on the central atom in the following molecules?

  a. CCl_4

  b. PCl_3

  c. CO_2

10 What is the strongest intermolecular force present in each of the following molecules?

  a. Cl_2

  b. H_2O

  c. HBr

Answers to Chapter 10 Quiz

(1) The formula of methane is CH_4. This means the total number of valence electrons (N) needed is 8 for carbon plus 2×4 for hydrogen giving you an N of 16. The total number of valence electrons available are 4 for carbon plus 1×4 for hydrogen giving you an A of 8. Solving for S doing N – A $(16-8)$ gives you an S of 8, meaning there are 8 shared electrons. Divide that by 2 (8/2) to show you there are 4 bonds, meaning you draw carbon in the middle with 4 hydrogens bonded to it as shown in the following figure.

$$
\begin{array}{c}
\text{H} \\
| \\
\text{H} - \text{C} - \text{H} \\
| \\
\text{H}
\end{array}
$$

(2) The formula of methane is CO_2. This means the total number of valence electrons (N) needed is 8 for carbon plus 2×8 for oxygen, giving you an N of 24. The total number of valence electrons available are 4 for carbon plus 2×6 for hydrogen giving you an A of 16. Solving for S doing N – A $(16-8)$ gives you an S of 8 meaning there are 8 shared electrons. Divide that by 2 (8/2) to show you there are 4 bonds. In this case, that means you need to draw carbon bonded to each oxygen twice using a double bond between each C and O as shown in the following figure.

$$ \text{O} = \text{C} = \text{O} $$

(3) The formula you have is PCl_3. In this case P will be the central atom. This means the total number of valence electrons (N) needed is 8 for phosphorous plus 3×8 for chlorine giving you an N of 32. The total number of valence electrons available are 5 for phosphorous plus 3×7 for hydrogen, giving you an A of 26. Solving for S doing N – A $(32-26)$ gives you an S of 6, meaning there are 6 shared electrons. Divide that by 2 (6/2) to show you there are 3 bonds. In this case that means you need to draw phosphorous bonded to each chlorine one time. Once you have done this you will then fill in with 3 electron pairs around each chlorine. Finally, you will add a lone pair of electrons on the central atom of phosphorous, as shown in the following figure.

$$
\begin{array}{c}
:\ddot{\text{C}}\text{l} - \ddot{\text{P}} - \ddot{\text{C}}\text{l}: \\
| \\
:\ddot{\text{C}}\text{l}:
\end{array}
$$

(4) The formula you have is H_2S. In this case S will be the central atom. This means the total number of valence electrons (N) needed is 8 for sulfur plus 2×1 for hydrogen, giving you an N of 10. The total number of valence electrons available are 6 for phosphorous plus 2×1 for hydrogen, giving you an A of 6. Solving for S doing N – A $(10 - 6)$ gives you an S of 4, meaning there are 6 shared electrons. Divide that by 2 $(4/2)$ to show you there 2 bonds. In this case that means you need to draw sulfur bonded to each hydrogen one time. For the hydrogens this fills their outer energy level, as each shared bond counts for 2 electrons and hydrogen only wants 2 electrons to begin with. This leaves you with 4 electrons to put on the sulfur. Add 2 lone pairs of electrons to sulfur, as shown in the following figure.

$$\overset{\displaystyle ..}{\underset{\displaystyle |}{\ddot{S}}} - H$$
$$H$$

(5) To determine whether a bond is nonpolar covalent, polar covalent, or ionic, you need to compare the electronegativities of the 2 atoms in the bond. If the electronegativity difference is 0 to 0.2 the bond is said to be nonpolar covalent, 0.3 to 1.4 is a polar covalent bond, and anything greater than 1.5 is ionic.

a. **Polar covalent bond.** Hydrogen has an electronegativity of 2.1 and fluorine has an electronegativity of 3.0. 3.0 – 2.1 = 0.9. This makes HCl a polar covalent bond.

b. **Ionic bond.** Sodium has an electronegativity of 0.9 and bromine has an electronegativity of 2.8. The electronegativity difference between the two is 2.8 – 0.9 = 1.9. This makes the bond ionic.

c. **Polar covalent bond.** Carbon has an electronegativity of 2.5 and oxygen has an electronegativity of 3.5. The electronegativity difference between the two is 3.5 – 2.5 = 1. This makes the bond polar covalent.

(6) a. The hydrogen has the partial positive charge and chlorine has the partial negative charge. Chlorine is far more electronegative than hydrogen this leads to the electrons in the shared covalent bond being far more likely to be found around chlorine than hydrogen. Having the extra electron more of the time leads to an uneven sharing of the electrons. Since chlorine tends to have the extra shared electron more it is said to be partially negative because a negative electron is around it enough to make it somewhat negative but not a strong or permanent negative charge. The same but opposite can be said for hydrogen. Since hydrogen has a lower electronegativity than chlorine it is less likely to have the shared electron, meaning it has a partially positive charge due to the lack of an electron in the presence of the proton that makes up hydrogen's nucleus.

b. Nitrogen is bonded to itself in this example. That means that both atoms in this molecule have the same electronegativity. Neither atom has a stronger desire for electrons than the other so there is no uneven sharing of electrons. The molecule is nonpolar and does not have a partial positive or negative charge due to this.

c. Sulfur is more electronegative than hydrogen. This means that sulfur will have the partial negative charge on it as the shared electrons are more likely to be found around the sulfur atom. The two hydrogens will each have a partial negative charge due to the shared electrons being more likely to be found with the sulfur atom.

7. To determine the polar character of different bonds you need to examine the electronegativity difference between the two atoms found in the bonds. Methane's bonds involve a carbon bonding to hydrogen atoms. Carbon has an electronegativity value of 2.5 and hydrogen has an electronegativity value of 2.1. The electronegativity difference between the two (2.5 – 2.1) is 0.4. Carbon tetrachloride's bonds involve carbon bonding to chlorines. Carbon has an electronegativity value of 2.5 and chlorine has an electronegativity value of 3.0. The difference here is (3.0 – 2.5) 0.5. So the bonds that make up carbon tetrachloride have a slightly greater polar character due to having a slightly greater electronegativity difference. The difference itself though is very, very small, though, so in reality there is actually minimal difference between the two.

8. The VSEPR shape of each of the following molecules is based on the Lewis structure. You need to determine the total number of electron bonding pairs and then also determine the number of nonbonding electron pairs. Once you know this you are able to identify the VSEPR shape.

   a. **Tetrahedral.** CCl_4 has a Lewis structure that contains 4 bonds and no lone pairs of electrons on the central atom. This corresponds to 4 total electron pairs and 4 bonding pairs of electrons. If you look at Table 10–1 you can see this indicates a shape of tetrahedral. This is a very common shape and one that you can easily recognize without giving it too much thought. If you see an atom with 4 bonds on it and no lone pairs of electrons around it, the shape will always be tetrahedral.

   b. **Trigonal pyramidal.** PCl_3 has a Lewis structure that contains 3 bonds and 1 lone pair of electrons on the central atom. This corresponds to having 4 total electron pairs around the central atom and 3 bonding pairs of electrons. If you look at Table 10–1 you can see this indicates a shape of trigonal pyramidal.

   c. **Bent, V shape.** H_2S has a Lewis structure that contains 2 bonds and 2 lone pair of electrons on the central atom. This corresponds to having 4 total electron pairs around the central atom and 2 bonding pairs of electrons. If you look at Table 10–1 you can see this indicates a bent, V shape.

9. The simplest way to think about hybridization of an atom is based on the molecular geometry of the molecule and the total number of electron domains found around the central atom. Remember that you are simply looking at the number of electron domains around the atom. It does not matter if the domain has a single, double, or triple bond or if it is simply a loan pair of electrons. All that matters is whether there are electrons present in that domain.

   a. **SP_3.** CCl_4 has 4 bonds around the atom corresponding to 4 bonding domains around the central atom. These four bonds give it a molecular geometry of tetrahedral. An atom with a tetrahedral geometry has a hybridization of SP^3.

   b. **SP_3.** PCl_3 has 3 bonds around the central atom and 1 lone pair of electrons. This corresponds to 4 electron domains around the central atom. Any atom with 4 electron domains will have a hybridization of SP^3. This makes the hybridization of PCl_3 the same as the hybridization of CCl_4 from the previous question. Even though they have different shapes, the hybridization is the same on both molecules because they have the same total number of electron domains.

   c. **SP.** CO_2 has 2 double bonds around the central atom of carbon and no lone pairs of electrons. Despite the fact that each of these double bonds contains 4 electrons, they only count as 1 electron domain each. This means that the carbon atom in CO_2 only has 2 electron domains around it. This gives it a hybridization of SP.

(10) a. **London/dispersion forces.** Cl_2 is a nonpolar molecule since chlorine and chlorine both have the exact same desire for electrons since they have the same electronegativity. This means the only intermolecular force present in chlorine molecules are London/dispersion forces.

b. **Hydrogen bonding.** The strongest intermolecular force present in a group of water molecules is hydrogen bonding. Hydrogen bonded to oxygen (or fluorine/nitrogen) results in a very strong partial positive and negative charge, creating a very strong dipole that is called hydrogen bonding. Hydrogen bonding is the strongest of the possible intermolecular forces.

c. **Dipole dipole.** HBr is a polar molecule since hydrogen and bromine have a difference in electronegativity, meaning the electrons in their bonds are not shared equally between one another. This uneven sharing creates a dipole in the molecule, resulting in dipole dipole interactions taking place between HBr molecules. London dispersion forces are also present, but these are much weaker than dipole dipole interactions.

IN THIS CHAPTER

» **Crafting names and formulas for ionic and molecular compounds**

» **Handling polyatomic ions**

» **Talking about acids**

» **Using a quick and easy scheme for naming any compound**

» **Introducing alkane naming conventions**

Chapter **11**

Naming Compounds and Writing Formulas

C hemists give compounds very specific names. Sometimes these names seem overly specific. For example, what's the point of referring to "dihydrogen monoxide" when you can simply say "water"? First, some people simply believe this kind of thing sounds cool. Beyond dubious notions of coolness, however, lies a more important reason: Chemical names clue you in to chemical structures — but only if you know the code. Fortunately, the code, as you find out in this chapter, is pretty straightforward — no advanced cryptology required. (Which is also fortunate, because putting chemists and cryptologists in the same room could result in the kind of party you don't want to admit to having attended.)

Labeling Ionic Compounds and Writing Their Formulas

Chapter 9 discusses the way that *anions* (atoms with negative charge) and *cations* (atoms with positive charge) attract one another to form ionic bonds. Ionic bonds hold together ionic compounds. The anions and cations in a given ionic compound are important factors in how you name and write the formula of that compound.

Determining formulas of ionic compounds

Writing out the formulas of ionic compounds is a rather straightforward affair. For example, imagine you react magnesium with bromine. To begin, make sure you have an ionic compound in the first place. The simplest and fastest way to check this is to see whether you have a metal and a nonmetal bonding together. In this case, magnesium is a metal and bromine is a nonmetal (you can tell by their locations on the periodic table), so you have an ionic compound. You can now determine the correct formula and write it out.

Start by putting the magnesium and bromine atoms side by side, with the metal (Mg) on the left, and then adding their charges. Figure 11-1 shows this process. (Forget about the criss-crossing lines for now. Well, if you're really curious, check out the next section.)

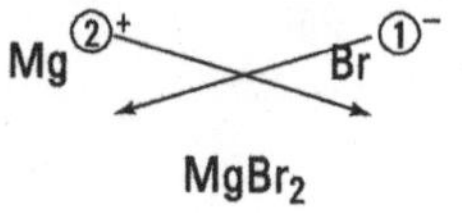

FIGURE 11-1:
Figuring the formula of magnesium bromide.

$$Mg^{2+} \diagdown \diagup Br^{1-}$$
$$MgBr_2$$

The electron configurations for magnesium and bromine are

Magnesium (Mg) $1s^2 2s^2 2p^6 3s^2$
Bromine (Br) $1s^2 2s^2 2p^6 3s^2 3p^6 4s^2 3d^{10} 4p^5$

Magnesium, an alkaline earth metal, has two valence electrons that it loses to form a cation with a +2 charge. The electron configuration for the magnesium cation is

Mg^{2+} $1s^2 2s^2 2p^6$

Bromine, a halogen, has seven valence electrons, so it gains one to complete its octet (eight valence electrons) and forms the bromide anion with a –1 charge. The electron configuration for the bromide anion is

Br^{1-} $1s^2 2s^2 2p^6 3s^2 3p^6 4s^2 3d^{10} 4p^6$

Note that if the anion simply has 1 unit of charge, positive or negative, you normally don't write the 1; you just use the plus or minus symbol, with the 1 being understood. But for the example of the bromide ion, we use the 1.

The compound must be neutral; it must have the same number of positive and negative charges so that, overall, it has a zero charge. The magnesium ion has a +2 charge, so it requires two bromide anions, each with a single negative charge, to balance the two positive charges of magnesium. So the formula of the compound that results from reacting magnesium with bromine is $MgBr_2$.

Applying the crisscross rule

A quick way to determine the formula of an ionic compound is to use the *crisscross rule*. The crisscross rule uses the ionic charges of the ions to predict the formula of the ionic compound. It doesn't work all the time, but it's a good way of checking your result using the method in the preceding section.

To see how this rule works, go back to Figure 11-1. Take the numerical value of the metal ion's superscript (forget about the charge symbol) and move it to the bottom right-hand side of the nonmetal's symbol — as a subscript. Then take the numerical value of the nonmetal's superscript and make it the subscript of the metal. (Note that if the numerical value is 1, it's just understood and not shown.) So in this example, you make magnesium's 2 a subscript of bromine and make bromine's 1 a subscript of magnesium (but because it's 1, you don't show it), and you get the formula $MgBr_2$.

So what happens if you react aluminum and oxygen? Figure 11-2 shows the crisscross rule used for this reaction.

FIGURE 11-2: Figuring the formula of aluminum oxide.

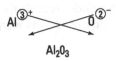

Al_2O_3

WARNING

The crisscross rule works very well, but you have to be careful if both ions have the same numeral in the superscript. Suppose that you want to write the compound formed when calcium reacts with oxygen. Magnesium, an alkaline earth metal, forms a 2+ cation, and oxygen forms a 2– anion. So you might predict that the formula is

Mg_2O_2

But this formula is incorrect. After you use the crisscross rule, you need to reduce all the subscripts by a common factor, if possible. In this case, you divide each subscript by 2 to get the correct formula:

MgO

Writing the names of ionic compounds

When you speak to another person, you probably don't stop to analyze his or her name and consider why it's significant or what meaning it might have beyond helping you identify that person. However, the names of chemical compounds give you all sorts of information. Rules govern how an ionic compound is named and what information that name conveys.

Before you write the name of an ionic compound from its formula, make sure the compound is actually ionic (if you end up with two nonmetals stuck together, skip to the section on naming covalent compounds, later in this chapter). To start, you pair the name of the cation with the name of the anion and then change the ending of the anion's name to -*ide*. The cation always precedes the anion in the full name. For example, the chemical name of NaCl (a compound made up of one sodium atom and one chlorine atom) is *sodium chloride*.

Of course, sodium chloride is more commonly known as table salt. Many compounds have such so-called *common names.* Common names aren't wrong, but they're less informative than chemical names. The name sodium chloride, properly decoded, tells you that you're dealing with a one-to-one ionic compound composed of sodium and chlorine. The name *table salt* just tells you one possible (albeit tasty) use for the compound.

Cations and anions combine in very predictable ways within ionic compounds, always acting to neutralize overall charge. Therefore, the name of an ionic compound implies more than just the identity of the atoms that make it up. It also helps you determine the correct chemical formula, which tells you the ratio in which the elements combine. Consider these two examples, both of which involve lithium:

>> In the compound lithium fluoride (LiF), lithium and fluorine combine in a one-to-one ratio because lithium's +1 charge and fluorine's –1 charge cancel one another (neutralize) perfectly. By itself, the name *lithium fluoride* tells you only that the compound is made up of a lithium cation and a fluoride anion, but by comparing their charges, you can see that Li^+ and F^- neutralize each other in the one-to-one compound LiF.

>> If lithium combines with oxygen to form an ionic compound, then two lithium ions, each with +1 charge, are required to neutralize the –2 charge of the oxide anion. So the name *lithium oxide* implies the formula Li_2O, because a two-to-one ratio of lithium cation to oxide anion is necessary to produce a neutral compound.

REMEMBER

Use the name of the ionic compound to identify the ions you're dealing with, and then combine those ions in the simplest way that results in a neutral compound.

Using a compound's name to identify the ions can be tricky when the cation is a metal. All Group B transition metals — with the exceptions of silver (which is always found as Ag^+) and zinc (always Zn^{2+}) — as well as several Group A elements on the right-hand side of the periodic table can take on a variety of charges. (Chapter 5 describes the periodic table.) Chemists use an additional naming device, the Roman numeral, to identify the charge state of the cation. A Roman numeral placed within parentheses after the name of the cation lists the positive charge of that cation. For example, copper(I) is copper with a +1 charge, and copper(II) is copper with a +2 charge. You may think this kind of distinction is simply chemical hair-splitting; who cares whether you're dealing with iron(II) bromide ($FeBr_2$) or iron(III) bromide ($FeBr_3$)? The difference matters because ionic compounds with different formulas (even those containing the same types of elements) can have very different properties.

REMEMBER

Use Roman numerals only for metals whose charge you don't know. Don't use Roman numerals if you know the charge of the metallic cation, including alkali metals such as lithium (Li^+) and sodium (Na^+), as well as alkaline earth metals such as calcium (Ca^{2+}) and magnesium (Mg^{2+}).

The trickiness of metals with variable charges also applies when you translate a chemical formula into the corresponding chemical name. Here are the two secrets to keep in mind:

>> The name implies only the charge of the ions that make up an ionic compound.

>> The charges of the ions determine the ratio in which they combine.

For example, you must do a little bit of sleuthing before assigning a name to the formula CrO. You know that oxygen brings a −2 charge to an ionic compound. Because O^{2-} combines with chromium in a one-to-one ratio within CrO, you know that chromium here must have the equal and opposite charge of +2. In this way, CrO can be electrically neutral (uncharged). So the chromium cation in CrO is Cr^{2+}, and the compound name is chromium(II) oxide. Simply calling the compound *chromium oxide* doesn't clearly identify the precise chromium cation in play. For example, chromium can assume a +3 charge. When Cr^{3+} combines with an oxide anion (which has a −2 charge), a different ion ratio is necessary to produce a neutral compound: Cr_2O_3, which is chromium(III) oxide. This equating of total positive and negative charges in an atom is called *balancing the charge*. After you've balanced the charges in the atom's formula, you can drop the charges on the individual ions.

Q. What is the name of Fe_2O_3?

EXAMPLE **A.** Iron(III) oxide. The initial name is easy to come up with. Simply identify Fe as iron and change the ending of *oxygen* to *-ide* to get *iron oxide*. However, iron is a metal that has more than one possible charge, so you need a Roman numeral to indicate the charge of iron. Because you know that oxygen has a charge of −2 in ionic compounds and you have three oxygen atoms present, the total negative charge is −6. Two iron atoms are present, so each atom must have a charge of +3 for a total charge of +6 to balance out the oxygen's negative charge. To indicate this +3 charge, you add (III) after the iron.

Q. What is the formula for the compound tin(IV) fluoride?

A. SnF_4. The Roman numeral within parentheses tells you that you're dealing with Sn^{4+}. Because fluorine is a halogen, it always has a charge of −1 in ionic compounds, which means that four fluoride anions are necessary to cancel the four positive charges of a single tin cation. Therefore, the compound is SnF_4.

YOUR TURN

1 Name the following compounds:

a. MgF_2

b. LiBr

c. Cs_2O

d. CaS

e. KCl

f. Sr_3P_2

g. Na_2S

 2 Name the following compounds that contain elements with variable charge. Don't forget to use Roman numerals!

a. FeF_2

b. CuBr

c. SnI_4

d. Mn_2O_3

e. Fe_2O_3

f. PbS_2

g. $CuCl_2$

3 Translate the following names into chemical formulas:

    a. Iron(III) oxide

    b. Beryllium chloride

    c. Tin(II) sulfide

    d. Potassium iodide

    e. Lithium phosphide

    f. Strontium oxide

    g. Iron(III) sulfide

Getting a Grip on Ionic Compounds with Polyatomic Ions

A confession: Not all ions are of single atoms. To get a full education in chemical nomenclature, you must grapple with an irksome and all-too-common group of molecules called the *polyatomic ions*, which are made up of groups of atoms. Polyatomic ions, like single-element ions, tend to quickly combine with other ions to neutralize their charges. Unfortunately, you can't use any simple, periodic trend–type rules to figure out the charge of a polyatomic ion. You must — gulp — memorize them.

Table 11-1 summarizes the most common polyatomic ions, grouping them by charge.

REMEMBER Notice in Table 11-1 that all the common polyatomic ions except ammonium have a negative charge ranging between –1 and –3. You also see a number of *-ite/-ate* pairs, such as chlorite and chlorate, phosphite and phosphate, and nitrite and nitrate. If you look closely at these pairs, you notice that the only difference between them is the number of oxygen atoms in each ion. Specifically, the *-ate* ion always has one more oxygen atom than the *-ite* ion but has the same overall charge.

TIP To complicate your life further, polyatomic ions sometimes occur multiple times within the same ionic compound. How do you specify that your compound has two sulfate ions in a way that makes visual sense? Put the entire polyatomic ion formula in parentheses and then add a subscript outside the parentheses to indicate how many such ions you have, as in $\left(SO_4^{2-}\right)_2$.

TABLE 11-1 Common Polyatomic Ions

–1 Charge	–2 Charge	–3 Charge	+1 Charge
Dihydrogen phosphate $(H_2PO_4^-)$	Hydrogen phosphate (HPO_4^{2-})	Phosphite (PO_3^{3-})	Ammonium (NH_4^+)
Acetate $(C_2H_3O_2^-)$	Oxalate $(C_2O_4^{2-})$	Phosphate (PO_4^{3-})	
Hydrogen sulfite (HSO_3^-)	Sulfite (SO_3^{2-})		
Hydrogen sulfate (HSO_4^-)	Sulfate (SO_4^{2-})		
Hydrogen carbonate (HCO_3^-)	Carbonate (CO_3^{2-})		
Nitrite (NO_2^-)	Chromate (CrO_4^{2-})		
Nitrate (NO_3^-)	Dichromate $(Cr_2O_7^{2-})$		
Cyanide (CN^-)	Silicate (SiO_3^{2-})		
Hydroxide (OH^-)			
Permanganate (MnO_4^-)			
Hypochlorite (ClO^-)			
Chlorite (ClO_2^-)			
Chlorate (ClO_3^-)			
Perchlorate (ClO_4^-)			

When you write a chemical formula that involves polyatomic ions, you treat them just like other ions. You still need to balance charges to form a neutral atom. We're sorry to report that when you're converting from a formula to a name, you can't use any simple rule for naming polyatomic ions. You just have to memorize the entire table of polyatomic ions and their charges.

When you're tasked with writing the name of a formula containing a polyatomic ion, you follow all the same naming rules as listed previously except for one very simple change: You don't change the ending of any polyatomic ion. You leave it exactly as it written.

Q. Write the name of the formula $LiNO_3$.

EXAMPLE **A.** Lithium nitrate. Because lithium is an alkali metal and the charge of alkali metals in Group IA is always +1, you don't need to use Roman numerals to indicate the charge of lithium. You simply write *lithium* and then the name of the polyatomic ion, which is *nitrate*. You don't change the ending of the polyatomic ion name.

Q. Write the formula for the compound barium chlorite.

A. $Ba(ClO_2)_2$. Barium is an alkaline earth metal (Group IIA) and thus has a charge of +2. You should recognize chlorite as the name of a polyatomic ion. In fact, any anion name that doesn't end in –ide should scream polyatomic ion to you. As Table 11-1 shows, chlorite is ClO_2^-, which reveals that the chlorite ion has a –1 charge. Two chlorite ions are necessary to neutralize the +2 charge of a single barium cation, so the chemical formula is $Ba(ClO_2)_2$.

4 Name the following compounds that contain polyatomic ions:

a. $Mg_3(PO_4)_2$

b. $Pb(C_2H_3O_2)_2$

c. $Cr(NO_2)_3$

d. $(NH_4)_2C_2O_4$

e. $KMnO_4$

5 Write the formula for the following compounds that contain polyatomic ions:

a. Potassium sulfate

b. Lead(II) dichromate

c. Ammonium chloride

d. Sodium hydroxide

e. Chromium(III) carbonate

Naming Covalent Compounds and Writing Their Formulas

Nonmetals tend to form covalent bonds with one another (see Chapter 9 for details). Compounds made up of nonmetals held together by one or more covalent bonds are called *covalent (or molecular) compounds*.

Predicting how the atoms within molecules will bond with one another is a tricky endeavor because two nonmetals often can combine in multiple ratios. Carbon and oxygen, for example, can combine in a one-to-two ratio to form CO_2 (carbon dioxide), a harmless gas you emit every time you exhale. Alternatively, the same two elements can combine in a one-to-one ratio to form CO (carbon monoxide), a poisonous gas. Clearly, having names that distinguish between these (and other) covalent compounds is useful.

Covalent compound names clearly specify how many of each type of atom participate in the compound. Table 11-2 lists the prefixes used to do so.

You can attach the prefixes in Table 11-2 to any of the elements in a covalent compound, as exemplified by SO_3 (sulfur trioxide) and N_2O (dinitrogen monoxide). The second element in each compound receives the *-ide* suffix, as in ionic compounds (which we discuss earlier in this chapter). In the case of covalent compounds, where cations or anions aren't involved, the more electronegative element (in other words, the element that's closer to the upper right-hand corner of the periodic table) tends to be named second.

TABLE 11-2 Common Prefixes for Binary Covalent Compounds

Number of Atoms	Prefix
1	mono-
2	di-
3	tri-
4	tetra-
5	penta-
6	hexa-
7	hepta-
8	octa-
9	nona-
10	deca-

Take a look at the following examples to see how to use the prefixes when naming binary covalent compounds (we've bolded the prefixes for you):

CO_2	carbon dioxide
P_4O_{10}	**tetra**phosphorus **dec**oxide (Chemists try to avoid putting an *a* and an *o* together with the oxide name, as in dec**ao**xide, so they normally drop the *a* from the prefix.)
SO_3	sulfur **tri**oxide
N_2O_4	**di**nitrogen **tetr**oxide

Note that the absence of a prefix from the first named element in a covalent compound implies that there's only one atom of that element. In other words, the prefix *mono-* is unnecessary for the first element only. You still have to attach a *mono-* prefix, when appropriate, to the names of subsequent elements.

Writing a formula for a covalent compound with a given name is much simpler than writing a formula for an ionic compound. In a covalent compound, the ratio in which the two elements combine is built into the name itself, and you don't need to worry about balancing charges. For example, the prefixes in the name *dihydrogen monoxide* imply that the chemical formula contains two hydrogens and one oxygen (H_2O).

Translating a formula into a name is equally simple. All you need to do is convert the subscripts into prefixes and attach them to the names of the elements that make up the compound. For example, for the compound N_2O_4, you simply attach the prefix *di-* to *nitrogen* to indicate the two nitrogen atoms and *tetra-* to *oxygen* to indicate the four oxygen atoms, giving you *dinitrogen tetroxide*.

REMEMBER

Hydrogen is located on the far left of the periodic table, but it's actually a nonmetal. In keeping with this hydrogenic craziness, hydrogen can appear as either the first or second element in a *binary* (two-element) molecular compound, as shown by dihydrogen monosulfide (H_2S) and phosphorus trihydride (PH_3).

Q. What are the names of the compounds N_2O, SF_6, and Cl_2O_8?

EXAMPLE **A.** Dinitrogen monoxide, sulfur hexafluoride, and dichlorine octoxide. Notice that none of these compounds contain any metals, which means that they're most certainly covalent compounds. The first compound contains two nitrogen atoms and one oxygen atom, so it's called *dinitrogen monoxide.* The second compound contains one sulfur and six fluorines. Because sulfur is the first named element, you don't need to include a *mono–* prefix. You simply name the compound *sulfur hexafluoride* (rather than *monosulfur hexafluoride*). Using the same methods, the third compound is named *dichlorine octoxide.*

Q. What is the formula of dicarbon tetrahydride?

A. C_2H_4. The prefixes in the name indicate the compound is covalent, so you don't need to worry about ionic charges. Just identify the element and the number of atoms based on the numerical prefix and then write it down. In this case, you have two carbons indicated by the *di–* and four hydrogens indicated by the *tetra–*.

YOUR TURN

6 Write the proper names for the following compounds:

a. N_2H_4

b. H_2S

c. NO

d. CBr_4

e. C_2H_6

f. C_2H_4

g. SO_2

7 Write the proper formulas for the following compounds:

a. Silicon difluoride

b. Nitrogen trifluoride

c. Disulfur decafluoride

d. Diphosphorus trichloride

e. Iodine heptafluoride

f. Xenon hexafluoride

g. Phosphorus pentoxide

Addressing Acids

The chemical compounds known as *acids* have their own special naming system, but writing the names and formulas of acids really isn't that big of a deal. As long as you pay attention to the details, you won't have any problem naming acidic compounds. You more than likely already know the names of several very common acids, though perhaps you don't know their formulas. Table 11-3 lists several of the most common acids and their formulas. (Flip to Chapter 20 for full details on acids.)

TABLE 11-3 Common Acids

Name	Formula
Acetic acid	$HC_2H_3O_2$
Carbonic acid	H_2CO_3
Hydrochloric acid	HCl
Nitric acid	HNO_3
Phosphoric acid	H_3PO_4
Sulfuric acid	H_2SO_4

The two most common types of acids you encounter in a basic chemistry class are binary acids and oxy-acids:

>> **Binary acids:** You can easily recognize a binary acid when you see hydrogen bonded to a nonmetallic element or polyatomic ion without oxygen present.

>> **Oxy-acids:** Oxy-acids contain hydrogen bonded to a polyatomic ion containing oxygen.

To name a binary acid (no oxygen), use the following steps:

1. **Write the prefix *hydro-* at the beginning of the name.**

   Say you begin with the acid HCl. No oxygens are present, so the name starts with *hydro-*.

2. **Write the name of the anion.**

   If the name of the anion begins with a vowel, you drop the *o* in *hydro-* to avoid having two vowels next to each other. The anion in this case is chlorine, so you write *hydro-chlorine*.

3. **Change the ending of the anion name to *-ic* and add *acid* to the end of the name.**

   So the name becomes *hydrochloric acid*.

The steps for naming an oxy-acid (in which oxygen is present) are a little different:

1. **Write the root name of the polyatomic anion.**

   Suppose you're naming HNO_3. Oxygen is present, so you don't begin by writing *hydro-*. Instead, simply write *nitrate*, because the anion is NO_3.

2. **If the polyatomic ion name ends in *-ate*, change the ending to *-ic*; if the** polyatomic ion name ends in *-ite*, change the ending to *-ous*.

   NO_3 is nitrate. It ends in *-ate*, so change the ending to *-ic*, giving you *nitric*.

3. **Put *acid* at the end of the name.**

   The name is *nitric acid*.

Q. What is the name of HBr?

EXAMPLE **A.** Hydrobromic acid. HBr is a binary acid; it doesn't contain oxygen, so it isn't an oxy-acid. You begin by writing *hydro-*. Next, change the ending of the anion name, bromine, to *-ic* and write *bromic* after *hydro-*. Write *acid* at the end, and you're done!

Q. What is the name of H_2SO_4?

A. Sulfuric acid. H_2SO_4 is an oxy-acid because it contains oxygen. You begin by identifying SO_4 as sulfate. Sulfate ends in *-ate*, so you change the ending to *-ic*. You then add *acid* to the end of the name. (Don't worry about the fact that two hydrogen atoms are present at the beginning of the formula. Those two hydrogens are just in place to balance out the –2 charge of SO_4 with two positive charges. Each hydrogen ion has a charge of +1, for a total charge of +2.)

YOUR TURN

⑧ Write the proper names for the following binary acids:

   a. HF

   b. HI

   c. HCl

   d. HCN

 ⑨ Write the proper names for the following oxy-acids:

   a. H_2CO_3

   b. H_2SO_3

   c. $HClO_2$

   d. HNO_3

Mixing the Rules for Naming and Formula Writing

Have all the naming rules left you confused and frustrated? Do you feel like rebelling against the chemical conspiracy and liberating all chemicals from the oppressive confines of their formal names? All right, calm down. If you take a step back to look at the big picture, you see that the naming system is actually pretty logical and straightforward. It gives you a lot of valuable information when dealing with chemical compounds. The best way to really conquer the world of naming chemical compounds is to practice everything at once.

In short, you can think of this section as the big game. You've practiced and you've worked; you've put blood, sweat, and tears into these compounds. Now you're ready to show the world how it's done.

In reality, you aren't given nice subheadings telling you which type of compound a chemical is. You need to figure it out yourself so you know which set of naming rules to use. For a lot of people, this is where naming compounds gets a bit confusing.

Fortunately, you can follow a series of rules that will help you figure out how exactly to write out a name or a formula until you get comfortable enough to just do it on your own. As with anything in chemistry, practice makes perfect. With that in mind, here's a simple series of questions to help you write the names of all the wonderful types of chemicals in the previous sections correctly and accurately:

1. **Does the formula begin with an *H*?** If so, use the rules presented earlier in "Addressing Acids." Be sure to identify whether the compound is a binary acid or an oxy-acid. If the compound doesn't begin with an *H*, move along to Question 2.

2. **Does the formula contain a metal (not hydrogen)?** If there's no metal, you're naming a molecular (covalent) compound, so you need to use the prefixes in Table 11-2. Be sure to change the ending of the second element to *-ide*. If there is a metal, then you're dealing with an ionic compound, so proceed to Question 3.

3. **Is the cation a transition metal (Group B) or a metal with a variable charge?** If the cation is a Group B metal (or other metal of variable charge, like tin), you need to use Roman numerals to specify its charge. See the earlier section "Labeling Ionic Compounds and Writing Their Formulas" for details. If the cation isn't a transition metal and you know the charge, you don't need to specify the charge with Roman numerals.

4. **Is the anion a polyatomic ion?** If so, you have to recognize it as such and have its name memorized (or easily accessible in a nifty table such as Table 11-1). If the anion isn't a polyatomic ion, you use an *-ide* ending.

Q. Write the name of C_2H_7.

EXAMPLE **A.** Dicarbon heptahydride. Carbon is a nonmetal, as is hydrogen, so this compound is covalent (molecular). That means you need to use prefixes when naming the formula. The compound has two carbons, so the carbon prefix is *di-*; it has seven hydrogens, so the hydrogen prefix is *hepta-*. You end the name by changing the ending of *hydrogen* to *-ide*.

Q. Write the formula of lead(IV) sulfate.

A. $Pb(SO_4)_2$. The cation lead is a metal with a variable charge, as indicated by the Roman numeral IV in parentheses. This classification means that lead has a charge of +4. SO_4 is the polyatomic ion sulfate with a charge of -2. To balance out the charges, you need two sulfate ions for each lead ion. Thus, the formula has one Pb ion with a total charge of +4 and two sulfate ions with a total charge of -4. To indicate the need for two polyatomic ions, you put parentheses around the sulfate ion and write the 2 as a subscript outside the parentheses.

10 Name each of the following compounds:

a. $PbCrO_4$

b. Mg_3P_2

c. $SrSiO_4$

d. H_2SO_4

e. Na_2S

f. B_3Se_2

g. HgF_2

h. $Ba_3(PO_4)_2$

11 Translate the following names into chemical formulas:

a. Barium hydroxide

b. Tin(IV) bromide

c. Sodium sulfate

d. Phosphorus triiodide

e. Magnesium permanganate

f. Acetic acid

g. Nitrogen dihydride

h. Iron(II) chromate

Beyond the Basics: Naming Organic Carbon Chains

One of the most common molecules studied in organic chemistry is the hydrocarbon. *Hydrocarbons* are compounds composed of carbon and hydrogen. The simplest of the hydrocarbons fall into the category of alkanes. *Alkanes* are chains of carbon molecules connected by single covalent bonds. Chapter 9 describes how single covalent bonds result when atoms share pairs of valence electrons. Because a carbon atom has four valence electrons, it's eager to donate those valence electrons to covalent bonds so it can receive four donated electrons in turn, filling carbon's valence shell. In other words, carbon really likes to form four bonds. In alkanes, each of these four is a single bond with a different partner.

As the name *hydrocarbon* suggests, these partners may be hydrogen or carbon. The simplest of the alkanes, called *continuous* or *straight-chain alkanes*, consist of one straight chain of carbon atoms linked with single bonds. Hydrogen atoms fill all the remaining bonds. Other types of alkanes include closed circles and branched chains, but we discuss the straight-chain alkanes here because they make clear the basic strategy for naming hydrocarbons. From the standpoint of naming, the hydrogen atoms in a hydrocarbon are more or less "filler atoms." Alkanes' names are based on the largest number of consecutively bonded carbon atoms, so the name of a hydrocarbon tells you about that molecule's structure.

To name a straight-chain alkane, simply match the appropriate chemical prefix with the suffix *-ane.* The prefixes, which relate to the number of carbons in the continuous chain, are listed in Table 11-4.

TABLE 11-4 Carbon Prefixes

Number of Carbons	Prefix	Chemical Formula	Alkane
1	meth-	CH_4	methane
2	eth-	C_2H_6	ethane
3	prop-	C_3H_8	propane
4	but-	C_4H_{10}	butane
5	pent-	C_5H_{12}	pentane
6	hex-	C_6H_{14}	hexane
7	hept-	C_7H_{16}	heptane
8	oct-	C_8H_{18}	octane
9	non-	C_9H_{20}	nonane
10	dec-	$C_{10}H_{22}$	decane

The naming method in Table 11-4 tells you how many carbons are in the chain. Because you know that each carbon has four bonds and because you're fiendishly clever, you can deduce the number of hydrogen atoms in the molecule as well. Consider the carbon structure of pentane, for example, shown in Figure 11-3.

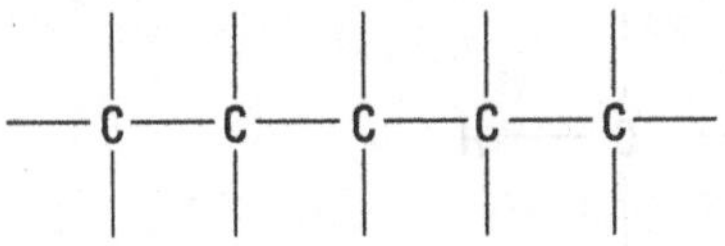

FIGURE 11-3:
Pentane's carbon skeleton.

Only four carbon-carbon bonds are required to produce the five-carbon chain of pentane. This situation leaves many bonds open — two for each interior carbon and three for each of the *terminal* carbons (the ones on either end of the chain). These open bonds are satisfied by carbon-hydrogen bonds, thereby forming a hydrocarbon, as shown in Figure 11-4.

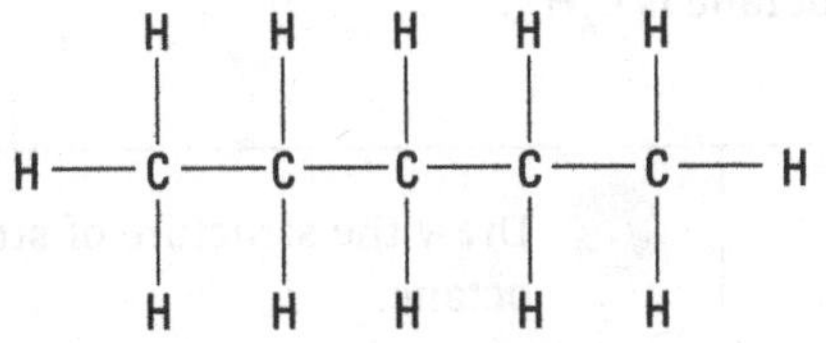

FIGURE 11-4:
Pentane's hydrocarbon structure.

If you add up the hydrogen atoms in Figure 11-4, you get 12. Therefore, pentane contains 5 carbon atoms and 12 hydrogen atoms.

REMEMBER The more complicated the organic molecule, the more important it is that you draw the molecular structure so you can visualize the molecule. In the case of straight-chain alkanes, the simplest of all organic molecules, you can remember a convenient formula for calculating the number of hydrogen atoms in the alkane without actually drawing the chain:

Number of hydrogen atoms $= (2 \times$ Number of carbon atoms$) + 2$

TIP

You can refer to the same molecule in a number of different ways. For example, you can refer to pentane by its name (ahem . . . *pentane*); by its molecular formula, C_5H_{12}; or by the complete structure in Figure 11-4. Clearly, these names include different levels of structural detail. A *condensed structural formula* is another naming method, one that straddles the divide between a molecular formula and a complete structure. For pentane, the condensed structural formula is $CH_3CH_2CH_2CH_2CH_3$. This kind of formula assumes that you understand how straight–chain alkanes are put together. Here's the lowdown:

>> Carbons on the end of a chain are bonded to only one other carbon, so they have three additional bonds that are filled by hydrogen and are labeled as CH_3 in a condensed formula.

>> Interior carbons are bonded to two neighboring carbons and have only two hydrogen bonds, so they're labeled as CH_2.

Your chemistry teacher will probably require you to draw structures of alkanes when given their names and require you to name alkanes when given their structures. If your teacher fails to make such requests, ask to see his credentials. You may be dealing with an impostor.

EXAMPLE

Q. What is the name of the following structure, and what is its molecular formula?

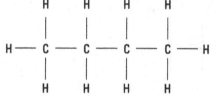

A. Butane; C_4H_{10}. First, count the number of carbons in the continuous chain. Four carbons are in the chain, and Table 11-4 helpfully points out that four-carbon chains earn the prefix *but-*. What's more, this molecule is an alkane (because it contains only single bonds), so it receives the suffix *-ane*. So what you've got is butane. With four carbon atoms in a straight chain, you need ten hydrogen atoms to satisfy all the carbon bonds, so the molecular formula of butane is C_4H_{10}.

YOUR TURN

12 What is the name of the following structure, and what is its molecular formula?

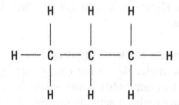

13 Draw the structure of straight–chain octane.

Practice Questions Answers and Explanations

(1) These compounds are all ordinary ionic compounds, so you simply need to pair the cation name with the anion name and change the anion name's ending to *-ide.*

  a. Magnesium fluoride

  b. Lithium bromide

  c. Cesium oxide

  d. Calcium sulfide

  e. Potassium chloride

  f. Strontium phosphide

  g. Sodium sulfide

(2) As the problem states, all the cations here are ones that can have varying amounts of positive charge, so you need to decipher their charges.

  a. **Iron(II) fluoride.** The fluoride ion has a charge of –1. Because two fluorides are present here, the single iron ion must have a +2 charge.

  b. **Copper(I) bromide.** The –1 charge of the bromide must be balanced by a +1 charge.

  c. **Tin(IV) iodide.** The four iodide anions each have a –1 charge, so the tin cation must have a charge of +4.

  d. **Manganese(III) oxide.** The three oxide anions here each have a charge of –2, giving an overall charge of –6. The manganese cations must carry a total charge of +6, split between the two cations, so you must be dealing with Mn^{3+}.

  e. **Iron(III) oxide.** There are two irons with an unknown charge and three oxygens, each with a known charge of –2. This means each iron must have a +3 charge.

  f. **Lead(IV) sulfide.** Each sulfur atom has a known charge of –2. Two lead atoms are present, so each individual atom must have a charge of +2 to cancel out the total charge of –4 from the two sulfur atoms.

  g. **Copper(II) chloride.** In this case, each chlorine atom has a known charge of –1, producing a total charge of –2. You have one copper atom, so that single copper must have a charge of +2 to cancel out the negative charges of the chlorine atoms.

(3) The names translate into the following chemical formulas:

  a. **Fe_2O_3.** Because the name specifies that you're dealing with Fe^{3+}, and because oxygen is always O^{2-}, you simply balance your charges to get Fe_2O_3.

  b. **$BeCl_2$.** Beryllium is an alkaline earth metal with a charge of +2, while chlorine is a halogen with a charge of –1.

  c. **SnS.** Because the name specifies that you're dealing with Sn^{2+}, and because sulfur is always S^{2-}, you simply balance your charges to get SnS.

  d. **KI.** Potassium is an alkali metal with a charge of +1, and iodine is a halogen with a charge of –1.

e. **Li₃P.** Each lithium atom has a +1 charge, and each phosphide atom has a charge of –3. To balance out the –3 charge, you need three lithium atoms to create a +3 charge.

f. **SrO.** In this case, no subscripts are needed. Strontium has a charge of +2, and oxygen has a charge of –2.

g. **Fe₂S₃.** Each iron atom has a charge of +3, as indicated by the Roman numeral. Each sulfur atom has a charge of –2. Thus, you need two irons and three sulfurs so the charges of +6 and –6 balance out.

(4) Look up (or recall) the polyatomic ions in each compound, and specify the charge of the cation if it's a metal that can take on different charges.

a. **Magnesium phosphate.** Magnesium is the cation here, and the anion is $PO_4{}^{3-}$. Table 11-1 tells you that this ion is the polyatomic ion phosphate (not to be confused with phosphite, $PO_3{}^{3-}$).

b. **Lead(II) acetate.** Acetate has a –1 charge, and because two acetate ions are necessary to balance out the charge on the lead cation, you must be dealing with Pb^{2+}.

c. **Chromium(III) nitrite.** Nitrite has a –1 charge, and because three of them are necessary to balance out the charge on the chromium cation, you must be dealing with Cr^{3+}.

d. **Ammonium oxalate.** Here, both the cation and the anion are polyatomic ions — annoying but true.

e. **Potassium permanganate.** Potassium is the cation here, so all you need to do is look up the anion $MnO_4{}^-$ to find that it's called *permanganate*.

(5) First, look up (or better, recall from memory) the charge of the polyatomic ion or ions, and then use subscripts as necessary to balance charges.

a. **K₂SO₄.** Potassium has a charge of +1, and sulfate has a charge of –2. To balance these charges, you must double the number of potassiums. This balances the charges at +2 and –2.

b. **PbCr₂O₇.** Here, the Roman numeral indicates that lead has a +2 charge. Dichromate is a polyatomic ion with a charge of –2, so you need only one of each to balance the charges.

c. **NH₄Cl.** Ammonium is a +1 polyatomic ion, and chlorine is a –1 ion. These charges are balanced, so you need one of each.

d. **NaOH.** Sodium has a charge of +1, and the hydroxide polyatomic ion has a charge of –1, so you need one of each.

e. **Cr₂(CO₃)₃.** The Roman numeral indicates that chromium has a +3 charge. Carbonate is a polyatomic ion with a charge of –2. To balance this formula, you must come up with a multiple of 2 and 3. In this case, the least common multiple is 6, so you double the +3 and triple the –2. This results in two chromiums and three carbonates.

(6) Using Table 11-2, translate the subscripts into prefixes. Omit the prefix *mono-* on the first named element in a compound where applicable.

a. Dinitrogen tetrahydride

b. Dihydrogen monosulfide

c. Nitrogen monoxide

d. Carbon tetrabromide

e. Dicarbon hexahydride

f. Dicarbon tetrahydride

g. Sulfur dioxide

7. Using Table 11-2, translate the prefixes into subscripts. If the first named element lacks a prefix, assume that only one such atom exists per molecule.

a. SiF_2

b. NF_3

c. S_2F_{10}

d. P_2Cl_3

e. IF_7

f. XeF_6

g. PO_5

8. Identify the anion for each binary acid. Add *hydro-* to the beginning of it and change the ending of it to *-ic* (if necessary, drop the *o* in *hydro-* to avoid having two vowels next to each other). Then write *acid* at the end.

a. Hydrofluoric acid

b. Hydroiodic acid

c. Hydrochloric acid

d. Hydrocyanic acid

9. Identify the anion for each oxy-acid from the polyatomic ion chart in Table 11-1. If the poly-atomic ion ends in *-ate*, change the ending to *-ic*. If the polyatomic ion ends in *-ite*, change the ending to *-ous*. Write *acid* at the end of the name. (Don't begin with *hydro-*! These aren't binary acids.)

a. Carbonic acid

b. Sulfurous acid

c. Chlorous acid

d. Nitric acid

10. Use the steps presented earlier in this section to guide yourself to a name from a formula.

a. **Lead(II) chromate.** Lead is a Group B element, and $CrO_4{}^{2-}$ is a polyatomic ion. Therefore, you need to determine the charge on the lead and specify that charge with a Roman numeral. If you haven't memorized the name of $CrO_4{}^{2-}$, find it in Table 11-1. Because chromate combines with lead in a one-to-one ratio, you know that the charge on the lead must be +2.

b. **Magnesium phosphide.** This compound is a simple ionic compound because Mg is a non–Group B metal and because P isn't a polyatomic ion.

c. **Strontium silicate.** Sr is neither a Group B element nor a nonmetal, but SiO_3 forms a polyatomic ion.

d. **Sulfuric acid**. The H at the beginning of this formula is your clue that this compound is an acid. The H is then followed by the polyatomic ion sulfate. Because sulfate (SO_4^{2-}) ends in *-ate*, you change the ending to *-ic*, adding *ur* to make the name sound better. You then write *acid* at the end.

e. **Sodium sulfide**. Simply name the cation and change the ending of the (polyatomic) anion to *-ide*.

f. **Triboron diselenide**. Both boron and selenium are nonmetals, so this compound is a molecular compound. Therefore, you name it by using prefixes.

g. **Mercury(II) fluoride**. Here you have an ionic compound with a Group B metal, which means you need to use a Roman numeral. The charge on the mercury atom must be +2 because it combines with two fluoride anions, each of which must have a –1 charge.

h. **Barium phosphate**. This compound is an ionic compound that contains a polyatomic ion, phosphate.

(11) Reverse your naming rules to deduce the chemical formula of each compound.

a. $Ba(OH)_2$. Barium is an ordinary metal, and hydroxide is a polyatomic ion with a charge of –1.

b. $SnBr_4$. The name indicates that you're dealing with Sn^{4+}. The bromide ion has a charge of –1, so you need four of them to balance the charge of a single tin cation.

c. Na_2SO_4. Sodium is a simple alkali metal, and sulfate is a polyatomic ion with a charge of –2.

d. PI_3. The prefixes indicate that this compound is a molecular compound containing a single phosphorus atom and three iodine atoms.

e. $Mg(MnO_4)_2$. Magnesium is a metal, and permanganate is a polyatomic ion with a charge of –1.

f. $HC_2H_3O_2$. Because this compound is an acid, you know the formula starts with an *H*. *Acetic* is nowhere on the periodic table, so you must assume it's a polyatomic ion. To figure out which one, change the *-ic* ending back to *-ate* to get *acetate*. Acetate has a charge of –1, and hydrogen has a charge of +1, so you need one of each to balance the charges.

g. NH_2. The *di-* prefix and the fact that both elements are nonmetals indicate that this compound is a molecular compound. Nitrogen, the first named element, lacks a prefix, so there must be only one nitrogen per molecule. The *di-* prefix indicates two hydrogen atoms per molecule.

h. $FeCrO_4$. The name tells you that you're dealing with Fe^{2+} and the polyatomic ion chromate, which has a charge of –2. A one-to-one ratio is sufficient to neutralize overall charge.

(12) **The structure is propane; its molecular formula is C_3H_8.** The figure shows a three-carbon chain with only single bonds. Therefore, it's propane. Its molecular formula is C_3H_8.

(13) The *oct–* prefix here tells you that this alkane is eight carbons long. Draw eight linked carbons and fill in the empty bonds with hydrogen. Your structure should look like one of these:

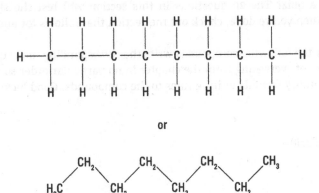

or

If you're ready to test your skills a bit more, take the following chapter quiz that incorporates all the chapter topics.

Whaddya Know? Chapter 11 Quiz

Ready for a quiz? The 20 questions in this section will test the skills you learned in this chapter. When you're done, check out the section that follows for answers and explanations.

For each of the following compounds write the corresponding name or formula correctly. This is a mixture of everything from the chapter in no particular order so make sure you apply the correct naming / formula writing rules to the compounds. Good luck!

1. P_2H_6

2. Calcium Chloride

3. HBr

4. $PbSO_4$

5. $Pb_3(PO_4)_4$

6. Dicarbon pentoxide

7. $HClO_3$

8. nitrous acid

9. Sodium cyanide

10. NH_4Cl

11. $Ba(ClO_2)_2$

12. Potassium nitrate

13. Hydrofluoric acid

14. V_2O_5

15. Aluminum Hydroxide

16. Sulfur heptoxide

17. Iron (II) Oxide

18. $Sr(NO_3)_2$

19. C_4H_{10}

20. C_6H_{14}

Answers to Chapter 11 Quiz

1. **Diphosphorous hexahydride.** This is a covalently bonded compound as these are two non-metals bonded together. That means you are not dealing with ionic charges. Instead simply use the correct naming prefixes to represent the number of each atom present. There are 2 phosphorous so you use the prefix di. There are 6 hydrogen so you use the prefix hexa.

2. $CaCl_2$. This is an ionic compound because you have a metal and a nonmetal bonding together. You must balance the ionic charge of this compound for it to be accurate. Calcium is 2+ and chlorine is 1–. To balance this you must have 2 chlorines for each calcium.

3. **Hydrobromic acid.** This formula represents an acid. You can usually identify this by the formula starting with a hydrogen atom. There are no oxygens present so it is not an oxy acid. Instead it is a binary acid because there are only 2 elements present, H and Br. To name it correctly you add hydro to the beginning and then change the ending of the second element to –*ic*. Don't forget to write acid at the end either.

4. **Lead (II) sulfate.** This is an ionic compound so you do not use prefixes when writing the name. Instead you simply write the name of the cation, lead, and then write the name of the anion, sulfate. In this case, since sulfate is a polyatomic ion you don't change the ending of the anion. You aren't done yet though since lead is a transition metal and is capable of having several different charges. You need to indicate the charge of this lead by including a Roman numeral. Sulfate is a 2– ion and you only have one of them. That means the single lead atom must be a 2+ charge to balance out the charge on this compound.

5. **Lead (IV) phosphate.** This is an ionic compound so you do not use prefixes when writing the name. Instead you simply write the name of the cation, lead, and then write the name of the anion, phosphate. Since phosphate is a polyatomic ion you don't change the ending of the anion. You aren't done yet though, since lead is a transition metal and is capable of having several different charges. You need to indicate the charge of this lead by including a roman numeral. Phosphate is a 3– ion and you have 4 of them. That gives a total negative charge of 12–. You have 3 leads so each lead must be a 4+ charge to equal a total charge of 12+ so you use a Roman numeral of (IV).

6. C_2O_5. This is a covalently bonded molecule since you have to nonmetals, you can also identify this easily as there are prefixes given to you in the name. Simply write the number of atoms for each element that matches the corresponding prefix. di=2 and pent=5.

7. **Chloric acid.** This is an acid as indicated by the first element being a hydrogen. In addition, it is an oxy acid because there is oxygen present in the formula. You do not use hydro in naming this because it is not a binary acid. Instead just write the name of the polyatomic ion, chlorate, but change the ending of that ion to –*ic* because the polyatomic ion ends in –*ate*. Remember, –*ate* gets change to –*ic* and –*ite* gets changed to –*ous*. Don't forget to write "acid" at the end.

8. HNO_2. Since there is no hydro present at the beginning of this name you know it is an oxy acid. The –*ous* ending tells you the polyatomic ion for it ends in –*ite*. This corresponds to nitrite. So you begin your acid with hydrogen and then write NO_2. Both are 1+ and 1– respectively so you only need one of each ion.

9) **NaCN.** This is an ionic compound because you have a metal and a nonmetal bonding together. You must balance the ionic charge of this compound for it to be accurate. Be sure to recognize that cyanide is a polyatomic ion. Sodium is 1+ and cyanide is 1– so you need one of each ion in the formula.

10) **Ammonium chloride.** This is an ionic compound so you do not use prefixes when writing the name. Instead you simply write the name of the cation, ammonium, and then write the name of the anion, chloride. There is no need for a Roman numeral because the charge of ammonium is always 1+.

11) **Barium chlorite.** This is an ionic compound so you do not use prefixes when writing the name. Instead you simply write the name of the cation, barium, and then write the name of the anion, chlorite. There is no need for a Roman numeral because the charge of barium is always 2+. You do not need to indicate the subscript 2 that is written on chlorite in the formula.

12) **KNO_3.** This is an ionic compound because you have a metal and a nonmetal bonding together. You must balance the ionic charge of this compound for it to be accurate. Be sure to recognize that nitrate, NO_3, is a polyatomic ion. Potassium is 1+ and nitrate is 1– so you need one of each ion in the formula.

13) **HF.** Hydrofluoric acid is a binary acid because it begins with hydro-. To write the formula correctly you need to add H to the beginning of the formula. Once you do that you can change the ending of –ic back to –ide and see that you need the fluoride ion in the compound. Be sure to write the formula correctly related to charge. H is 1+ and F is 1– so you need one of each to balance the charge.

14) **Vanadium (V) oxide.** This is an ionic compound so you do not use prefixes when writing the name. Instead you simply write the name of the cation, vanadium, and then write the name of the anion, oxide. You also change the ending of the anion, oxygen, to oxide. You aren't done yet though since vanadium is a transition metal and is capable of having several different charges. You need to indicate the charge of this vanadium by including a Roman numeral. Oxide is a 2– ion and you have 5 of them. That gives a total negative charge of 10–. You have 2 vanadiums so each vanadium must be a 5+ charge to equal a total charge of 10+ so you use a Roman numeral of (V).

15) **$Al(OH)_3$.** This is an ionic compound because you have a metal and a nonmetal bonding together. You must balance the ionic charge of this compound for it to be accurate. Be sure to recognize that hydroxide is a polyatomic ion. Aluminum is 3+ and hydroxide is 1– so you need 3 hydroxides to balance out the 3+ of aluminum. Be sure to use parenthesis around the hydroxide. If you wrote your answer as OH_3 that is incorrect because there needs to be 3 hydroxide ions, not just 3 hydrogens.

16) **SO_7.** This is a covalently bonded molecule since you have to nonmetals, you can also identify this easily as there are prefixes given to you in the name. Simply write the number of atoms for each element that matches the corresponding prefix. In the case of sulfur there is no prefix given. Since no prefix is given you can assume that there is only 1 of that atom present. This would normally be written as mono but the prefix mono- is generally dropped from the first element in a compound. Mono=1, hept=7.

17 **FeO.** This is an ionic compound because you have a metal and a nonmetal bonding together. You must balance the ionic charge of this compound for it to be accurate. The Roman numeral II that is given after iron indicates that iron is a 2+ charge and you know that oxygen is a 2− charge. Based on them having the same but opposite charge, you only need one of each element. Be sure not to write this as Fe_2O_2. If you have a situation like that, you need to remember to simplify the subscripts to 1 and 1.

18 **Strontium nitrate.** This is an ionic compound so you do not use prefixes when writing the name. Instead you simply write the name of the cation, strontium, and then write the name of the anion, nitrate. Since nitrate is a polyatomic ion you do not change ending of it. There is no need for a Roman numeral because the charge of strontium is always 2+. You do not need to indicate the subscript 2 that is written on nitrate in the formula.

19 **Butane.** The easiest way to identify this is to see that you have a hydrocarbon of only carbon and hydrogen. Since there are 4 carbons and the number of hydrogen is $4 \times 2 + 2$ relative to the carbon, this makes it an alkane molecule. An alkane with 4 carbons is called butane.

20 **Hexane.** This is an organic carbon molecule. Since there are 6 carbons and the number of hydrogen is $6 \times 2 + 2$ relative to the carbon, this makes it an alkane molecule. An alkane with 6 carbons is called hexane.

4

Working with Reactions

In This Unit . . .

Chapter **12**

Understanding the Many Uses of the Mole

C hemists routinely deal with hunks of material containing trillions of trillions of atoms, but ridiculously large numbers can induce migraines. For this reason, chemists count particles (like atoms and molecules) in multiples of a quantity called the *mole*.

The mole is, without a doubt, the largest number you'll ever deal with on a regular basis. It represents 6.02×10^{23} particles. Chemists came up with this number to help work with the incredibly large number of atoms they deal with on a daily basis. Saying 2 moles of sodium chloride is far easier than saying 1.204×10^{24} atoms of sodium chloride.

The best way to understand the concept of the mole is to see it like any other representative quantity, such as a pair, a dozen, or a gross. For example, you know that 1 dozen of something is 12 of that same thing. Well, 1 mole of something is 6.02×10^{23} of that same thing. No different from a dozen — the mole is just much, much bigger. In this chapter, we explain what you need to know about moles.

REMEMBER

For many mole conversions, you need to look up atomic masses on the periodic table (see Chapter 5) or in the alphabetized table of the elements in Chapter 4. The atomic masses you see in different periodic tables may vary slightly, so for consistency, we've rounded all atomic mass values to two decimal places before plugging them into the equations. We round answers according to significant figure rules (see Chapter 1 for details).

The Mole Conversion Factor: Avogadro's Number

If 6.02×10^{23} strikes you as an unfathomably large number, then you're thinking about it correctly. It's larger, in fact, than the number of stars in the sky or the number of fish in the sea, and it's many, many times more than the number of people who've been born throughout all of human history. When you think about the number of particles in something as simple as, say, a cup of water, all your previous conceptions of "big numbers" are blown out of the water, as it were.

When written in longhand notation, it's

602,200,000,000,000,000,000,000

And *that* is why chemists like scientific notation.

The number 6.02×10^{23}, known as *Avogadro's number*, is named after the 19th-century Italian scientist Amedeo Avogadro. Posthumously, Avogadro really pulled one off in giving his name to this number, because he never actually thought of it. The real brain behind Avogadro's number was that of a French scientist named Jean Baptiste Perrin. Nearly 100 years after Avogadro had his final pasta, Perrin named the number after Avogadro as an homage. Ironically, this humble act of tribute has misdirected the resentment of countless hordes of high school chemistry students to Avogadro instead of Perrin.

REMEMBER

Avogadro's number is the conversion factor used to move between particle counts and numbers of moles. Notice that *mole* is abbreviated as *mol*.

$$\frac{1 \text{ mol}}{6.02 \times 10^{23} \text{ particles}}$$

If you had a mole of marshmallows, it would cover the United States to a depth of about 600 miles. A mole of rice grains would cover the land area of the world to a depth of about 75 meters. And a mole of moles . . . no, we don't even want to think about that!

The most common particles that you'll use with the mole are atoms and molecules. (You more than likely won't be doing calculations involving a mole of chocolate bars, though it is nice to imagine.) When encountering a problem that deals with a specific unit, like molecules, you just replace *particles* with the correct unit, as in the following example. Like all conversion factors, you can invert this fraction to move in the other direction, from moles to particles. (Flip to Chapter 2 for an introduction to conversion factors.)

EXAMPLE

Q. How many water molecules are in 1 tablespoon of water if the tablespoon holds 0.82 mol?

A. 4.9×10^{23}. To convert from moles to particles, use a conversion factor with moles in the denominator and particles in the numerator:

$$(0.82 \text{ mol } H_2O)\left(\frac{6.02 \times 10^{23} \text{ molecules } H_2O}{1 \text{ mol } H_2O}\right) = 4.9 \times 10^{23} \text{ molecules } H_2O$$

The units of moles in the numerator of the first term and the denominator of the second term cancel, so you're left with the final answer of 4.9×10^{23} molecules. (If you need tips on multiplying in scientific notation, check out Chapter 1.)

Q. How many moles are in 5.6×10^8 molecules of H_2O?

A. Set things up exactly as if you were converting from moles to particles, with one very big change: Flip the conversion factor to make sure your units will cancel out.

$$(5.6 \times 10^8 \text{ molecules } H_2O)\left(\frac{1 \text{ mol } H_2O}{6.02 \times 10^{23} \text{ molecules } H_2O}\right) = 9.3 \times 10^{-16} \text{ mol } H_2O$$

YOUR TURN

1 If you have 1.3 mol of sodium (Na), how many atoms of sodium are present?

2 If you have 7.9×10^{24} molecules of methane (CH_4), how many moles of methane are present?

Doing Mass and Volume Mole Conversions

Chemists always begin a discussion about moles with Avogadro's number. They do this for two reasons. First, it makes sense to start the discussion with the way the mole was originally defined. Second, Avogadro's number is sufficiently large to intimidate the unworthy.

Still, for all its importance and intimidating size, Avogadro's number quickly grows tedious in everyday use, because counting out very large numbers isn't very practical. Suppose that you have a job packing 1,000 nuts and 1,000 bolts in big bags, and you get paid for each bag you fill. What's the most efficient and quickest way of counting out nuts and bolts? Determine the mass of 100, or even 10, of each, and then figure out the mass of 1,000 of each. Fill up the bag with nuts until it equals the mass you figured for 1,000 nuts. After you have the correct number of nuts, use the same process to fill the bag with bolts. In other words, count by massing; that's one of the most efficient ways of counting large numbers of objects and it is exactly what we do with elements and compounds.

Perhaps even more interesting than Avogadro's number is the fact that 1 mole of an individual element turns out to possess exactly its atomic mass's worth of grams (see Chapter 4 for more about atomic mass). In other words, 1 mole of lithium atoms has a mass of about 6.94 grams, and 1 mole of helium atoms has a mass of about 4.00 grams. The same is true no matter where you wander through the corridors of the periodic table. The number listed as the atomic mass of an element also equals that element's *molar mass*. So in short, 1 mole of an individual element is equal to its atomic mass in grams.

Of course, chemistry involves the making and breaking of bonds (as you find out in Chapter 8), so talk of individual atoms only goes so far. How lucky, then, that calculating the molecular mass of a complex molecule is essentially no different from finding the mass per mole of individual elements. For example, a molecule of glucose ($C_6H_{12}O_6$) is assembled from 6 carbon atoms, 12 hydrogen atoms, and 6 oxygen atoms. To calculate the number of grams per mole of a complex molecule (such as glucose), simply do the following:

1. **Multiply the number of atoms per mole of the first element by its atomic mass.**

   In glucose, the first element is carbon. You'd multiply its atomic mass, 12.01, by the number of atoms, 6, for a total of 72.06.

2. **Multiply the number of atoms per mole of the second element by its atomic mass. Keep going until you've covered all the elements in the molecule.**

   Multiply hydrogen's atomic mass of 1.01 by the number of hydrogen atoms, 12, for a total of 12.12.

   The third element in glucose is oxygen, so you multiply 16.00, the atomic mass, by 6, the number of oxygen atoms, for a total of 96.00.

3. **Finally, add the masses together. The units that you use for molar mass is grams/mol.**

   Here are all the calculations for this example:

$$\left(12.01\frac{g}{mol}\right)(6\,atoms\ carbon) = 72.06\,g/mol\ C$$

$$\left(1.01\frac{g}{mol}\right)(12\,atoms\ hydrogen) = 12.12\,g/mol\ H$$

$$\left(16.00\frac{g}{mol}\right)(6\,atoms\ oxygen) = 96.00\,g/mol\ O$$

$$72.06\,g/mol\ C + 12.12\,g/mol\ H + 96.00\,g/mol\ O = 180.18\,g/mol\,C_6H_{12}O_6$$

This kind of quantity, called the *gram molecular mass*, is exceptionally convenient for chemists, who are much more inclined to measure the mass of a substance than to count all the individual particles that make up a sample.

WARNING

Chemists may distinguish between the molar masses of pure elements, molecular compounds, and ionic compounds by referring to them as the *gram atomic mass*, *gram molecular mass*, and *gram formula mass*, respectively. Don't be fooled! The basic concept behind each term is the same: molar mass.

REMEMBER

It's all very good to find the mass of a solid or liquid and then go about calculating the number of moles in that sample. But what about gases? Let's not engage in phase discrimination; gases are made of matter, too, and their moles have the right to stand and be counted. Fortunately, there's a convenient way to convert between the moles of gaseous particles and their volume. Unlike gram atomic/molecular/formula masses, this conversion factor is constant — no matter what kinds of molecules make up the gas. Every gas has a volume of 22.4 liters per mole, regardless of the size of the gaseous molecules.

Before you start your hooray-chemistry-is-finally-getting-simple dance, understand that certain conditions apply to this conversion factor. For example, it's true only at *standard temperature and pressure* (STP), or 0°C and 1 atmosphere. Also, the figure of 22.4 L/mol applies only to the extent that a gas resembles an *ideal gas*, one whose particles have zero volume and neither attract nor repel one another. Ultimately, no gas is truly ideal, but many are so close to being so that the 22.4 L/mol conversion is very useful.

TIP

What if you want to convert between the volume of a gaseous substance and its mass, or between the mass of a substance and the number of particles it contains? You already have all the information you need! To make these kinds of conversions, simply build a chain of conversion factors, converting units step by step from the ones you have (say, liters) to the ones you want (say, grams). You'll find that your chain of conversion factors always includes central links featuring units of moles. You can think of the mole as a family member who passes on what you've said, loudly barking into the ear of your nearly deaf grandmother because you have laryngitis and can't speak any louder. Without such a central translator, your message would no doubt be misinterpreted. "Grandma, how was your day?" would be received as "Grandma, you want to eat clay?" So unless you're bent on force-feeding clay to your grandmother, do not attempt to convert directly from volume to mass, from mass to particles, or any other such shortcut. Use your translator, the mole.

EXAMPLE

Q. Convert 65 g of carbon dioxide (CO_2) to moles of carbon dioxide.

A. 1.5 mol CO_2. Before beginning this problem, you need to determine the molecular mass of CO_2. Do this by adding the atomic mass of 1 carbon atom $(1 \times 12.01 = 12.01)$ and the atomic mass of 2 oxygens $(2 \times 16.00 = 32.00)$ for a total of 44.01 g/mol. Then use this molecular mass to create a conversion factor and determine the moles. Be sure your units cancel out to ensure your conversion factor is set up correctly.

$$\left(65 \text{ g CO}_2\right)\left(\frac{1 \text{ mol CO}_2}{44.01 \text{ g CO}_2}\right) = 1.5 \text{ mol CO}_2$$

Q. Convert 26.7 g of carbon dioxide (CO_2) to liters of carbon dioxide gas.

A. 13.6 L CO_2. The first thing you should notice about this problem is that it requires two steps. You must first convert from grams of CO_2 to moles of CO_2 using the molecular mass of carbon dioxide. Find this molecular mass by adding the atomic mass of 1 carbon atom $(1 \times 12.01 = 12.01)$ and the atomic mass of 2 oxygens $(2 \times 16.00 = 32.00)$ for a total of 44.01 g/mol. After converting to moles, you convert to liters of CO_2 gas using the molar volume constant, 22.4 L/mol. However, instead of doing each of these steps in separate calculations, you can combine them so you do everything on one line. As always, be sure your units cancel out so that you end up with your goal, in this case liters of carbon dioxide.

$$(26.7\text{g } CO_2)\left(\frac{1 \text{ mol } CO_2}{44.01\text{g } CO_2}\right)\left(\frac{22.4 \text{ L } CO_2}{1 \text{ mol } CO_2}\right) = 13.6 \text{ L } CO_2$$

Q. How many molecules are there in 100.0 grams of carbon dioxide?

A. To solve this question, do the following:

Determine the molecular mass of CO_2 by looking at the periodic table. The periodic table shows that one carbon atom has a mass of 12.01 amu and one oxygen atom has a mass of 15.99 amu.

Figure the molecular mass. For this example, your calculations look like this:

$$(1\,C \times 12.01 \text{ g/mol } C) + (2\,O \times 16.00 \text{ g/mol } O) = 44.01 \text{ g/mol } CO_2$$

Set up your problem so that units cancel, and you get your answer:

$$(100.0 \text{ g } CO_2)\left(\frac{1 \text{ mol } CO_2}{44.01 \text{ g } CO_2}\right)\left(\frac{6.02 \times 10^{23} \text{ molecules of } CO_2}{1 \text{ mol } CO_2}\right) = 1.368 \times 10^{24} \text{ molecules of } CO_2$$

YOUR TURN

3 Do the conversion:

a. How many grams are present in 2.6 mol of lithium?

b. How many moles are present in 85.2 g of sodium chloride (NaCl)?

4 Do the conversion:

    a. How many liters of gas are present in 4.3 mol of oxygen gas (O_2)?

    b. How many moles of gas are present in 64.3 L of nitrogen gas (N_2)?

5 Do the conversion:

    a. How many liters of hydrogen gas (H_2) are present in 76.2 g of hydrogen gas?

    b. How many atoms are present in 10 g of lithium?

Determining Percent Composition

Chemists are often concerned with precisely what percentage of a compound's mass consists of one particular element. Lying awake at night, uttering prayers to Avogadro, they fret over this quantity, called *percent composition*. Calculating percent composition is trickier than you may think. Consider the following problem, for example.

The human body is composed of 60 to 70 percent water, and water contains twice as many hydrogen atoms as oxygen atoms. If two-thirds of every water molecule is hydrogen and if water makes up 60 percent of the body, it seems logical to conclude that hydrogen makes up 40 percent of the body. Yet hydrogen is only the third most abundant element in the body by mass. What gives?

Oxygen is 16 times more massive than hydrogen, so equating *atoms* of hydrogen and *atoms* of oxygen is a bit like equating a toddler to a sumo wrestler. When the doors of the elevator won't close, the sumo wrestler is the first one you should kick out, weep though he may.

Within a compound, it's important to sort out the atomic toddlers from the atomic sumo wrestlers. To do so, follow three simple steps:

1. **Calculate the molar mass of the compound, as we explain in the preceding section.**

   This value represents the total mass of the compound.

2. **Determine the total mass of each element in the compound, being sure to account for all atoms present.**

   For example, if you have three carbon atoms, be sure to multiply the mass of carbon (12.01 g/mol) by 3. These values represent the individual mass of each element in the compound.

3. **Use the following formula to find the percent composition for each element in the compound:**

$$\% \text{ composition of element} = \left(\frac{\text{Total mass of element}}{\text{Total mass of entire compound}} \right) \times 100$$

Q. Calculate the percent composition for each element in sodium sulfate, Na_2SO_4.

A. Na: 32.4%, S: 22.6%, O: 45.0%. To begin, determine the molecular mass of sodium sulfate. Do this by adding the masses of two sodium atoms plus one sulfur atom and four oxygen atoms together:

$$\left(22.99 \, \frac{g}{mol} \right) (2 \text{ atoms Na}) = 45.98 \text{ g/mol Na}$$

$$\left(32.06 \, \frac{g}{mol} \right) (1 \text{ atom S}) = 32.06 \text{ g/mol S}$$

$$\left(16.00 \, \frac{g}{mol} \right) (4 \text{ atoms O}) = 64.00 \text{ g/mol O}$$

45.98 g/mol Na + 32.06 g/mol S + 64.00 g/mol O = 142.04 g/mol Na_2SO_4

You've now also determined the total mass of each element present in the formula. Now all you need to do is plug each element's value into the percent composition formula and solve:

$$\% \text{ composition of Na} = \left(\frac{45.98 \text{ g Na}}{142.04 \text{ g Na}_2SO_4} \right) \times 100 = 32.4\% \text{ Na}$$

$$\% \text{ composition of S} = \left(\frac{32.06 \text{ g S}}{142.04 \text{ g Na}_2SO_4} \right) \times 100 = 22.6\% \text{ S}$$

$$\% \text{ composition of O} = \left(\frac{64.00 \text{ g O}}{142.04 \text{ gNa}_2SO_4} \right) \times 100 = 45.0\% \text{ O}$$

6 Calculate the percent composition of potassium chromate, K_2CrO_4.

 7 Calculate the percent composition of lithium phosphate, Li_3PO_4.

Calculating Empirical Formulas

What if you don't know the formula of a compound? Chemists sometimes find themselves in this disconcerting scenario. Instead of cursing Avogadro (or perhaps *after* doing so), they analyze samples of the frustrating unknown to identify the percent composition. From there, they calculate the ratios of different types of atoms in the compound. They express these ratios as an *empirical formula*, the lowest whole-number ratio of elements in a compound.

Here's how to find an empirical formula when given percent composition:

REMEMBER 1. **Assume that you have 100 grams of the unknown compound.**

The beauty of this little trick is that you conveniently gift yourself with the same number of grams of each elemental component as its contribution to the percent composition. For example, if you assume that you have 100 grams of a compound composed of 60.3% magnesium and 39.7% oxygen, you know that you have 60.3 grams of magnesium and 39.7 grams of oxygen. (The only time you don't do this is if the problem specifically gives you the mass of each element present in the unknown compound.)

2. **Convert the masses from Step 1 into moles using the molar mass.**

See the earlier section "Doing Mass and Volume Mole Conversions."

3. **Determine which element has the smallest mole value. Then divide all the mole values you calculated in Step 2 by this smallest value.**

This division yields the mole ratios of the elements of the compound.

4. **If any of your mole ratios aren't whole numbers, multiply all numbers by the smallest possible factor that produces whole-number mole ratios for all the elements.**

For example, if you have 1 nitrogen atom for every 0.5 oxygen atoms in a compound, the empirical formula is not $N_1O_{0.5}$. Such a formula casually suggests that an oxygen atom has been split, something that would create a small-scale nuclear explosion.

Though impressive sounding, this scenario is almost certainly false. Far more likely is that the atoms of nitrogen and oxygen are combining in a 1:0.5 *ratio* but do so in a larger but equivalent ratio of 2:1. The empirical formula is thus N_2O.

Sometimes you only have to double all the numbers in your mole ratio to get a whole-number ratio, as we've shown here. In other problems, figuring out which number to multiply by may take a bit more work.

Because the original percent composition data is typically experimental, expect to see a bit of error in the numbers. For example, 2.03 is probably within experimental error of 2, 2.99 is probably 3, and so on.

5. **Write the empirical formula by attaching these whole-number mole ratios as sub-scripts to the chemical symbols of the elements. Order the elements according to the general rules for naming ionic and molecular compounds.**

    We describe the naming rules in Chapter 6.

Now, keeping the preceding steps in mind, suppose you're working with some unknown compound in a lab and you determine that this particular compound has the following weight percentage of elements present:

26.4% Na

36.8% S

36.8% O

Because you're dealing with percentage data (amount per hundred), remember to assume you have 100 grams of the total compound when you begin solving the problem, then convert each of the mass values to moles.

$$\frac{26.4 \text{ g Na}}{1} \times \frac{1 \text{ mol Na}}{22.99 \text{ g}} = 1.15 \text{ mol Na}$$

$$\frac{36.8 \text{ g S}}{1} \times \frac{1 \text{ mol S}}{32.07 \text{ g}} = 1.15 \text{ mol S}$$

$$\frac{36.8 \text{ g O}}{1} \times \frac{1 \text{ mol O}}{16.00 \text{ g}} = 2.30 \text{ mol O}$$

Now take each mole value and divide it by the smallest mole value of the three elements in the question.

$$\frac{1.15 \text{ Na}}{1.15} = 1 \text{ Na} \qquad \frac{1.15 \text{ S}}{1.15} = 1 \text{ S} \qquad \frac{2.30 \text{ O}}{1.15} = 2 \text{ O}$$

When you have the answer to that, write the empirical formula: $NaSO_2$. (If a subscript is 1, it's not shown.)

Q. What is the empirical formula of a substance that is 40.0% carbon, 6.7% hydrogen, and 53.3% oxygen by mass?

EXAMPLE

A. CH_2O. For the sake of simplicity, assume that you have a total of 100 g of this mystery compound. Therefore, you have 40.0 g of carbon, 6.7 g of hydrogen, and 53.3 g of oxygen. Convert each of these masses to moles by using the gram atomic masses of C, H, and O:

$$(40.0 \text{ g C})\left(\frac{1 \text{ mol C}}{12.01 \text{ g C}}\right) = 3.33 \text{ mol C}$$

$$(6.7 \text{ g H})\left(\frac{1 \text{ mol H}}{1.01 \text{ g H}}\right) = 6.6 \text{ mol H}$$

$$(53.3 \text{ g O})\left(\frac{1 \text{ mol O}}{16.00 \text{ g O}}\right) = 3.33 \text{ mol O}$$

Notice that the carbon and oxygen mole numbers are the same, so you know the ratio of these two elements is 1:1 within the compound. Next, divide all the mole numbers by the smallest among them, which is 3.33. This division yields

$$\frac{3.33 \text{ mol C}}{3.33} = 1 \text{ mol C} \quad \frac{6.6 \text{ mol H}}{3.33} = 2 \text{ mol H} \quad \frac{3.33 \text{ mol O}}{3.33} = 1 \text{ mol O}$$

The compound has the empirical formula CH_2O. The actual number of atoms within each particle of the compound is some multiple of the numbers expressed in this formula.

YOUR
TURN

8 Calculate the empirical formula of a compound with a percent composition of 88.9% oxygen and 11.1% hydrogen.

9 Calculate the empirical formula of a compound with a percent composition of 40.0% sulfur and 60.0% oxygen.

Using Empirical Formulas to Find Molecular Formulas

Many compounds in nature — particularly compounds made of carbon, hydrogen, and oxygen — are composed of atoms that occur in numbers that are multiples of their empirical formula. In other words, their empirical formulas don't reflect the actual numbers of atoms within them; instead, they reflect only the ratios of those atoms. What a nuisance! Fortunately, this is an old nuisance, so chemists have devised a means to deal with it. To account for these annoying types of compounds, chemists are careful to differentiate between an empirical formula and a molecular formula. A *molecular formula* uses subscripts that report the actual number of each type of atom in a molecule of the compound (a *formula unit* accomplishes the same thing for ionic compounds).

Molecular formulas are associated with gram molecular masses that are simple whole-number multiples of the corresponding *empirical formula mass*. For example, a molecule with the empirical formula CH_2O has an empirical formula mass of about 30 g/mol (12 for the carbon + 2 for the two hydrogens + 16 for the oxygen). The molecule may have a molecular formula of CH_2O, $C_2H_4O_2$, $C_3H_6O_3$, or the like. As a result, the compound may have a gram molecular mass of 30 g/mol, 60 g/mol, 90 g/mol, or another multiple of 30 g/mol.

WARNING You can't calculate a molecular formula based on percent composition alone. If you attempt to do so, Avogadro and Perrin will rise from their graves, find you, and slap you 6.02×10^{23} times per cheek. You can clearly see the folly of such an approach by comparing formaldehyde with glucose. The two compounds have the same empirical formula, CH_2O, but different molecular formulas, CH_2O and $C_6H_{12}O_6$, respectively. Glucose is a simple sugar, the one made by photosynthesis and the one broken down during cellular respiration. You can dissolve it in your coffee with pleasant results. Formaldehyde is a carcinogenic component of smog. Solutions of formaldehyde have historically been used to embalm dead bodies. Dissolving formaldehyde in your coffee is not advised. In other words, molecular formulas differ from empirical formulas, and the difference is important in the real world.

REMEMBER To determine a molecular formula, you must know the gram formula mass of the compound as well as the empirical formula (or enough information to calculate it yourself from the percent composition; see the preceding section for details). With these tools in hand, calculating the molecular formula involves three steps:

1. **Calculate the empirical formula mass.**

2. **Divide the gram molecular mass by the empirical formula mass.**

3. **Multiply each of the subscripts within the empirical formula by the number calculated in Step 2.**

Q. What is the molecular formula of a compound that has a gram molecular mass of 34 g/mol and the empirical formula HO?

EXAMPLE **A.** H_2O_2. The empirical formula mass is 17.01 g/mol. You determine this number by finding the mass of HO (1 hydrogen atom and 1 oxygen atom).

$$H: 1.01 \cdot 1 = 1.01$$
$$O: 16.00 \cdot 1 = 16.00$$
$$1.01 + 16.00 = 17.01 \text{ g/mol HO}$$

Dividing the gram molecular mass by this value yields the following:

$$\frac{\text{Molecular formula mass}}{\text{Empirical formula mass}} = \frac{34 \, \frac{g}{mol}}{17.01 \, \frac{g}{mol}} = 2$$

Multiplying the subscripts within the empirical formula by this number gives you the molecular formula H_2O_2. This formula corresponds to the compound *hydrogen peroxide.*

YOUR TURN

 10 What is the molecular formula of a compound that has a gram formula mass of 78 g/mol and the empirical formula NaO?

11 A compound has a percent composition of 49.5% carbon, 5.2% hydrogen, 16.5% oxygen, and 28.8% nitrogen. The compound's gram molecular mass is 194.2 g/mol. What are the empirical and molecular formulas?

Practice Questions Answers and Explanations

(1) **7.8×10^{23} atoms Na.** To calculate the answer, make sure you use the correct conversion factor with Avogadro's number. The units of *moles* in the first term and in the denominator of the second term cancel, leaving you the answer in number of atoms:

$$(1.3 \text{ mol Na})\left(\frac{6.02 \times 10^{23} \text{ atoms Na}}{1 \text{mol Na}}\right) = 7.8 \times 10^{23} \text{ atoms Na}$$

(2) **13 mol CH_4.** To calculate the answer, use the correct conversion factor with Avogadro's number. In this case, you flip it so that moles are on top. The units of molecules in the first term and in the denominator of the second term cancel, giving you the answer in moles:

$$(7.9 \times 10^{24} \text{ molecules } CH_4)\left(\frac{1 \text{ mol } CH_4}{6.02 \times 10^{23} \text{ molecules } CH_4}\right) = 13 \text{ mol } CH_4$$

(3) Complete the conversions:

a. **18 g Li.** To solve this problem, you need to take your mole value of Li and multiply it by the molar mass of the element lithium (which is approximately 6.94, according to the periodic table). Ensure units cancel to make sure you set up your conversion factor correctly.

$$(2.6 \text{ mol Li})\left(\frac{6.94 \text{ g Li}}{1 \text{ mol Li}}\right) = 18 \text{ g Li}$$

b. **1.46 mol NaCl.** Determine the formula mass of NaCl. To do this, add the molar masses of 1 sodium (22.99 g) and 1 chlorine (35.45 g) together for a formula mass of 54.88 g. Then divide the given initial mass by the formula mass of NaCl:

$$(85.2 \text{ g NaCl})\left(\frac{1 \text{ mol NaCl}}{58.44 \text{ g NaCl}}\right) = 1.46 \text{ mol NaCl}$$

(4) Complete the conversions:

a. **96 L O_2.** To solve this problem, you need to take your mole value of O_2 and multiply it by the molar volume constant of 22.4 L/mol. Ensure units cancel to make sure you set up your conversion factor correctly.

$$(4.3 \text{ mol } O_2)\left(\frac{22.4 \text{ L } O_2}{1 \text{ mol } O_2}\right) = 96 \text{ L } O_2$$

b. **2.87 mol N_2.** Perform the same conversion as in part (a), except this time flip the conversion factor so that *liters* is on the bottom:

$$(64.3 \text{ L } N_2)\left(\frac{1 \text{ mol } N_2}{22.4 \text{ L } N_2}\right) = 2.87 \text{ mol } N_2$$

(5) Complete the conversions:

a. **845 L H₂.** This is a two-step problem. The first step requires converting from grams to moles of hydrogen. Then convert from moles to liters of hydrogen. To determine the molecular mass of hydrogen (H_2), add together the masses of 2 hydrogen atoms (1.01) for a total of 2.02.

$$\left(76.2 \text{ g H}_2\right)\left(\frac{1 \text{ mol H}_2}{2.02 \text{ g H}_2}\right)\left(\frac{22.4 \text{ L H}_2}{1 \text{ mol H}_2}\right) = 845 \text{ L H}_2$$

b. **8.7×10²³ atoms Li.** First convert grams of lithium to moles of lithium by dividing the initial value by the molar mass of lithium off the periodic table. Then multiply by Avogadro's number to determine the number of particles (atoms in this case).

$$\left(10 \text{ g Li}\right)\left(\frac{1 \text{ mol Li}}{6.94 \text{ g Li}}\right)\left(\frac{6.02 \times 10^{23} \text{ atoms Li}}{1 \text{ mol Li}}\right) = 8.7 \times 10^{23} \text{ atoms Li}$$

(6) **K: 40.3%, Cr: 26.8%, O: 33.0%.**

First, calculate the gram molecular mass of potassium chromate, which comes to 194.2 g/mol:

$$\left(39.10 \frac{\text{g}}{\text{mol}}\right)\left(2 \text{ atoms K}\right) = 78.20 \text{ g/mol K}$$

$$\left(52.00 \frac{\text{g}}{\text{mol}}\right)\left(1 \text{ atom Cr}\right) = 52.00 \text{ g/mol Cr}$$

$$\left(16.00 \frac{\text{g}}{\text{mol}}\right)\left(4 \text{ atoms O}\right) = 64.00 \text{ g/mol O}$$

$78.20 \text{ g/mol K} + 52.00 \text{ g/mol Cr} + 64.00 \text{ g/mol O} = 194.20 \text{ g/mol K}_2\text{CrO}_4$

In a 194.20 g sample, 78.20 g is potassium, 52.00 g is chromium, and 64.00 g is oxygen. Divide each of these masses by the gram molecular mass, and then multiply by 100 to get the percent composition:

$$\% \text{ composition of K} = \left(\frac{78.20 \text{ g Na}}{194.2 \text{ g K}_2\text{CrO}_4}\right) \times 100 = 40.3\% \text{K}$$

$$\% \text{ composition of Cr} = \left(\frac{52.00 \text{ g Cr}}{194.2 \text{ g K}_2\text{CrO}_4}\right) \times 100 = 26.8\% \text{Cr}$$

$$\% \text{ composition of O} = \left(\frac{64.00 \text{ g O}}{194.2 \text{ g K}_2\text{CrO}_4}\right) \times 100 = 33.0\% \text{O}$$

Note: If you rounded your percentages properly, they add to 100.1%, not 100%. If you do away with rounding, you get exactly 100%. Rounding is common practice, though, so don't be too worried if your answer is off by a tenth or two. However, if your percentages add up to anything beyond that — say, 198% — then you'll probably want to go back and check your work.

(7) **Li: 17.98% P: 26.75% O: 55.27%.**

First, calculate the formula mass of lithium phosphate:

$(6.94 \text{ g/mol})(3 \text{ atoms Li}) = 20.82 \text{ g/mol Li}$
$(30.97 \text{ g/mol})(1 \text{ atoms P}) = 30.97 \text{ g/mol P}$
$(16.0 \text{ g/mol})(4 \text{ atoms O}) = 64 \text{ g/mol O}$
$20.82 \text{ g/mol Li} + 30.97 \text{ g/mol P} + 64.0 \text{ g/mol O} = 115.79 \text{ g/mol Li}_3\text{PO}_4$

The gram molecular mass comes to 115.79 g/mol. In a 115.79 g sample, 20.82 g is lithium, 30.97 g is phosphorous, and 64.00 g is oxygen. Divide each of these masses by the gram molecular mass and then multiply by 100 to get the percent composition:

$$\% \text{ composition of Li} = \left(\frac{20.82 \text{ g Li}}{115.79 \text{ g Li}_3\text{PO}_4} \right) \times 100 = 17.98\% \text{ Li}$$

$$\% \text{ composition of P} = \left(\frac{30.97 \text{ g P}}{115.79 \text{ g Li}_3\text{PO}_4} \right) \times 100 = 26.75\% \text{ P}$$

$$\% \text{ composition of O} = \left(\frac{64.0 \text{ g O}}{115.79 \text{ g Li}_3\text{PO}_4} \right) \times 100 = 55.27\% \text{ O}$$

As always, you can check your answer by adding up your percentages to see whether they equal 100%. In this case, they add up to exactly 100%.

(8) **H$_2$O.** First, assume that you have 88.9 g of oxygen and 11.1 g of hydrogen in a 100 g sample. Then convert each of these masses into moles by using the gram atomic masses of oxygen and hydrogen:

$$(88.9 \text{ g O}) \left(\frac{1 \text{ mol O}}{16.00 \text{ g O}} \right) = 5.56 \text{ mol O}$$

$$(11.1 \text{ g H}) \left(\frac{1 \text{ mol H}}{1.01 \text{ g H}} \right) = 11.0 \text{ mol H}$$

Next, divide each of these mole quantities by the smallest among them, 5.56 mol:

$$\frac{5.56 \text{ mol O}}{5.56} = 1 \quad \frac{11.0 \text{ mol H}}{5.56} = 2$$

Attach these quotients as subscripts and list the atoms properly. This yields H$_2$O. The compound is water.

(9) **SO$_3$.** Following the same procedure as in Question 1, you calculate 1.25 mol of sulfur and 3.75 mol of oxygen by dividing each mass value by the molar mass of each element.

$$(40.0 \text{ g S}) \left(\frac{1 \text{ mol S}}{32.06 \text{ g S}} \right) = 1.25 \text{ mol S}$$

$$(60.0 \text{ g O}) \left(\frac{1 \text{ mol O}}{16.00 \text{ g O}} \right) = 3.75 \text{ mol O}$$

Dividing each of these quantities by 1.25 mol (the smallest quantity) yields 1.25/1.25 = 1 sulfur and 3.75/1.25 = 3 oxygen, or a mole ratio of 1 sulfur to 3 oxygens. The compound is SO$_3$, sulfur trioxide.

(10) **Na_2O_2.** First, find the empirical formula mass of NaO, which is 38.99 g/mol. You determine this by adding one Na (22.99) to one O (16.00). Then divide the gram formula mass of the mystery compound, 78 g/mol, by this empirical formula mass to obtain the quotient, 2. Multiply each of the subscripts within the empirical formula by this number to obtain Na_2O_2. You've just found the molecular formula for sodium peroxide.

(11) **$C_4H_5N_2O$ is the empirical formula; $C_8H_{10}N_4O_2$ is the molecular formula.** You're not directly given the empirical formula of this compound, but you *are* given the percent composition. Using the percent composition, you can calculate the empirical formula. To do so, assume that you have 100 g of the substance, giving you 49.5 g of carbon, 5.2 g of hydrogen, 16.5 g of oxygen, and 28.8 g of nitrogen. Then divide these masses by the atomic mass of each element, giving you the following numbers of moles:

$$\frac{49.5 \text{ g C}}{12.01 \text{ g/mol}} = 4.125 \text{ mol C}$$

$$\frac{5.2 \text{ g H}}{1.01 \text{ g/mol}} = 5.2 \text{ mol H}$$

$$\frac{16.5 \text{ g O}}{16.00 \text{ g/mol}} = 1.031 \text{ mol O}$$

$$\frac{28.8 \text{ g N}}{14.01 \text{ g/mol}} = 2.057 \text{ mol N}$$

Finally, divide each of these mole values by the lowest among them, 1.031. You get 4.0 mol carbon, 5.0 mol hydrogen, 1.0 mol oxygen, and 2.0 mol nitrogen, giving you the empirical formula $C_4H_5N_2O$.

The empirical formula mass is 97.1 g/mol, which you calculate by multiplying the number of atoms of each element in the compound by the element's atomic mass and adding them all up:

C: $12.01 \cdot 4 = 48.04$
H: $1.01 \cdot 5 = 5.05$
N: $4.01 \cdot 2 = 28.02$
O: $16.00 \cdot 1 = 16.00$
$48.04 + 5.05 + 28.02 + 16.00 = 97.1$ g/mol

Dividing the gram molecular mass you were given (194.2 g/mol) by this empirical formula mass yields the quotient, 2. Multiplying each of the subscripts in the empirical formula by 2 produces the molecular formula, $C_8H_{10}N_4O_2$. The common name for this culturally important compound is caffeine.

If you're ready to test your skills a bit more, take the following chapter quiz that incorporates all the chapter topics.

Whaddya Know? Chapter 12 Quiz

Ready for a quiz? The 15 questions in this section will test the skills you learned in this chapter. When you're done, check out the section that follows for answers and explanations.

1. What is the molar mass of $CaCl_2$?

2. What is the molar mass of $Al(NO_3)_3$?

3. How many grams are there in 3 moles of carbon?

4. Convert 56.4 grams of sodium chloride to moles of sodium chloride.

5. How many atoms are there in 8.5 moles of lithium?

6. What is the volume, in L, of 40.0 moles of hydrogen gas?

7. Convert 8.4×10^{23} molecules of carbon dioxide (CO_2) to moles of carbon dioxide.

8. Convert 334 dm^3 of nitrogen gas (N_2) to molecules of nitrogen gas.

9. Convert 45.0 grams of $Al(ClO_3)_3$ to formula units of $Al(ClO_3)_3$.

10. What is the percent composition of nitrate (NO_3)?

11. What is the percent composition of acetate ($C_2H_3O_2$)?

12. A substance is found and analyzed. The substance is determined to be composed of 36.5 grams of sodium (Na), 25.4 grams of sulfur (S), and 38.1 grams of oxygen (O). What is the empirical formula of this compound?

13. You are given a substance and are told this substance is 36.84% nitrogen and 63.16% oxygen. Determine the empirical formula of this compound.

14. You are given a compound that contains 43.64 grams of phosphorous and 56.36 grams of oxygen. The molecular mass of this compound is 283.9 amu (atomic mass units). Given this information, calculate the molecular formula of this compound.

15. Convert 98.2 grams of copper (Cu) to atoms of Cu.

Answers to Chapter 12 Quiz

(1) **110.98 g/mol $CaCl_2$.** The molar mass calcium chloride can be calculated by adding up the molar masses of each of the elements present in the formula of calcium chloride multiplied by the number of each atom present.

$$Ca\left(40.08\,\frac{g}{mol}\right)(1\text{ atom Ca}) = 40.08\text{ g/mol Ca}$$

$$Cl\left(35.45\,\frac{g}{mol}\right)(21\text{ atoms Cl}) = 70.90\text{ g/mol Cl}$$

$$40.08 + 70.09 = 110.98\text{ g/mol }CaCl_2$$

(2) **213.01 g/mol $Al(NO_3)_3$.** The molar mass aluminum nitrate can be calculated by adding up the molar masses of each of the elements present in the formula of aluminum nitrate multiplied by the number of each atom present. When you do this, make sure to account for the atoms inside the parentheses correctly. The 3 on the outside of NO_3 is distributed to the N and the O_3, giving you 3 nitrogen and 9 oxygen. The 3 does not go to the aluminum at all.

$$Al\left(26.98\,\frac{g}{mol}\right)(1\text{ atom Al}) = 26.98\text{ g/mol Al}$$

$$N\left(14.01\,\frac{g}{mol}\right)(3\text{ atoms N}) = 42.03\text{ g/mol N}$$

$$O\left(16.00\,\frac{g}{mol}\right)(9\text{ atoms O}) = 144\text{ g/mol O}$$

$$26.98 + 42.03 + 144 = 213.01\text{ g/mol }Al(NO_3)_3$$

(3) **36.03 g C.** To solve this problem, you need to convert 3 moles of carbon to the mass of carbon in grams. Doing this requires you to know the conversation factor between moles of carbon and the mass of carbon. This is determined by looking at the molar mass of carbon on the period table, which is 12.01 g/mol. The units are what show you the conversion: for every 12.01 g of carbon, you have the equivalent of 1 mole of carbon. This can be shown as: $\frac{12.01\text{ g C}}{1\text{ mol C}}$. Now use this conversion factor to solve your problem:

$$(3\text{ mol C})\left(\frac{12.01\text{ g C}}{1\text{ mol C}}\right) = 36.03\text{ g C}$$

(4) **0.97 mol NaCl.** Determine the molar mass of sodium chloride by adding together the atomic mass of sodium (22.99 g/mol) and chlorine (35.45 g/mol) from the periodic table: 58.44 g/mol. Use this molar mass to perform your conversion from grams to moles:

$$(56.4\text{ g NaCl})\left(\frac{1\text{ mol NaCl}}{58.44\text{ g NaCl}}\right) = 0.97\,mol\,NaCl$$

(5) **5.1×10^{24} mol of Li.** To solve this problem, you are going to convert from grams to atoms of lithium. In this case the periodic table is not used at all. You simply need to know that there are 6.02 x 10²³ atoms of lithium in 1 mole of lithium. Remember, this is the same for any atom, molecule, or formula unit. Avogadro's number is always the conversion between mole and a representative particle of matter.

$$(8.5 \text{ mol Li})\left(\frac{6.02 \times 10^{23} \text{ atom of Li}}{1 \text{ mol Li}} \right) = 5.1 \times 10^{24} \text{mol of Li}$$

(6) **896 L H_2.** In this problem you are going to convert from moles of hydrogen gas to a volume of hydrogen gas, liters in this case. To do this, you will use the molar volume constant of 22.4 L / 1 mol of gas:

$$40.0 \text{ mol H}_2 \left(\frac{22.4 \text{ L H}_2}{1 \text{ mol H}_2} \right) = 896 \text{ L H}_2$$

(7) **1.4 mol of CO_2.** This problem requires you to convert from molecules to moles. You will be using a conversion factor with Avogadro's number in relation to one mole of particles (molecules in this case) of 6.02×10^{23} molecules / 1 mol.

$$\left(8.4 \times 10^{23} \text{ molecules CO}_2\right)\left(\frac{1 \text{ mol CO}_2}{6.02 \times 10^{23} \text{ molecules CO}_2} \right) = 1.4 \text{ mol of CO}_2$$

(8) **8.98×10^{24} molecules N_2.** This is a two-step problem. Remember, when solving two-step problems it is always best to do it all on one line. Don't break it up into separate problems because this can lead to more chances for error. Reminder: The volume of the gas given in this problem is in dm³, which is the same as liters, so you do not need to do another conversion or anything else. You can simply use that number and then the standard molar volume constant. You will convert from a volume of a gas to moles of a gas using the molar volume constant of 22.4 dm³, then you will multiply by Avogadro's number to determine the molecules of that gas present.

$$\left(334 \text{ dm}^3 \text{ N}_2\right)\left(\frac{1 \text{ mol N}_2}{22.4 \text{ dm}^3 \text{ N}_2} \right)\left(\frac{6.02 \times 10^{23} \text{ molecules N}_2}{1 \text{ mol N}_2} \right) = 8.98 \times 10^{24} \text{ molecules N}_2$$

(9) This is a two-step problem as well. You will need to convert from grams of aluminum chlorate to moles of aluminum chlorate. To do this, you will need to know the molar mass of aluminum chlorate. Then from there you must convert to formula units of aluminum chlorate using Avogadro's number.

$$\text{Al}\left(26.98 \frac{g}{\text{mol}} \right)(1 \text{ atom Al}) = 26.98 \text{ g/mol Al}$$

$$\text{Cl}\left(35.45 \frac{g}{\text{mol}} \right)(3 \text{ atoms Cl}) = 106.35 \text{ g/mol N}$$

$$\text{O}\left(16.00 \frac{g}{\text{mol}} \right)(9 \text{ atoms O}) = 144 \text{ g/mol O}$$

$$26.98 + 106.35 + 144 = 277.33 \text{ g/mol Al}\left(\text{ClO}_3 \right)_3$$

$$\left(45.0 \text{ g Al}(ClO_3)_3\right)\left(\frac{1 \text{mol Al}(ClO_3)_3}{277.33 \text{ g Al}(ClO_3)_3}\right)\left(\frac{6.02\times10^{23} \text{ formula units Al}(ClO_3)_3}{1 \text{mol Al}(ClO_3)_3}\right)$$

$$= 9.77\times10^{22} \text{ formula units Al}(ClO_3)_3$$

10) You first need to determine the molar mass of NO_3. Once you have done this, you will use the individual masses of the elements and divide those by the total mass of nitrate and then multiply by 100.

$$N\left(14.01\frac{g}{mol}\right)(1 \text{ atom N}) = 14.01 \text{ g/mol N}$$

$$O\left(16.00\frac{g}{mol}\right)(3 \text{ atoms O}) = 48.00 \text{ g/mol O}$$

$$14.01 + 48.00 = 62.01 \text{ g/mol } NO_3$$

$$\% \text{ composition of N} = \left(\frac{14.01 \text{ g N}}{62.01 \text{ g } NO_3}\right)\times100 = 22.6\% \text{ N}$$

$$\% \text{ composition of O} = \left(\frac{48.00 \text{ g O}}{62.01 \text{ g } NO_3}\right)\times100 = 77.4\% \text{ O}$$

11) You first need to determine the molar mass of acetate. Once you have done this, you will use the individual masses of the elements and divide those by the total mass of acetate and then multiply by 100.

$$C\left(12.01\frac{g}{mol}\right)(2 \text{ atoms C}) = 24.02 \text{ g/mol C}$$

$$H\left(1.01\frac{g}{mol}\right)(3 \text{ atoms H}) = 3.03 \text{ g/mol H}$$

$$O\left(16.00\frac{g}{mol}\right)(2 \text{ atoms O}) = 32.00 \text{ g/mol O}$$

$$24.02 + 3.03 + 32.00 = 59.05 \text{ g/mol } C_2H_3O_2$$

$$\% \text{ composition of C} = \left(\frac{24.02 \text{ g C}}{59.05 \text{ g } C_2H_3O_2}\right)\times100 = 40.68\% \text{ C}$$

$$\% \text{ composition of H} = \left(\frac{3.03 \text{ g H}}{59.05 \text{ g } C_2H_3O_2}\right)\times100 = 5.13\% \text{ H}$$

$$\% \text{ composition of O} = \left(\frac{32.00 \text{ g O}}{59.05 \text{ g } C_2H_3O_2}\right)\times100 = 54.19\% \text{ O}$$

12) **Na_2SO_3.** To determine the empirical formula of the compound, you first need to take all of your mass values and convert them to moles. Once you do that, you will then determine the element with the smallest mole value and then divide all of the moles values by the smallest one. In this case, you don't need to perform the extra step of multiplying by the smallest possible factor.

$$(36.5 \text{ g Na})\left(\frac{1 \text{ mol Na}}{22.99 \text{ g Na}}\right) = 1.59 \text{ mol Na}$$

$$(25.4 \text{ g S})\left(\frac{1 \text{ mol S}}{32.07 \text{ g S}}\right) = 0.79 \text{ mol S}$$

$$(38.1 \text{ g O})\left(\frac{1 \text{ mol O}}{16.00 \text{ g O}}\right) = 2.38 \text{ mol O}$$

The smallest value after converting each of your starting masses to moles is 0.79 moles of sulfur. You now divide each of the three values by 0.79 to determine the number of atoms of each element in your empirical formula.

$$\frac{1.59 \text{ Na}}{0.79} = 2 \text{ Na} \quad \frac{0.79}{0.79} = 1 \text{ S} \quad \frac{2.38 \text{ O}}{0.79} = 3 \text{ O}$$

Finally, write the formula:

Na_2SO_3

13. **Na_2O_3.** To solve this problem, you need to determine the grams of each element present. In this case, you are not given grams but instead percentage. To make this simpler just assume you have 100 grams of the compound and apply your percentages to 100 grams. That means you have 36.84 grams of nitrogen and 63.16 grams of oxygen. Next, you convert your values to moles and then divide by the smallest value.

$$(36.84 \text{ g N})\left(\frac{1 \text{ mol N}}{14.01 \text{ g N}}\right) = 2.63 \text{ mol N}$$

$$(63.16 \text{ g O})\left(\frac{1 \text{ mol O}}{16.00 \text{ g O}}\right) = 3.95 \text{ mol O}$$

$$\frac{2.63 \text{ N}}{2.63} = 1 \text{ Na} \quad \frac{3.95 \text{ O}}{2.63} = 1.50 \text{ O}$$

In this case sodium is a nice whole number, but oxygen is 1.5, which is not a whole number. That means you will need to somehow turn these into whole numbers. To do this, multiply both values by the smallest number possible to make them both whole numbers. In this case that means you double them by multiplying by 2.

$$(1 \text{ Na})2 = 2 \text{ Na} \quad (1.5 \text{ O})2 = 3 \text{ O}$$

Your final formula is Na_2O_3.

14. **P_4O_{10}.** There are several steps you will follow to solve this problem. You first determine the empirical formula, then you determine the empirical formula mass; finally, once you know this, you divide the molecular mass by the empirical mass and multiply the empirical formula by your resulting answer.

First step, calculate empirical formula:

$$(43.64 \text{ g P})\left(\frac{1 \text{ mol N}}{30.97 \text{ g P}}\right) = 1.41 \text{ mol P}$$

$$(56.36 \text{ g O})\left(\frac{1 \text{ mol O}}{16.00 \text{ g O}}\right) = 3.52 \text{ mol O}$$

$$\frac{1.41 \text{ P}}{1.41} = 1 \text{ P} \quad \frac{3.52 \text{ O}}{1.41} = 2.50 \text{ O}$$

$$(1 \text{ P})2 = 2 \text{ P} \quad (2.5 \text{ O})2 = 5 \text{ O}$$

Empirical Formula:

P_2O_5

Second step, determine the empirical formula mass:

$$P\left(30.97\frac{g}{mol}\right)(2 \text{ atoms P}) = 61.94 \text{ g/mol P}$$

$$O\left(16.00\frac{g}{mol}\right)(5 \text{ atoms O}) = 80.00 \text{ g/mol O}$$

$$61.94 + 48.00 = 141.94 \text{ g/mol P}_2O_5$$

Final step, divide the empirical formula mass you just calculated by the molecular formula mass given in the problem:

$$\frac{283.9 \text{ molecular formula mass}}{141.94 \text{ empirical formula mass}} = 2$$

This means you will double the empirical formula from P_2O_5 to P_4O_{10}.

15) **9.30×10^{23} atom Cu.** This is a two-step problem that requires you to convert from grams of copper to moles of copper first, using copper's molar mass from the periodic table. Once you have this conversion done, you will then be converting from moles of copper to atoms of copper using Avogadro's number.

$$\left(98.2 \text{ g Cu}\right)\left(\frac{1 \text{ mol Cu}}{63.55 \text{ g Cu}}\right)\left(\frac{6.02 \times 10^{23} \text{ atom Cu}}{1 \text{ mol Cu}}\right) = 9.30 \times 10^{23} \text{ atom Cu}$$

Chapter **13**

Getting a Grip on Chemical Equations

Chapters 9, 10, and 11 focus on chemical compounds and the bonds that bind them. You can think of a compound in two ways:

» As the product of one chemical reaction

» As a starting material in another chemical reaction

In the end, chemistry is about action — about the breaking and making of bonds. Chemists describe action by using *chemical equations*, chemical symbols turned into sentences that say who reacted with whom and who remained when the smoke cleared. This chapter explains how to read, write, balance, and predict the products of these action-packed chemical sentences.

Translating Chemistry into Equations and Symbols

In a chemical reaction, substances (elements and/or compounds) are changed into other substances (compounds and/or elements). You can't change one element into another in a chemical reaction — that happens in nuclear reactions, as we describe in Chapter 8. Instead, you create new substances with chemical reactions.

REMEMBER

A number of clues show that a chemical reaction has taken place — something new is visibly produced, a gas is created, heat is given off or taken in, and so on. The chemical substances that are eventually changed are called the *reactants*, and the new substances that are formed are called the *products*. *Chemical equations* show the reactants and products as well as other factors such as energy changes, catalysts, and so on.

In general, all chemical equations are written in the basic form

Reactants → Products

where the arrow in the middle means *yields*. The basic idea is that the reactants react and the reaction produces products. By *reacting*, we simply mean that bonds within the reactants are broken, to be replaced by new and different bonds within the products.

Chemists fill chemical equations with symbols because they think it looks cool and, more importantly, because the symbols pack a lot of meaning into a small space. Table 13-1 summarizes the most important symbols you find in chemical equations.

After you understand how to interpret chemical symbols, the names of compounds (see Chapter 11), and the symbols in Table 13-1, you can understand almost anything. You're equipped, for example, to decode a chemical equation into an English sentence describing a reaction. Conversely, you can translate an English sentence into the chemical equation it describes. When you're fluent in this language, you regrettably won't be able to talk to the animals; you will, however, be able to describe their metabolism in great detail.

TABLE 13-1 Symbols Commonly Used in Chemical Equations

Symbol	Explanation
+	The plus sign separates two reactants or products.
→	The *yields* symbol separates the reactants from the products. The single arrowhead suggests the reaction occurs in only one direction.
↔	A two-way *yields* symbol means the reaction can occur reversibly, in both directions. You may also see this symbol written as two stacked arrows with opposing arrowheads.
(s)	A reactant or product followed by this symbol exists as a solid.
(l)	A reactant or product followed by this symbol exists as a liquid.
(g)	A reactant or product followed by this symbol exists as a gas.
(aq)	A reactant or product followed by this symbol exists in aqueous solution, dissolved in water.
Δ	This symbol, usually written above the *yields* symbol, signifies that heat is added to the reactants.
Ni, LiCl	Sometimes a chemical symbol (such as those for nickel or lithium chloride here) is written above the *yields* symbol. This means that the indicated chemical was added as a catalyst. Catalysts speed up reactions but do not otherwise participate in them.

For example, take a look at the reaction that occurs when you light your natural gas range in order to fry your breakfast eggs. Methane (natural gas) reacts with the oxygen in the atmosphere to produce carbon dioxide and water vapor. (If your burner isn't properly adjusted to produce that nice blue flame, you may also get a significant amount of carbon monoxide along with carbon dioxide. This is not a good thing!) The chemical equation that represents this reaction is written like this:

$$CH_4(g) + 2O_2(g) \rightarrow CO_2(g) + 2H_2O(g)$$

You can read the equation like this: One molecule of methane gas, $CH_4(g)$, reacts with two molecules of oxygen gas, $O_2(g)$, to form one molecule of carbon dioxide gas, $CO_2(g)$, and two molecules of water vapor, $H_2O(g)$. The 2 in front of the oxygen gas and the 2 in front of the water vapor are called the reaction *coefficients*. They indicate the number of each chemical species that reacts or is formed. You see how to figure out the value of the coefficients in the section "Balancing Chemical Equations," later in the chapter.

Methane and oxygen (oxygen is a diatomic [two-atom] element) are the reactants, and carbon dioxide and water are the products. All the reactants and products are gases (indicated by the *g*s in parentheses).

In this reaction, all reactants and products are invisible. The heat being given off is the clue that tells you a reaction is taking place. By the way, this is a good example of an *exothermic* reaction, a reaction in which heat is given off. A lot of reactions are exothermic. Some reactions, however, absorb energy rather than release it. These reactions are called *endothermic* reactions. Cooking involves a lot of endothermic reactions — frying those eggs, for example. You can't just break the shells and let the eggs lie on the pan and then expect the myriad chemical reactions to take place without heating the pan (except when you're outside in Texas during August; there, the sun will heat the pan just fine).

Thinking about cooking those eggs brings to mind another issue about exothermic reactions. You have to ignite the methane coming out of the burners with a match, lighter, pilot light, or built-in electric igniter. In other words, you have to put in a little energy to get the reaction going. The energy you have to supply to get a reaction going is called the *activation energy* of the reaction. (In the next section, we show you that there's also an activation energy associated with endothermic reactions, but it isn't nearly as obvious.)

Understanding How Reactions Occur

In order for a chemical reaction to take place, the reactants must collide. It's like playing pool. In order to drop the eight ball into the corner pocket, you have to hit it with the cue ball. This collision transfers *kinetic energy* (energy of motion) from one ball to the other, sending the second ball (hopefully) toward the pocket. Energy is required to break a bond between atoms, and energy is released when a bond is made. The *collision theory* states that the collision between the molecules can provide the energy needed to break the necessary bonds so that new bonds can be formed. The collision takes place at the right spot and transfers sufficient energy. The following sections provide three examples of what can happen during a collision.

One-step collision example

When you play pool, not every shot you make causes a ball to go into the pocket. Sometimes you don't hit the ball hard enough, and you don't transfer enough energy to get the ball to the pocket. This situation also occurs with molecular collisions and reactions. Sometimes, even if a collision takes place, not enough kinetic energy is available to be transferred — the molecules aren't moving fast enough. You can help the situation somewhat by heating the mixture of reactants. The temperature is a measure of the average kinetic energy of the molecules; raising the temperature increases the kinetic energy available to break bonds during collisions.

Sometimes, even if you hit the ball hard enough, it doesn't go into the pocket because you didn't hit it in the right spot. The same is true during a molecular collision. The molecules must collide in the right orientation, or hit at the right spot, in order for the reaction to occur.

Suppose you have an equation showing molecule A–B reacting with C to form C–A and B, like this:

$$A - B + C \rightarrow C - A + B$$

The way this equation is written, the reaction requires that reactant C collide with A–B on the A end of the molecule. (You know this because the product side shows C hooked up with A: C–A.) If it hits the B end, nothing happens. The A end of this hypothetical molecule is called the *reactive site*, the place on the molecule that the collision must take place in order for the reaction to occur. If C collides at the A end of the molecule, then enough energy may be transferred to break the A–B bond. After the A–B bond is broken, the C–A bond can be formed. The equation for this reaction process can be shown in this way (we show the breaking of the A–B bond and the forming of the C–A bond as squiggly lines):

$$C - A \sim B \rightarrow C \sim A + B$$

So in order for this reaction to occur, a collision between C and A–B must occur at the reactive site. The collision between C and A–B has to transfer enough energy to break the A–B bond, allowing the C–A bond to form.

If instead of having a simple A–B molecule, you have a large complex molecule, like a protein or a polymer, then the likelihood of C colliding at the reactive site is much smaller. You may have a lot of collisions but few at the reactive site. This reaction will probably be much slower than the simple case.

Note that this example is a simple one. We've assumed that only one collision is needed, making this a one-step reaction. Many reactions are one-step, but many others require several steps before the reactants become the final products. In the process, several compounds may be formed that react with each other to give the final products. These compounds are called *intermediates.* They're shown in the reaction *mechanism*, the series of steps that the reaction goes through in going from reactants to products. But in this chapter, we keep it simple and pretty much limit our discussion to one-step reactions.

Considering an exothermic example

Imagine that the hypothetical reaction $A - B + C \rightarrow C - A + B$ is *exothermic* — a reaction in which thermal energy is given off (released) when going from reactants to products. The reactants start off at a higher energy state than the products, so energy is released in going from reactants to products. Figure 13-1 shows an energy diagram of this reaction.

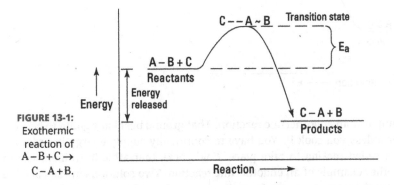

FIGURE 13-1:
Exothermic
reaction of
$A - B + C \rightarrow$
$C - A + B$.

In Figure 13-1, E_a is the activation energy for the reaction — the energy that you have to put in to get the reaction going. We show the collision of C and A–B with the breaking of the A–B bond and the forming of the C–A bond at the top of an activation energy hill. This grouping of reactants at the top of the activation energy hill is sometimes called the *transition state* of the reaction. As we show in Figure 13-1, the difference in the energy level of the reactants and the energy level of the products is the amount of energy (heat) that is released in the reaction.

Some reactions give off energy but not thermal energy. An example is light sticks. You mix two chemical solutions by flexing the light stick, and the resulting product glows — it gives off light but not heat. Another example is fireflies, which mix two chemicals in their bodies and give off light. These reactions that give off energy are *exergonic*. If that energy is in the form of heat, the reaction is subclassified as exothermic.

Looking at an endothermic example

Suppose that the hypothetical reaction $A - B + C \rightarrow C - A + B$ is *endothermic* — a reaction in which heat is absorbed — so the reactants are at a lower energy state than the products. Figure 13-2 shows an energy diagram of this reaction.

Just as in the exothermic-reaction energy diagram shown in Figure 13-1, this diagram shows that an activation energy is associated with the reaction (represented by E_a). In going from reactants to products, you have to put in more energy initially to get the reaction started, and then you get some of that energy back out as the reaction proceeds. Notice that the transition state appears at the top of the activation energy hill — just like in the exothermic-reaction energy diagram. But although both endothermic and exothermic reactions require activation energy, exothermic reactions release thermal energy, and endothermic reactions absorb it.

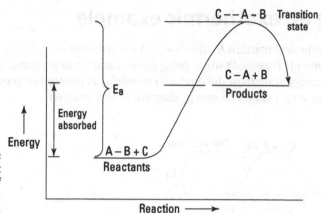

FIGURE 13-2:
Endothermic
reaction of
$A - B + C \rightarrow$
$C - A + B$.

Cooking is a great example of an endothermic reaction. That ground beef isn't going to become a delicious hamburger unless you cook it. You have to continually supply energy in order for the chemical reactions called *cooking* to take place. Cold packs that athletic trainers use to treat injuries offer another example of an endothermic reaction. Two solutions in the pack are mixed, and the pack absorbs thermal energy from the surroundings. The surroundings therefore become colder.

Other reactions may absorb energy but not necessarily heat. For example, some reactions absorb light energy in order to react. The general term that chemists use to describe reactions that absorb energy (heat or otherwise) is *endergonic*. Endothermic reactions are a subset of endergonic reactions.

Q. Write out the chemical equation for the following sentence:

EXAMPLE Solid iron(III) oxide reacts with gaseous carbon monoxide to produce solid iron and gaseous carbon dioxide.

A. $Fe_2O_3(s) + CO(g) \rightarrow Fe(s) + CO_2(g)$. First, convert each formula into the written name for the compound (check out Chapter 11 for details). Next, annotate the physical state of the compound if it's provided. Then group the compounds into "reactant" and "product" categories. Things that react are reactants, and things that are produced are products. List the reactants on the left side of a reaction arrow, separating each pair with a plus sign. Do the same for the products, but list them on the right side of the reaction arrow.

Q. Write a sentence that describes the following chemical reaction:

$H_2O(l) + N_2O_3(g) \xrightarrow{\Delta} HNO_2(aq)$

"Liquid water is heated with gaseous dinitrogen trioxide to produce an aqueous solution of nitrous acid." First, figure out the names of the compounds. Next, note their states (liquid, solid, gas, or aqueous solution). Then observe what the compounds are actually doing — combining, decomposing, combusting, and so on. Finally, assemble all these observations into a sentence. Many sentence variations are correct, as long as they include these elements.

YOUR TURN

1 Write chemical equations for the following reactions:

(a) Solid magnesium is heated with gaseous oxygen to form solid magnesium oxide.

(b) Solid diboron trioxide reacts with solid magnesium to make solid boron and solid magnesium oxide.

2 Write sentences describing the reactions summarized by the following chemical equations:

(a) $S(s) + O_2(g) \rightarrow SO_2(g)$

(b) $H_2(g) + O_2(g) \xrightarrow{\text{Pt}} H_2O(l)$

3 Write out the chemical reaction for the following statement, being sure to include the phase of each reactant and product:

One way to test the salinity of a water sample is to add a few drops of silver nitrate solution with a known concentration. As the solutions of sodium chloride and silver nitrate mix, a precipitate of silver chloride forms, and sodium nitrate is left in solution.

Balancing Chemical Equations

The equations you read and write in the preceding section are *skeleton equations*, and they're perfectly adequate for a qualitative description of the reaction: What are the reactants, and what are the products? But if you look closely, you'll see that those equations just don't add up. As written, the mass of 1 mole of each of the reactants doesn't equal the mass of 1 mole of each of the products (see Chapter 12 for details on moles). The skeleton equations break the *law of conservation of mass*, which states that all the mass present at the beginning of a reaction must be present at the end. To be quantitatively accurate, these equations must be *balanced* so the masses of reactants and products are equal.

REMEMBER

To balance an equation, you use *coefficients* to alter the number of moles of reactants and/or products so the mass on one side of the equation equals the mass on the other side. A *coefficient* is simply a number that precedes the symbol of an element or compound, multiplying the number of moles of that *entire* compound within the equation. Coefficients are different from *subscripts*, which multiply the number of atoms or groups within a compound. Consider the following:

$$4Cu(NO_3)_2$$

The number 4 that precedes the compound is a coefficient, indicating 4 moles of copper(II) nitrate. The subscripted 3 and 2 within the compound indicate that each nitrate contains three oxygen atoms and that there are two nitrate groups per copper ion. Coefficients and subscripts multiply to yield the total numbers of each atom present in the formula. In the preceding example, there are 4 atoms of copper present, 8 atoms of nitrogen, and 24 atoms of oxygen.

WARNING

When you balance an equation, you change *only the coefficients.* Changing subscripts alters the chemical compounds themselves, and you can't do that. If your pencil were equipped with an electrical shocking device, that device would activate the moment you attempted to change a subscript while balancing an equation.

Getting the lowdown on balancing equations

Now you can look at balancing an equation in a bit more detail. One interesting reaction, the *Haber process,* is a method for preparing ammonia (NH_3) by reacting nitrogen gas with hydrogen gas. This reaction is worth looking at because it helps feed the world. The ammonia that's produced is used to produce ammonium nitrate and ammonium phosphate, both of which are synthetic fertilizers that have allowed the increased production of food. Here's the equation:

$$N_2(g) + H_2(g) \rightarrow NH_3(g)$$

This equation shows you what happens in the reaction, but it doesn't show you how much of each element you need to produce the ammonia. To find out how much of each element you need, you have to balance the equation; that is, make sure that the number of atoms on the left side of the equation equals the number of atoms on the right.

TIP

In most cases, waiting until the end to balance hydrogen atoms or oxygen atoms is a good idea based on many years of experience. Balance the other atoms first.

So in this example, you need to balance the nitrogen atoms first. You have 2 nitrogen atoms on the left side of the arrow (reactant side) and only 1 nitrogen atom on the right side (product side). In order to balance the nitrogen atoms, use a coefficient of 2 in front of the ammonia on the right.

$$N_2(g) + H_2(g) \rightarrow 2NH_3(g)$$

Now you have 2 nitrogen atoms on the left and 2 nitrogen atoms on the right.

Next, tackle the hydrogen atoms. You have 2 hydrogen atoms on the left and 6 hydrogen atoms on the right (2 NH_3 molecules, each with 3 hydrogen atoms, for a total of 6 hydrogen atoms). So put a 3 in front of the H_2 on the left, giving you

$$N_2(g) + 3H_2(g) \rightarrow 2NH_3(g)$$

That should do it. Check to be sure: You have 2 nitrogen atoms on the left and 2 nitrogen atoms on the right. You have 6 hydrogen atoms on the left ($3 \times 2 = 6$) and 6 hydrogen atoms on the right ($2 \times 3 = 6$). The equation is balanced. You can read the equation this way: 1 nitrogen molecule reacts with 3 hydrogen molecules to yield 2 ammonia molecules.

This equation also balances with coefficients of 2, 6, and 4 instead of 1, 3, and 2. In fact, any multiple of 1, 3, and 2 balances the equation, but chemists always show the lowest whole-number ratio. If you ever end up with a bunch of even-number coefficients, divide them all by 2 or 4 or whatever to show the equation in the lowest whole-number ratio.

Walking through the steps of balancing equations

Now here's another example that breaks equation-balancing down into a simple step-by-step process. An interesting equation — one that many, many people in this world see every day — is the burning or combustion of butane in oxygen (see the next section for details on combustion). This reaction occurs whenever someone lights a butane lighter.

1. **Start with the unbalanced reaction written in a molecular equation.**

   The equation of burning butane with excess oxygen available is

   $$C_4H_{10}(g) + O_2(g) \rightarrow CO_2(g) + H_2O(g)$$

2. **Balance the carbon atoms first.**

   You want to wait until the end to balance hydrogen and oxygen atoms.

   REMEMBER

   You have 4 carbon atoms on the left and 1 carbon atom on the right, so put a coefficient of 4 in front of the carbon dioxide:

   $$C_4H_{10}(g) + O_2(g) \rightarrow 4CO_2(g) + H_2O(g)$$

3. **When all non-hydrogen and non-oxygen atoms are balanced, balance the hydrogen atoms.**

   Carbon is the only other atom in this example, so you can move on to hydrogen now. You have 10 hydrogen atoms on the left and 2 hydrogen atoms on the right, so use a coefficient of 5 in front of the water on the right:

   $$C_4H_{10}(g) + O_2(g) \rightarrow 4CO_2(g) + 5H_2O(g)$$

4. **Balance the oxygen atoms.**

   You have 2 oxygen atoms on the left and a total of 13 oxygen atoms on the right [$(4 \times 2) + (5 \times 1) = 13$]. What can you multiply 2 with in order for it to equal 13? How about 6.5?

   $$C_4H_{10}(g) + 6.5O_2(g) \rightarrow 4CO_2(g) + 5H_2O(g)$$

5. **Multiple all coefficients in the equation to get the lowest whole-number ratio of coefficients.**

For this example, multiply the entire equation by 2 (just the coefficients, please) in order to generate whole numbers:

$$[C_4H_{10}(g) + 6.5O_2(g) \rightarrow 4CO_2(g) + 5H_2O(g)] \times 2$$

Multiplying every coefficient by 2 (don't touch the subscripts!) gives you

$$2C_4H_{10}(g) + 13O_2(g) \rightarrow 8CO_2(g) + 10H_2O(g)$$

6. **Check the atom count on both sides of the equation to ensure that the equation is balanced and the coefficients are in the lowest whole-number ratio.**

Most simple reactions can be balanced in this fashion. Just keep everything organized and be neat when writing out your equations and coefficients. Sometimes all you need to do is add a single coefficient, perhaps a 2 or a 3, to what seems like a monstrous problem, and you're done. On the other hand, those small equations that look like they'll be simple may end up being some of the toughest ones. Stay focused and organized, and you'll be fine.

EXAMPLE

Q. Balance the following equation:

$$Na(s) + Cl_2(g) \rightarrow NaCl(s)$$

A. $2Na(s) + Cl_2(g) \rightarrow 2NaCl(s)$. You can't change the subscripted 2 in the chlorine gas reactant, so you must add a coefficient of 2 to the sodium chloride product. This change requires you to balance the sodium reactant with another coefficient of 2.

Q. Write and balance the following reaction:

Solid zinc reacts with hydrochloric acid (HCl) to produce aqueous zinc chloride and hydrogen gas.

A. $Zn(s) + 2HCl(aq) \rightarrow ZnCl_2(aq) + H_2(g)$

First write out the skeleton equation, making sure you use the correct formulas for the reactants and products. If any part of that is incorrect, then balancing the equation will be a fruitless endeavor. The reactants are pretty straightforward. Zinc (Zn) is a solid element, as the problem tells you, and you're given the formula of hydrochloric acid. Each zinc atom has a charge of +2, so you need two –1 chlorine atoms to make zinc chloride a neutral compound; therefore, zinc chloride has 2 chlorines. Hydrogen gas is written with a subscript of 2 because hydrogen is a diatomic element when found as a gas.

The skeleton equation shows 2 Cl atoms and 2 H atoms on the product side of the reaction but only 1 of each on the reactant side. Thankfully, H and Cl are a compound on the reactant side, so you only need to add a 2 in front of HCl to balance the equation; everything else is equal.

4 Balance the following reactions:

YOUR TURN

(a) $N_2(g) + H_2(g) \rightarrow NH_3(g)$

(b) $C_3H_8(g) + O_2(g) \rightarrow CO_2(g) + H_2O(l)$

(c) $Al(s) + O_2(g) \rightarrow Al_2O_3(s)$

(d) $AgNO_3(aq) + Cu(s) \rightarrow Cu(NO_3)_2(aq) + Ag(s)$

---

5 Balance the following reactions:

(a) $Ag_2SO_4(aq) + AlCl_3(aq) \rightarrow AgCl(s) + Al_2(SO_4)_3(aq)$

(b) $CH_4(g) + Cl_2(g) \rightarrow CH_2Cl_2(l) + HCl(g)$

(c) $Cu(s) + HNO_3(aq) \rightarrow Cu(NO_3)_2(aq) + NO(g) + H_2O(l)$

(d) $HCl(aq) + Ca(OH)_2(aq) \rightarrow CaCl_2(s) + H_2O(l)$

---

6 Balance the following reactions:

(a) $H_2C_2O_4(aq) + KOH(aq) \rightarrow K_2C_2O_4(aq) + H_2O(l)$

(b) $P(s) + Br_2(l) \rightarrow PBr_3(g)$

(c) $Pb(NO_3)_2(aq) + KI(aq) \rightarrow KNO_3(aq) + PbI_2(s)$

(d) $Zn(s) + AgNO_3(aq) \rightarrow Zn(NO_3)_2(aq) + Ag(s)$

Recognizing Reactions and Predicting Products

You can't begin to wrap your brain around the unimaginably large number of possible chemical reactions. That so many reactions can occur is a good thing, because they make things like life, the universe, and everything possible. From the perspective of a mere human brain trying to grok all these reactions, we have another bit of good news: A few categories of reactions pop up over and over again. After you see the very basic patterns in these categories, you'll be able to make sense of the majority of reactions out there.

The following sections describe five types of reactions that you'd do well to recognize (notice how their names tell you what happens in each reaction). By recognizing the patterns of these types of reactions, you can often predict reaction products when given only a set of reactants. (*Note:* Figuring out the formulas of products often requires you to apply knowledge about how ionic and molecular compounds are put together. To review these concepts, see Chapter 11.) Here are the five types of reactions you'll examine next:

>> Combination/Synthesis

>> Decomposition

>> Single displacement

>> Double displacement

>> Combustion

Combination (synthesis)

In *combination* (sometimes called *synthesis*), two or more reactants combine to form a single product, following the general pattern

$$A + B \rightarrow AB$$

For example,

$$2Na(s) + Cl_2(g) \rightarrow 2NaCl(s)$$

The combining of elements to form compounds (like NaCl) is a particularly common kind of combination reaction. Here's another example:

$$2Ca(s) + O_2(g) \rightarrow 2CaO(s)$$

Compounds can also combine to form new compounds, such as in the combination of sodium oxide with water to form sodium hydroxide:

$$Na_2O(s) + H_2O(l) \rightarrow 2NaOH(aq)$$

Note that, depending on conditions or the relative amounts of the reactants, more than one product can be formed in a combination reaction. Take the burning of coal, for example. If an excess of oxygen is present, the product is carbon dioxide. But if a limited amount of oxygen is available, the product is carbon monoxide:

$$2C(s) + O_2(g) \rightarrow 2CO_2(g) \qquad \text{(limited oxygen)}$$

Decomposition

In *decomposition*, a single reactant breaks down (decomposes) into two or more products, following the general pattern

$$AB \rightarrow A + B$$

For example,

$$2H_2O(l) \rightarrow 2H_2(g) + O_2(g)$$

Notice that combination and decomposition reactions are the same reaction in opposite directions.

TIP

Many decomposition reactions produce gaseous products, such as in the decomposition of carbonic acid into water and carbon dioxide:

$$H_2CO_3(aq) \rightarrow H_2O(l) + CO_2(g)$$

Single displacement (single replacement)

In a *single displacement* reaction (sometimes referred to as a *single replacement* reaction), a single, more reactive element or group replaces a less reactive element or group, following the general pattern

$$A + BC \rightarrow AC + B$$

For example,

$$Zn(s) + CuSO_4(aq) \rightarrow ZnSO_4(aq) + Cu(s)$$

Single displacement reactions in which metals replace other metals are especially common. Not all single displacement reactions occur as written, though. You sometimes need to refer to a chart called the *activity series* to determine whether such a reaction will take place. Table 13-2 presents the activity series. To determine whether a single displacement reaction will occur, compare the two metals in the reaction:

TIP

>> If the metal that is single and not bonded is higher on the activity series than the metal that is bonded in the compound, the reaction will take place.

>> If the metal by itself is not higher on the series, then no reaction will take place.

Take a quick look at the example equation for single displacement again:

$$Zn(s) + CuSO_4(aq) \rightarrow ZnSO_4(aq) + Cu(s)$$

The notation (*aq*) indicates that the compound is dissolved in water — in an aqueous solution. Because zinc replaces copper in this case, it's said to be more active. If you place a piece of copper in a zinc sulfate solution, nothing happens because zinc is more active than copper, which you can tell from Table 13-2.

TABLE 13-2 Activity Series

Metal	Notes
Lithium	Most-reactive metals; react with cold water to form hydroxide and hydrogen gas.
Potassium	
Strontium	
Calcium	
Sodium	
Magnesium	React with hot water/acid to form oxides and hydrogen gas.
Aluminum	
Zinc	
Chromium	
Iron	Replace hydrogen ion from dilute strong acids.
Cadmium	
Cobalt	
Nickel	
Tin	
Lead	
Hydrogen	Nonmetal, listed in reactive order.
Antimony	Combine directly with oxygen to form oxides.
Arsenic	
Bismuth	
Copper	
Mercury	Least-reactive metals; often found as free metals; oxides decompose easily.
Silver	
Palladium	
Platinum	
Gold	

Double displacement (double replacement)

In single displacement reactions, only one chemical species is displaced. In *double displacement reactions* (sometimes called *double replacement*), or *metathesis reactions,* two species (normally ions) are displaced. Most of the time, reactions of this type occur in a solution, and either an insoluble solid (precipitation reactions) or water (neutralization reactions) is formed.

Precipitation reactions

If you mix a solution of potassium chloride and a solution of silver nitrate, a white insoluble solid is formed in the resulting solution. The formation of an insoluble solid in a solution is called *precipitation*. Here are the molecular, ionic, and net ionic equations for this double displacement reaction:

$$KCl\,(aq) + AgNO_3(aq) \rightarrow AgCl(s) + KNO_3(aq) \qquad \text{(molecular)}$$

$$K^+(aq) + Cl^-(aq) + Ag^+(aq) + NO_3^-(aq) \rightarrow AgCl(s) + K^-(aq) + NO_3^-(aq) \qquad \text{(ionic)}$$

$$Cl^-(aq) + Ag^+ \rightarrow AgCl(s) \qquad \text{(net ionic)}$$

The white insoluble solid that's formed is silver chloride. You can drop out the potassium cation and nitrate anion spectator ions, because they don't change during the reaction and are found on both sides of the equation in an identical form.

To write these equations, you have to know something about the solubility of ionic compounds. Don't fret. Here you go: If a compound is soluble, it will remain in its free ion form, but if it's insoluble, it will precipitate (form a solid). Table 13-3 gives the solubility of selected ionic compounds.

TABLE 13-3 Solubility of Selected Ionic Compounds

Water Soluble	Water Insoluble
All chlorides, bromides, iodides	Except those of Ag^+, Pb^{2+}, Hg_2^{2+}
All compounds of NH_4^+	Oxides
All compounds of alkali metals	Sulfides
All acetates	Most phosphates
All nitrates	Most hydroxides
All chlorates	
All sulfates	Except $PbSO_4$, $BaSO_4$, and $SrSO_4$

To use Table 13-3, take the cation of one reactant and combine it with the anion of the other reactant, and vice versa (keeping the neutrality of the compounds). This allows you to predict the possible products of the reaction. Then look up the solubility of the possible products in the table. If the compound is insoluble, it precipitates. If it's soluble, it remains in solution.

Neutralization reactions

The other type of double displacement reaction is the reaction between an acid and a base. This double displacement reaction, called a *neutralization reaction*, forms water. Take a look at the mixing solutions of sulfuric acid (auto battery acid, H_2SO_4) and sodium hydroxide (lye, NaOH). Here are the molecular, ionic, and net ionic equations for this reaction:

$$H_2SO_4(aq) + 2NaOH(aq) \rightarrow Na_2SO_4(aq) + 2H_2O(l) \qquad \text{(molecular)}$$

$$2H^+(aq) + SO_4^{2-}(aq) + 2Na^+(aq) + 2OH^-(aq) \rightarrow 2Na^+(aq) + SO_4^{2-}(aq) + 2H_2O(l) \qquad \text{(ionic)}$$

$$2H^+(aq) + 2OH^-(aq) \rightarrow 2H_2O(l) \text{ or } H^+(aq) + OH^-(aq) \rightarrow H_2O(l)$$
$$\text{(net ionic)}$$

For more on net ionic equations, see the upcoming section "Canceling Spectator Ions: Net Ionic Equations." You can find more about acid-base reactions in Chapter 20.

Combustion

Oxygen is always a reactant in *combustion reactions*, which often release heat and light as they occur. Combustion reactions frequently involve hydrocarbon reactants (like propane, $C_3H_8(g)$, the gas used to fire up backyard grills) and yield carbon dioxide and water as products. For example,

$$C_3H_8(g) + 5O_2(g) \rightarrow 3CO_2(g) + 4H_2O(l)$$

Combustion reactions also include combination reactions between elements and oxygen, such as

$$S(s) + O_2(g) \rightarrow SO_2(g)$$

TIP

If the reactants include oxygen (O_2) and a hydrocarbon or an element, you're probably dealing with a combustion reaction. If the products are carbon dioxide and water, you're almost certainly dealing with a combustion reaction.

EXAMPLE

Q. Predict and balance the following reaction:

$$Be(s) + O_2 \rightarrow$$

A. $2Be(s) + O_2 \rightarrow 2BeO(s)$. Although beryllium isn't on the metal activity series, you can make a pretty good prediction that this is a combination/combustion reaction. Why? First, you have two reactants. A single reactant would imply decomposition. The beryllium reactant is an element, not a compound, so you can rule out double displacement. The metal element reactant might make you consider single displacement, but there's no metal in oxygen (the other reactant), so there's no obvious displacement partner. So the most likely reaction is combination. When elements combine with oxygen, that's also a combustion reaction.

YOUR TURN

7 Complete and balance the following reactions:

(a) $C(s) + O_2(g) \rightarrow$

(b) $Mg(s) + I_2(g) \rightarrow$

(c) $Sr(s) + I_2(g) \rightarrow$

8 Complete and balance the following reactions:

(a) $HI(g) \rightarrow$

(b) $H_2O_2(l) \rightarrow H_2O(l) +$

(c) $NaCl(s) \rightarrow$

9 Complete and balance the following reactions:

(a) $Zn(s) + H_2SO_4(aq) \rightarrow$

(b) $Al(s) + HCl(aq) \rightarrow$

(c) $Li(s) + H_2O(l) \rightarrow$

10 Complete and balance the following reactions:

(a) $Ca(OH)_2(aq) + HCl(aq) \rightarrow$

(b) $HNO_3(aq) + NaOH(aq) \rightarrow$

(c) $FeS(s) + H_2SO_4(aq) \rightarrow$

Canceling Spectator Ions: Net Ionic Equations

Chemistry is often conducted in aqueous solutions. Soluble ionic compounds dissolve into their component ions, and these ions can react to form new products. In these kinds of reactions, sometimes only the cation or anion of a dissolved compound reacts. The other ion merely watches the whole affair, twiddling its charged thumbs in electrostatic boredom. These uninvolved ions are called *spectator ions*.

REMEMBER Because spectator ions don't actually participate in the chemistry of a reaction, you don't need to include them in a chemical equation. Doing so leads to a needlessly complicated reaction equation, so chemists prefer to write *net ionic equations,* which omit the spectator ions. A net ionic equation doesn't include every component that may be present in a given beaker. Rather, it includes only those components that actually react.

TIP Here's a simple recipe for making net ionic equations of your own:

1. **Examine the starting equation to determine which ionic compounds are dissolved, as indicated by the (aq) symbol following the compound name.**

   $Zn(s) + HCl(aq) \rightarrow ZnCl_2(aq) + H_2(g)$

2. **Rewrite the equation, explicitly separating dissolved ionic compounds into their component ions.**

   $Zn(s) + H^+(aq) + Cl^-(aq) \rightarrow Zn^{2+}(aq) + 2Cl^-(aq) + H_2(g)$

   Polyatomic ions don't break apart in solution, so be sure to familiarize yourself with the common ones (see Chapter 11 for details).

3. **Compare the reactant and product sides of the rewritten reaction and cross out the spectator ions.**

Any dissolved ions that appear in the same form on both sides are spectator ions. Cross out the spectator ions to produce a net reaction. If all reactants and products cross out, then no reaction will occur.

$$Zn(s) + H^+(aq) + \cancel{Cl^-(aq)} \rightarrow Zn^{2+}(aq) + \cancel{2Cl^-(aq)} + H_2(g)$$

The net reaction is

$$Zn(s) + H^+(aq) \rightarrow Zn^{2+}(aq) + H_2(g)$$

As written, the preceding reaction is imbalanced with respect to the number of hydrogen atoms and the amount of positive charge.

4. **Balance the net reaction for mass and charge.**

$$Zn(s) + 2H^+(aq) \rightarrow Zn^{2+}(aq) + H_2(g)$$

TIP

If you want, you can balance the equation for mass and charge first (at Step 1). That way, when you cross out spectator ions at Step 3, you cross out equivalent numbers of ions. Either method produces the same net ionic equation in the end. Some people prefer to balance the starting reaction equation, but others prefer to balance the net reaction because it's a simpler equation.

Q. Generate a balanced net ionic equation for the following reaction:

EXAMPLE

$$CaCO_3(s) + 2HCl(aq) \rightarrow CaCl_2(aq) + H_2O(l) + CO_2(g)$$

A. $CaCO_3(s) + 2H^+(aq) \rightarrow Ca^{2+}(aq) + H_2O(l) + CO_2(g)$. Because HCl and $CaCl_2$ are listed as aqueous (aq), rewrite the equation, explicitly separating those compounds into their ionic components:

$$CaCO_3(s) + 2H^+(aq) + 2Cl^-(aq) \rightarrow Ca^{2+}(aq) + 2Cl^-(aq) + H_2O(l) + CO_2(g)$$

Next, cross out any components that appear in the same form on both sides of the equation. In this case, the chloride ions (Cl^-) are crossed out:

$$CaCO_3(s) + 2H^+(aq) + \cancel{2Cl^-(aq)} \rightarrow Ca^{2+}(aq) + \cancel{2Cl^-(aq)} + H_2O(l) + CO_2(g)$$

This leaves the net reaction:

$$CaCO_3(s) + 2H^+(aq) \rightarrow Ca^{2+}(aq) + H_2O(l) + CO_2(g)$$

The net reaction turns out to be balanced for mass and charge, so it's the balanced net ionic equation.

YOUR TURN

11 Generate balanced net ionic equations for the following reactions:

(a) $LiOH(aq) + HI(aq) \rightarrow H_2O(l) + LiI(aq)$

(b) $AgNO_3(aq) + NaCl(aq) \rightarrow AgCl(s) + NaNO_3(aq)$

(c) $Pb(NO_3)_2(aq) + H_2SO_4(aq) \rightarrow PbSO_4(s) + HNO_3(aq)$

12 Generate balanced net ionic equations for the following reactions:

(a) $HCl(aq) + ZnS(aq) \rightarrow H_2S(g) + ZnCl_2(aq)$

(b) $Ca(OH)_2(aq) + H_3PO_4(aq) \rightarrow Ca_3(PO_4)_2(aq) + H_2O(l)$

(c) $(NH_4)_2S(aq) + Co(NO_3)_2(aq) \rightarrow CoS(s) + NH_4NO_3(aq)$

Practice Questions Answers and Explanations

(1) Don't forget the symbols indicating state and whether any heat or catalysts were added.

  a. $Mg(s) + O_2(g) \xrightarrow{\Delta} MgO(s)$

  b. $B_2O_3(s) + Mg(s) \rightarrow B(s) + MgO(s)$

(2) Here are the sentences describing the provided reactions:

  a. Solid sulfur and gaseous oxygen react to produce the gas sulfur dioxide.

  b. Hydrogen gas reacts with oxygen gas in the presence of platinum to produce liquid water.

(3) $AgNO_3(aq) + NaCl(aq) \rightarrow AgCl(s) + NaNO_3(aq)$

The key to understanding and writing this reaction is to focus on what we've given you in the problem and nothing more. Two reactants are clearly mentioned, sodium chloride and silver nitrate. You know they're reactants because the problem states that they mix to produce new products, and they're both written as *aqueous* because they're both said to be in solution. You can determine the products by looking at what forms according to the problem, with *forms* being the key word. Silver chloride is the first product mentioned, and it's written as a solid because the problem identifies it as a *precipitate*, which is a solid in solution formed by a chemical reaction. The second product mentioned is sodium nitrate, which is aqueous because the problem identifies it as still in solution.

(4) Here are the balanced reactions:

  a. $N_2(g) + 3H_2(g) \rightarrow 2NH_3(g)$

  b. $C_3H_8(g) + 5O_2(g) \rightarrow 3CO_2(g) + 4H_2O(l)$

  c. $4Al(s) + 3O_2(g) \rightarrow 2Al_2O_3(s)$

  d. $2AgNO_3(aq) + Cu(s) \rightarrow Cu(NO_3)_2(aq) + 2Ag(s)$

(5) More balanced reactions:

  a. $3Ag_2SO_4(aq) + 2AlCl_3(aq) \rightarrow 6AgCl(s) + Al_2(SO_4)_3(aq)$

  b. $CH_4(g) + 2Cl_2(g) \rightarrow CH_2Cl_2(l) + 2HCl(g)$

  c. $3Cu(s) + 8HNO_3(aq) \rightarrow 3Cu(NO_3)_2(aq) + 2NO(g) + 4H_2O(l)$

  d. $2HCl(aq) + Ca(OH)_2(aq) \rightarrow CaCl_2(s) + 2H_2O(l)$

(6) Still more balanced reactions:

  a. $H_2C_2O_4(aq) + 2KOH(aq) \rightarrow K_2C_2O_4(aq) + 2H_2O(l)$

  b. $2P(s) + 3Br_2(l) \rightarrow 2PBr_3(g)$

  c. $Pb(NO_3)_2(aq) + 2KI(aq) \rightarrow 2KNO_3(aq) + PbI_2(s)$

  d. $Zn(s) + 2AgNO_3(aq) \rightarrow Zn(NO_3)_2(aq) + 2Ag(s)$

(7) After completing the reaction, forgetting to balance it is easy. Of course, proper balancing means that you have to pay attention to the amount of each atom in the product compounds. These reactions are all combination reactions:

a. $C(s) + O_2(g) \rightarrow CO_2(g)$

b. $Mg(s) + I_2(g) \rightarrow MgI_2(s)$

c. $Sr(s) + I_2(g) \rightarrow SrI_2(s)$

(8) Here are some more completed, balanced reactions. All of these are decomposition reactions:

a. $2HI(g) \rightarrow H_2 + I_2$

b. $2H_2O_2(l) \rightarrow 2H_2O(l) + O_2(g)$

c. $2NaCl(s) \rightarrow 2Na(s) + Cl_2(g)$

(9) Here are the completed, balanced versions of a series of single displacement reactions:

a. $Zn(s) + H_2SO_4(aq) \rightarrow ZnSO_4(aq) + H_2(g)$

b. $2Al(s) + 6HCl(aq) \rightarrow 2AlCl_3(aq) + 3H_2(g)$

c. $2Li(s) + 2H_2O(l) \rightarrow 2LiOH(aq) + H_2(g)$

(10) This final set, completed and balanced, is composed of double displacement reactions:

a. $Ca(OH)_2(aq) + 2HCl(aq) \rightarrow CaCl_2(aq) + 2H_2O(l)$

b. $HNO_3(aq) + NaOH(aq) \rightarrow H_2O(l) + NaNO_3(aq)$

c. $FeS(s) + H_2SO_4(aq) \rightarrow H_2S(g) + FeSO_4(aq)$

(11) The following answers show the original reaction, the expanded reaction, the expanded reaction with spectator ions crossed out, and the final balanced reaction:

a.
$$LiOH(aq) + HI(aq) \rightarrow H_2O(l) + LiI(aq)$$
$$Li^+(aq) + OH^-(aq) + H^+(aq) + I^-(aq) \rightarrow H_2O(l) + Li^+(aq) + I^-(aq)$$
$$\cancel{Li^+(aq)} + OH^-(aq) + H^+(aq) + \cancel{I^-(aq)} \rightarrow H_2O(l) + \cancel{Li^+(aq)} + \cancel{I^-(aq)}$$
$$OH^-(aq) + H^+(aq) \rightarrow H_2O(l)$$

b.
$$AgNO_3(aq) + NaCl(aq) \rightarrow AgCl(s) + NaNO_3(aq)$$
$$Ag^+(aq) + NO_3^-(aq) + Na^+(aq) + Cl^-(aq) \rightarrow AgCl(s) + Na^+(aq) + NO_3^-(aq)$$
$$Ag^+(aq) + \cancel{NO_3^-(aq)} + \cancel{Na^+(aq)} + Cl^-(aq) \rightarrow AgCl(s) + \cancel{Na^+(aq)} + \cancel{NO_3^-(aq)}$$
$$Ag^+(aq) + Cl^-(aq) \rightarrow AgCl(s)$$

c.
$$Pb(NO_3)_2(aq) + H_2SO_4(aq) \rightarrow PbSO_4(s) + 2HNO_3(aq)$$
$$Pb^{2+}(aq) + 2NO_3^-(aq) + 2H^+(aq) + SO_4^{2-}(aq) \rightarrow PbSO_4(s) + 2H^+(aq) + 2NO_3^-(aq)$$
$$Pb^{2+}(aq) + \cancel{2NO_3^-(aq)} + \cancel{2H^+(aq)} + SO_4^{2-}(aq) \rightarrow PbSO_4(s) + \cancel{2H^+(aq)} + \cancel{2NO_3^-(aq)}$$
$$Pb^{2+}(aq) + SO_4^{2-}(aq) \rightarrow PbSO_4(s)$$

(12) Here are the answers for the second batch of net ionic equation questions, again showing the original reaction, the expanded reaction, the expanded reaction with spectator ions crossed out, and the final balanced reaction:

a.

$$2HCl(aq) + ZnS(aq) \rightarrow H_2S(g) + ZnCl_2(aq)$$
$$2H^+(aq) + 2Cl^-(aq) + Zn^{2+}(aq) + S^{2-}(aq) \rightarrow H_2S(g) + Zn^{2+}(aq) + 2Cl^-(aq)$$
$$2H^+(aq) + \cancel{2Cl^-(aq)} + \cancel{Zn^{2+}(aq)} + S^{2-}(aq) \rightarrow H_2S(g) + \cancel{Zn^{2+}(aq)} + \cancel{2Cl^-(aq)}$$
$$2H^+(aq) + S^{2-}(aq) \rightarrow H_2S(g)$$

b.

$$3Ca(OH)_2(aq) + 2H_3PO_4(aq) \rightarrow Ca_3(PO_4)_2(aq) + 6H_2O(l)$$
$$3Ca^{2+}(aq) + 6OH^-(aq) + 6H^+(aq) + 2PO_4^{2-}(aq) \rightarrow 3Ca^{2+}(aq) + 2PO_4^{2-}(aq) + H_2O(l)$$
$$\cancel{3Ca^{2+}(aq)} + 6OH^-(aq) + 6H^+(aq) + \cancel{2PO_4^{2-}(aq)} \rightarrow \cancel{3Ca^{2+}(aq)} + \cancel{2PO_4^{2-}(aq)} + H_2O(l)$$
$$6OH^-(aq) + 6H^+(aq) \rightarrow 6H_2O(l)$$
$$OH^-(aq) + H^+(aq) \rightarrow H_2O(l)$$

c.

$$(NH_4)_2S(aq) + Co(NO_3)_2(aq) \rightarrow CoS(s) + 2NH_4NO_3(aq)$$
$$2NH_4^+(aq) + S^{2-}(aq) + Co^{2+}(aq) + 2NO_3^-(aq) \rightarrow CoS(s) + 2NH_4^+(aq) + 2NO_3^-(aq)$$
$$2NH_4^+(aq) + S^{2-}(aq) + Co^{2+}(aq) + \cancel{2NO_3^-(aq)} \rightarrow CoS(s) + \cancel{2NH_4^+(aq)} + \cancel{2NO_3^-(aq)}$$
$$S^{2-}(aq) + Co^{2+}(aq) \rightarrow CoS(s)$$

If you're ready to test your skills a bit more, take the following chapter quiz that incorporates all the chapter topics.

Whaddya Know? Chapter 13 Quiz

Ready for a quiz? The 10 questions in this section will test the skills you learned in this chapter. When you're done, check out the section that follows for answers and explanations.

1 Write the chemical reaction described below and balance it:

You react a solid piece of magnesium with aqueous hydrochloric acid to produce hydrogen gas and aqueous magnesium chloride.

2 Write the chemical reaction described below and balance it:

You burn methane gas (CH_4) in the presence of oxygen gas. This produces gaseous carbon dioxide and gaseous water vapor.

3 Balance the following reaction and identify the type of reaction shown:

$P_4 + O_2 \rightarrow P_2O_3$

4 Balance the following reaction and identify the type of reaction shown:

$MgI_2 + Mn(SO_3)_2 \rightarrow MgSO_3 + MnI_4$

5 Balance the following reaction and identify the type of reaction shown:

$Na_3PO_4 + KOH \rightarrow NaOH + K_3PO_4$

6 Balance the following reaction and identify the type of reaction shown:

$C_6H_{12} + O_2 \rightarrow CO_2 + H_2O$

7 Predict the following reaction and balance it correctly:

$Ag + CuSO_4 \rightarrow$

8 Predict the following reaction and balance it correctly:

$C_4H_8 + O_2 \rightarrow$

9 Predict the following reaction and balance it correctly:

$Li + CuCl_2 \rightarrow$

10 Predict the following reaction and balance it correctly:

$Al + O_2 \rightarrow$

11. Predict the following reaction and balance it correctly:

$$NO_2 \rightarrow$$

12. Predict the following reaction and balance it correctly:

$$AgNO_2 + BaSO_4 \rightarrow$$

13. Identify whether the following compounds are soluble (aqueous) or insoluble (solid):

   a. NaCl

   b. AgCl

   c. $CaSO_4$

   d. $BaSO_4$

14. Identify the phases of each reactant and product along with the precipitate, if there is one, in the following chemical reaction:

$$Pb(NO_3)_2 + Na_2SO_4 \rightarrow PbSO_4 + 2NaNO_3$$

15. Write the net ionic reaction of the following reaction shown here. Balance it and include the phases of the reactants and products. Identify the precipitate, if there is one, in the net ionic reaction:

$$CaCl_2 + NaOH \rightarrow$$

Answers to Chapter 13 Quiz

(1) $Mg(s) + 2HCl(aq) \rightarrow H_2(g) + MgCl_2(aq)$

When writing out an equation make sure to pay attention to the phases given. Hydrogen has a 2 subscript because when it is found by itself in a reaction as a gas it is always diatomic. To balance the reaction you need to add a 2 in front of HCl.

(2) $CH_4(g) + 2O_2(g) \rightarrow CO_2(g) + 2H_2O(g)$

When writing out an equation, make sure you always get the phases correct by referencing the information given to you in the problem. Oxygen has a subscript of 2 because it is a diatomic element. The coefficients of 2 were added in front of oxygen and water to balance the reaction.

(3) Here is the balanced reaction:

$P_4 + 3O_2 \rightarrow 2P_2O_3$

This is a **synthesis reaction**. You can tell this because there are two reactants but only one product.

(4) Here is the balanced reaction:

$2MgI_2 + Mn(SO_3)_2 \rightarrow 2MgSO_3 + MnI_4$

This is a **double displacement reaction.** You can tell this because there are two different elements that are being displaced. In this case both compounds present in the reactants have an ion that is replaced.

(5) Here is the balanced reaction:

$Na_3PO_4 + 3KOH \rightarrow 3NaOH + K_3PO_4$

This is a **double displacement reaction** for the same reasons as described in problem 4.

(6) Here is the balanced reaction:

$C_6H_{12} + 9O_2 \rightarrow 6CO_2 + 6H_2O$

This is a combustion reaction. You should be able to be able to identify this based on the reactants being a hydrocarbon plus oxygen and the reaction products being carbon dioxide and water. Those are all signs of a combustion of a reaction.

(7) **No reaction takes place.** This is a single displacement reaction. You identify that by the presence of a single element (Ag) being combined with an ionic compound ($CuSO_4$). To determine if a single displacement reaction happens you must look at the activity series and compare the elements of like charge. In this case, silver and copper are both positive cations so you should see which of them is more active using the activity series. Since copper, the element in the compound, is higher on the chart than silver, the single element by itself, no reaction happens. You don't need to do anything beyond that.

(8) Here is the complete, balanced reaction:

$$C_4H_8 + 6O_2 \rightarrow 4CO_2 + 4H_2O$$

This reaction is a combustion reaction. You can identify that without the carbon dioxide and water products by recognizing that you have a hydrocarbon C_4H_8 combining with oxygen gas. Any time you have reactants like that, you have a combustion reaction. The products of a combustion reaction are always carbon dioxide and water. Remember, it doesn't matter what order you put them in though. You can write $CO_2 + H_2O$ or $H_2O + CO_2$. Either one is fine. Once you've predicted this correctly, you just need to add the correct coefficients to balance it.

(9) Here is the complete, balanced reaction:

$$2Li + CuCl_2 \rightarrow Cu + 2LiCl$$

This is a single displacement reaction. You identify that by the presence of a single element (Li) being combined with an ionic compound ($CuCl_2$). To determine if a single displacement reaction happens you must look at the activity series and compare the elements of like charge. In this case, lithium and copper are both positive cations so you should see which of them is more active using the activity series. Lithium is at the very top of the activity series, while copper is almost all the way at the bottom. This means that the lithium will replace copper. You must then write the correct ionic formula of lithium chloride. Since lithium is a 1+ charge and chloride is a 1- charge, you do not need subscripts for the formula. You then balance the reaction.

(10) Here is the complete, balanced reaction:

$$4Al + 3O_2 \rightarrow 2Al_2O_3$$

This is a synthesis/combination reaction. You can identify this by seeing there are two elemental reactants. The only thing they can do is combine with one another. This combination creates the ionic compound of aluminum oxide. You must make sure to write it correctly by balancing the 3+ charge of aluminum and the 2- charge of oxide. Once you have done that you must balance the reaction.

(11) Here is the complete, balanced reaction:

$$2NO_2 \rightarrow N_2 + 2O_2$$

This is a decomposition reaction. This can be determined by there being only one reactant. The only possible option for this reactant is for it to break apart. The products of nitrogen and oxygen are both diatomic molecules and need to have a subscript of 2 written after each one. Once you do this, you simply balance the reaction and you are done.

(12) Here is the complete, balanced reaction:

$$2AgNO_2 + BaSO_4 \rightarrow Ag_2SO_4 + Ba(NO_2)_2$$

This is a double displacement reaction. You can identify this when you see two ionic compounds as reactants. Those reactants will have their positive cations switch with one another. Once this switch occurs you must write the correct ionic formulas that balance the charge of the positive and negative ion for each of the new compounds, as shown in the reaction and then balance it.

(13) Solubility for each of the following:

    a. soluble (aq)

    b. insoluble (s)

    c. soluble (aq)

    d. insoluble (s)

(14) Here is the reaction with the phases written in for each reactant and product

$$Pb(NO_3)_2(aq) + Na_2SO_4(aq) \rightarrow PbSO_4(s) + 2NaNO_3(aq)$$

The precipitate in the reaction is the product that is insoluble, meaning it is a solid. The solid product in this reaction is **$PbSO_4$**, making it the precipitate.

(15) The net ionic reaction for the question is:

$$Ca^{2+}(aq) + 2OH^-(aq) \rightarrow Ca(OH)_2(s)$$

The precipitate is **$Ca(OH)_2$ (s)**.

IN THIS CHAPTER

» **Using balanced chemical equations to perform mole-to-mole conversions**

» **Expanding mole-to-mole conversions to perform calculations with mass, particles, and volume**

» **Figuring out what happens when one reactant runs out before the others**

» **Using percent yield to determine the efficiency of reactions**

Chapter **14**

Putting Stoichiometry to Work

S toichiometry. Such a complicated word for such a simple idea. The Greek roots of the word mean "measuring elements," which doesn't sound nearly as intimidating. Moreover, the ancient Greeks couldn't tell an ionic bond from an Ionic column, so just how technical and scary could stoichiometry really be? Simply stated, *stoichiometry* is the quantitative relationship between components of chemical substances. In compound formulas and reaction equations, you express stoichiometry by using subscripted numbers and coefficients.

If you arrived at this chapter by first wandering through Chapters 12 and 13, then you've already had breakfast, lunch, and an afternoon snack with stoichiometry. If you bypassed the aforementioned chapters, then you haven't eaten all day. Either way, it's time for dinner. Please pass the coefficients.

REMEMBER

We've rounded molar masses to the hundredths place before doing the calculations. Answers have been rounded according to the rules for significant figures (see Chapter 1 for details).

Using Mole-Mole Conversions from Balanced Equations

Mass and energy are conserved. It's the law. Unfortunately, this means that there's no such thing as a free lunch or any other type of free meal. Ever. On the other hand, the conservation of mass makes it possible to predict how chemical reactions will turn out.

REMEMBER

Chapter 13 describes why chemical reaction equations should be balanced for equal mass in reactants and products. You balance an equation by adjusting the coefficients that precede reactant and product compounds within the equation. Balancing equations can seem like a chore, like taking out the trash. But a balanced equation is far better than any collection of coffee grounds and orange peels, because such an equation is a useful tool. After you've got a balanced equation, you can use the coefficients to build *mole-mole conversion factors*. These kinds of conversion factors tell you how much of any given product you get by reacting any given amount of reactant. This is one of those calculations that makes chemists particularly useful, so they needn't get by on looks and charm alone. (For more about moles, see Chapter 12.)

Consider the following balanced equation for generating ammonia from nitrogen and hydrogen gases:

$$N_2(g) + 3H_2(g) \rightarrow 2NH_3(g)$$

Industrial chemists around the globe perform this reaction, humorlessly fixating on how much ammonia product they'll end up with at the end of the day. (In fact, clever methods for improving the rate and yield of this reaction garnered Nobel Prizes for two German gentlemen, Fritz Haber and Karl Bosch.) In any event, how are chemists to judge how closely their reactions have approached completion? The heart of the answer lies in a balanced equation and the mole-mole conversion factors that spring from it.

For every mole of nitrogen reactant, a chemist expects 2 moles of ammonia product. Similarly, for every 3 moles of hydrogen reactant, the chemist expects 2 moles of ammonia product. These expectations are based on the coefficients of the balanced equation and are expressed as mole-mole conversion factors, as shown in Figure 14-1.

FIGURE 14-1:
Building mole-mole conversion factors from a balanced equation.

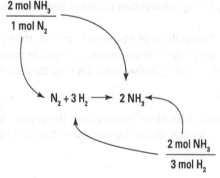

Q. How many moles of ammonia can be expected from the reaction of 278 mol of N_2 gas?

EXAMPLE **A.** 556 mol of ammonia. Begin with your known quantity, the 278 mol of nitrogen that's to be reacted. Multiply that quantity by the mole–mole conversion factor that relates moles of nitrogen to moles of ammonia. Write the conversion factor so that *mol NH₃* is on top and *mol N₂* is on the bottom. That way, the *mol N₂* units cancel, leaving you with the desired units, *mol NH₃*. The numbers you put in front of the units for the conversion factor come directly from the coefficients in the balanced chemical equation.

$$(278 \text{ mol } N_2)\left(\frac{2 \text{ mol } NH_3}{1 \text{ mol } N_2}\right) = 556 \text{ mol } NH_3$$

YOUR TURN

1 One source of hydrogen gas is the electrolysis of water, in which electricity is passed through water to break hydrogen–oxygen bonds, yielding hydrogen and oxygen gases:

$$2H_2O(l) \rightarrow 2H_2(g) + O_2(g)$$

(a) How many moles of hydrogen gas result from the electrolysis of 78.4 mol of water?

(b) How many moles of water are required to produce 905 mol of hydrogen?

(c) Running the electrolysis reaction in reverse constitutes the combustion of hydrogen. How many moles of oxygen are required to combust 84.6 mol of hydrogen?

---

2 Aluminum reacts with copper(II) sulfate to produce aluminum sulfate and copper, as summarized by this skeleton equation:

$$Al(s) + CuSO_4(aq) \rightarrow Al_2(SO_4)_3(aq) + Cu(s)$$

(a) Balance the equation.

(b) How many moles of aluminum are needed to react with 10.38 mol of copper(II) sulfate?

(c) How many moles of copper are produced if 2.08 mol of copper(II) sulfate react with aluminum?

(d) How many moles of copper(II) sulfate are needed to produce 0.96 mol of aluminum sulfate?

(e) How many moles of aluminum are needed to produce 20.01 mol of copper?

Solid iron reacts with solid sulfur to form iron(III) sulfide:

$$Fe(s) + S(s) \rightarrow Fe_2S_3(s)$$

(a) Balance the equation.

(b) How many moles of sulfur are needed to react with 6.2 mol of iron?

(c) How many moles of iron(III) sulfide are produced from 10.6 mol of iron?

(d) How many moles of iron(III) sulfide are produced from 3.5 mol of sulfur?

---

4 Ethane combusts to form carbon dioxide and water:

$$C_2H_6(g) + O_2 \rightarrow CO_2(g) + H_2O(g)$$

(a) Balance the equation.

(b) How many moles of carbon dioxide does 15.4 mol of ethane produce?

(c) How many moles of ethane does it take to produce 293 mol of water?

(d) How many moles of oxygen are required to combust 0.178 mol of ethane?

Putting Moles at the Center: Conversions Involving Particles, Volumes, and Masses

The mole is the beating heart of stoichiometry, the central unit through which other quantities flow. Real-life chemists don't have magic mole vision, however. A chemist can't look at a pile of potassium chloride crystals, squint her eyes, and proclaim, "That's 0.539 moles of salt." Real *reagents* (reactants) tend to be measured in units of mass or volume. Real products are measured in the same way. So you need to be able to use mole-mass, mole-volume, and mole-particle conversion factors to translate between these different dialects of counting. Figure 14-2 summarizes the interrelationship among all these things and serves as a flowchart for problem-solving. All roads lead to and from the mole.

REMEMBER

If you look at Figure 14-2, you can see that it isn't possible to convert directly between the mass of one substance and the mass of another substance. You must convert to moles and then use the mole-mole conversion factor before converting to the mass of a new substance. The same can be said for conversions from the particles or volume of one substance to that of another substance. The mole is always the intermediary you use for the conversion.

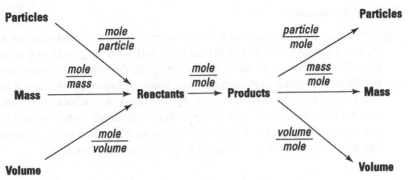

FIGURE 14-2: A problem-solving flowchart showing the use of mole-mole, mole-mass, mole-volume, and mole-particle conversion factors.

The challenging part of stoichiometry is not the calculations themselves but the work and critical thinking required to set up the problem in the first place. If you begin with the mass of one substance and want to convert to the volume of a new substance in a reaction, Figure 14-2 can show you the exact path to take. Sometimes actually tracing the path with your finger will help you commit the relationships to memory.

TIP

When performing stoichiometry calculations, save any rounding for the end. Set everything up, and then get your calculator out. When you have a string of conversion factors, you're far better off calculating everything in one step instead of breaking a problem into multiple steps where you type a calculation into your calculator, hit *enter*, round the answer, and then move on to the next step. Keep it as simple as you can, do everything in the same calculation on the same line. All that rounding leads to a greater degree of error. Don't round anything until you are done.

Another thing to pay attention to is a very common error that can happen when doing stoichiometry calculations involving Avogadro's number (6.02×10^{23}). If you are converting from particles (atoms/molecules/formula units) to moles, you are going to be dividing by Avogadro's number. Make sure you use parenthesis around this value when you type it into your calculator. Not doing this can lead to you getting an answer that appears correct but has the wrong exponent value for your calculations. This advice really applies whenever you are dividing something by a number in scientific notation. It is commonly encountered doing stoichiometry and mole conversion problems, so it is best to be aware of it when working through these. A simple suggestion in this case is that if you get an exponent that is somewhere in the 10^{40}s range, you did something in the problem wrong. This is a commonly seen answer that results when you don't use the order of operations correctly when solving these problems.

EXAMPLE

Q. Calcium carbonate decomposes to produce solid calcium oxide and carbon dioxide gas according to the following reaction. Answer each part of the question, assuming that 10.0 g of calcium carbonate decomposes.

$$CaCO_3(s) \rightarrow CaO(s) + CO_2(g)$$

(a) How many grams of calcium oxide are produced?

(b) At standard temperature and pressure (STP), how many liters of carbon dioxide are produced?

A. Here are the answers:

(a) 5.60 g CaO. First, convert 10.0 g of calcium carbonate to moles of calcium carbonate by using the molar mass of calcium carbonate (100.09 g/mol) as a conversion factor (see Chapter 7 for more about molar mass). To determine the grams of calcium oxide produced, you must then convert from moles of calcium carbonate to moles of calcium oxide. Keep in mind that you take the number of moles for the mole-mole conversion from the coefficients in the balanced chemical equation (see Chapter 8). Then convert from moles of calcium oxide to grams of calcium oxide by using the molar mass of calcium oxide (56.08 g/mol) as a conversion factor:

$$\left(10.0\text{g CaCO}_3\right)\left(\frac{1\text{ mol CaCO}_3}{100.09\text{g CaCO}_3}\right)\left(\frac{1\text{ mol CaO}}{1\text{ mol CaCO}_3}\right)\left(\frac{56.08\text{ g CaO}}{1\text{ mol CaO}}\right)=5.60\text{ g CaO}$$

(b) 2.24 L CO_2. To determine the liters of carbon dioxide produced, follow the initial mass-mole conversion with a mole-mole conversion to find the moles of carbon dioxide produced. Then convert from moles of carbon dioxide to liters by using the fact that at STP, each mole of gas occupies 22.4 L:

$$\left(10.0\text{ g CaCO}_3\right)\left(\frac{1\text{ mol CaCO}_3}{100.09\text{ g CaCO}_3}\right)\left(\frac{1\text{ mol CO}_2}{1\text{ mol CaCO}_3}\right)\left(\frac{22.4\text{ LCO}_2}{1\text{ mol CO}_2}\right)=2.24\text{ L CO}_2$$

Notice how both calculations require you to first convert to moles and then perform a mole-mole conversion using stoichiometry from the reaction equation. Then you convert to the desired units. Both solutions consist of a chain of conversion factors, each factor bringing the units one step closer to those needed in the answer.

YOUR TURN

5 Hydrogen peroxide decomposes into oxygen gas and liquid water:

$$H_2O_2(l) \rightarrow H_2O(l) + O_2(g)$$

(a) Balance the equation.

(b) How many grams of water are produced when 2.94×10^{24} molecules of hydrogen peroxide decompose? (Remember that molecules are particles.)

(c) What is the volume of oxygen produced at STP when 32.9 g of hydrogen peroxide decomposes?

Dinitrogen trioxide gas reacts with liquid water to produce nitrous acid:

$$N_2O_3(g) + H_2O(l) \rightarrow HNO_2(aq)$$

(a) Balance the equation.

(b) At STP, how many liters of dinitrogen trioxide react to produce 36.98 g of dissolved nitrous acid?

(c) How many molecules of water react with 17.3 L of dinitrogen trioxide at STP?

---

7 Lead(II) chloride reacts with chlorine to produce lead(IV) chloride: (Assume the reaction occurs at STP.)

$$PbCl_2(s) + Cl_2(g) \rightarrow PbCl_4(l)$$

(a) What volume of chlorine reacts to convert 50.0 g of lead(II) chloride entirely into product?

(b) How many formula units (particles of an ionic compound) of lead(IV) chloride result from the reaction of 13.71 g of lead(II) chloride?

(c) How many grams of lead(II) chloride react to produce 84.8 g of lead(IV) chloride?

Calculating Limiting Reactants

In real-life chemical reactions, not all of the reactants present convert into product. That would be perfect and convenient. Does that sound like real life to you? More typically, one reagent, another word for reactant, is completely used up, and others are left in excess, perhaps to react another day.

To understand how limiting reactants work, imagine you want a sandwich — a nice, tasty ham sandwich. If you want to get all nerdy about things (which we do), you can write a chemical reaction describing how a ham sandwich is made, focusing on two specific things: the ingredients needed and the ratio in which you combine these ingredients. Now, we make a pretty basic ham sandwich: two pieces of bread combine with one thick slice of ham and one slice of cheese to make one ham sandwich. You can see the equation here:

2 pieces of bread + 1 ham slice + 1 cheese slice → 1 ham sandwich

Suppose we check our supplies and find that we have 12 pieces of bread, 5 slices of ham, and 10 slices of cheese. How many sandwiches can we make? We can make 5, of course. We have enough bread for 6 sandwiches, enough ham for 5, and enough cheese for 10. But we're going to run out of ham first — we'll have bread and cheese left over. And the ingredient we run out of first really limits the amount of product (sandwiches) we'll be able to make; this component can be called the *limiting ingredient* (though in a real chemical reaction, we call this the *limiting reactant* or *limiting reagent*).

In any chemical reaction, you can simply pick one reactant as a candidate for the limiting reactant, calculate how many moles of that reagent you have, and then calculate how many grams of the other reactant you'd need to react both to completion. You'll discover one of two things. Either you have an excess of the first reactant, or you have an excess of the second reactant. The one you have in excess is the *excess reactant.* The one that isn't in excess is the *limiting reactant.* When you know which reactant is limiting, you can use that info to solve any stoichiometry problems involving that reaction. Why? You must use the limiting reactant for calculations because it runs out first and governs how much of the products you'll create in the reaction. If you used the excess for a calculation, you'd end up with a much larger amount of product than you should.

Q. Ammonia reacts with oxygen to produce nitrogen monoxide and liquid water:

EXAMPLE

$$NH_3(g) + O_2(g) \rightarrow NO(g) + H_2O(l)$$

(a) Balance the equation (as we explain in Chapter 13).

(b) Determine the limiting reagent if 100 g of ammonia and 100 g of oxygen are present at the beginning of the reaction.

(c) What is the excess reagent, and how many grams of the excess reagent will remain when the reaction reaches completion?

(d) How many grams of nitrogen monoxide will be produced if the reaction goes to completion?

(e) How many grams of water will be produced if the reaction goes to completion?

A. Here's the solution:

(a) Before doing anything else, you must have a balanced reaction equation. Don't waste good thought on an unbalanced equation. The balanced form of the given equation is

$$4NH_3(g) + 5O_2(g) \rightarrow 4NO(g) + 6H_2O(l)$$

(b) Oxygen is the limiting reagent. Two candidates, NH_3 and O_2, vie for the status of limiting reagent. You start with 100 g of each, which corresponds to some number of moles of each. Furthermore, you can tell from the coefficients in the balanced equation this reaction requires 4 mol of ammonia for every 5 mol of oxygen gas.

To find the limiting reactant, you simply need to perform a mass-to-mass (gram-to-gram) calculation from one reactant to the other (see Chapter 12 for more about molar mass). This allows you to see which reactant runs out first. You can start with either reactant and convert to mass of the other. In this example, we start with ammonia:

$$\left(100 \text{ g NH}_3\right)\left(\frac{1 \text{ mol NH}_3}{17.04 \text{ g NH}_3}\right)\left(\frac{5 \text{ mol O}_2}{4 \text{ mol NH}_3}\right)\left(\frac{32.00 \text{ g O}_2}{1 \text{ mol O}_2}\right) = 235 \text{ g O}_2$$

The calculation reveals that you'd need 235 g of oxygen gas to completely react with 100 g of ammonia. But you have only 100 g of oxygen. You'll run out of oxygen before you run out of ammonia, so oxygen is the limiting reagent.

(c) The excess reagent is ammonia, and 57.5 g of ammonia will remain when the reaction reaches completion. To calculate how many grams of ammonia will be left at the end of the reaction, assume that all 100 g of oxygen react:

$$\left(100 \text{ g O}_2\right)\left(\frac{1 \text{ mol O}_2}{32.00 \text{ g O}_2}\right)\left(\frac{4 \text{ mol NH}_3}{5 \text{ mol O}_2}\right)\left(\frac{17.04 \text{ g NH}_3}{1 \text{ mol NH}_3}\right) = 42.5 \text{ g NH}_3$$

This calculation shows that 42.5 g of the original 100 g of ammonia will react before the limiting reagent is expended. So 57.5 g of ammonia will remain in excess (just subtract 42.5 from 100).

(d) 75 g of nitrogen monoxide will be produced. This problem asks how much of a product is produced. For this calculation, you must begin with the limiting reactant. To determine the grams of nitrogen monoxide that are generated by the complete reaction of oxygen, start with the assumption that all 100 g of the oxygen react:

$$\left(100 \text{ g O}_2\right)\left(\frac{1 \text{ mol O}_2}{32.00 \text{ g O}_2}\right)\left(\frac{4 \text{ mol NO}}{5 \text{ mol O}_2}\right)\left(\frac{30.01 \text{ g NO}}{1 \text{ mol NO}}\right) = 75.0 \text{ g NO}$$

(e) 67.5 g of water will be produced. Again, assume that all 100 g of the oxygen react in order to determine how many grams of water are produced:

$$\left(100 \text{ g O}_2\right)\left(\frac{1 \text{ mol O}_2}{32.00 \text{ g O}_2}\right)\left(\frac{6 \text{ mol H}_2\text{O}}{5 \text{ mol O}_2}\right)\left(\frac{18.02 \text{ g H}_2\text{O}}{1 \text{ mol H}_2\text{O}}\right) = 67.5 \text{ g H}_2\text{O}$$

YOUR TURN

8   Iron(III) oxide reacts with carbon monoxide to produce iron and carbon dioxide:

$$\text{Fe}_2\text{O}_3(s) + 3\text{CO}(g) \rightarrow 2\text{Fe}(s) + 3\text{CO}_2(g)$$

(a) What is the limiting reagent if 50 g of iron(III) oxide and 67 g of carbon monoxide are present at the beginning of the reaction?

(b) What is the excess reagent, and how many grams of it will remain after the reaction proceeds to completion?

(c) How many grams of each product should be expected if the reaction goes to completion?

Solid sodium reacts (violently) with water to produce sodium hydroxide and hydrogen gas:

$$2Na(s) + 2H_2O(l) \rightarrow 2NaOH(aq) + H_2(g)$$

(a) What is the limiting reagent if 25 g of sodium and 40.2 g of water are present at the beginning of the reaction?

(b) What is the excess reagent, and how many grams of it will remain after the reaction has gone to completion?

(c) How many grams of sodium hydroxide and how many liters of hydrogen gas (at STP) should be expected if the reaction goes to completion?

---

10 Aluminum reacts with chlorine gas to produce aluminum chloride:

$$2Al(s) + 3Cl_2(g) \rightarrow 2AlCl_3(s)$$

(a) What is the limiting reagent if 29.3 g of aluminum and 34.6 L of chlorine gas (at STP) are present at the beginning of the reaction?

(b) What is the excess reagent, and how many grams (or liters at STP) of it will remain after the reaction has gone to completion?

(c) How many grams of aluminum chloride should be expected if the reaction goes to completion?

Counting Your Chickens after They've Hatched: Percent Yield Calculations

In a way, reactants have it easy. Maybe they'll make something of themselves and actually react. Or maybe they'll just lean against the inside of the beaker, scrolling through their feed and sipping a caramel macchiato.

REMEMBER

Chemists don't have it so easy. Someone is paying them to do reactions. That someone doesn't have time or money for excuses about loitering reactants. So you, as a fresh-faced chemist, have to be concerned with just how completely your reactants react to form products. To compare the amount of product obtained from a reaction with the amount that should have been obtained, chemists use *percent yield*. You determine percent yield with the following formula:

$$Percent\ Yield = \left(\frac{Actual\ Yield}{Theoretical\ Yield} \right) \times 100$$

Lovely, but what is an actual yield, and what is a theoretical yield? An *actual yield* is, well, the amount of product actually produced by the reaction in a lab or as told to you in the chemistry problem. A *theoretical yield* is the amount of product that could've been produced had everything gone perfectly, as described by theory if every single atom of reactants worked together perfectly. The theoretical yield is what you calculate when you do a calculation on paper (like in this chapter) or before you do a reaction in a lab.

REMEMBER

The actual yield will always be less than the theoretical yield because no chemical reaction ever reaches 100 percent completion. In a lab setting, there's always some amount of error, whether it's big or small.

EXAMPLE

Q. Calculate the percent yield of sodium sulfate in the following scenario: 32.18 g of sulfuric acid reacts with excess sodium hydroxide to produce 37.91 g of sodium sulfate.

$$H_2SO_4(aq) + 2NaOH(aq) \rightarrow 2H_2O(l) + Na_2SO_4(aq)$$

A. 81.37% is the percent yield. The question clearly notes that sodium hydroxide is the excess reagent. (*Tip:* You always can ignore a reactant if the problem says it's in excess. That's like a big this-one-isn't-important sign in the problem.) So sulfuric acid is the limiting reagent and is the reagent you should use to calculate the theoretical yield:

$$\left(32.18\ g\ H_2SO_4 \right) \left(\frac{1\ mol\ H_2SO_4}{98.08\ g\ H_2SO_4} \right) \left(\frac{1\ mol\ Na_2SO_4}{1\ mol\ H_2SO_4} \right) \left(\frac{142.04\ g\ Na_2SO_4}{1\ mol\ Na_2SO_4} \right) = 46.60\ g\ Na_2SO_4$$

Theory predicts that 46.60 g of sodium sulfate product is possible if the reaction proceeds perfectly and to completion. But the question states that the actual yield is only 37.91 g of sodium sulfate. With these two pieces of information, you can calculate the percent yield using the formula given at the start of this section:

$$Percent\ Yield = \left(\frac{37.91\ g}{46.60\ g} \right) \times 100 = 81.37\%$$

Sulfur dioxide reacts with liquid water to produce sulfurous acid:

$$SO_2(g) + H_2O(l) \rightarrow H_2SO_3(aq)$$

(a) What is the percent yield if 19.07 g of sulfur dioxide reacts with excess water to produce 21.61 g of sulfurous acid?

(b) When 8.11 g of water reacts with excess sulfur dioxide, 27.59 g of sulfurous acid is produced. What is the percent yield?

 Liquid hydrazine is a component of some rocket fuels. Hydrazine combusts to produce nitrogen gas and water:

$$N_2H_4(l) + O_2(g) \rightarrow N_2(g) + 2H_2O(g)$$

(a) If the percent yield of a combustion reaction in the presence of 23.4 g of N_2H_4 (and excess oxygen) is 98%, how many liters of nitrogen (at STP) are produced?

(b) What is the percent yield if 84.8 g of N_2H_4 reacts with 54.7 g of oxygen gas to produce 51.33 g of water?

Practice Questions Answers and Explanations

1 In the following equations, keep in mind that the number you use for moles is the coefficient for the compound taken from the balanced chemical reaction:

a. **78.4 mol H_2**

$$\left(78.4 \text{ mol } H_2O\right)\left(\frac{2 \text{ mol } H_2}{2 \text{ mol } H_2O}\right) = 78.4 \text{ mol } H_2$$

b. **905 mol H_2O**

$$\left(905 \text{ mol } H_2\right)\left(\frac{2 \text{ mol } H_2O}{2 \text{ mol } H_2}\right) = 905 \text{ mol } H_2O$$

c. **42.3 mol O_2**

$$\left(84.6 \text{ mol } H_2\right)\left(\frac{1 \text{ mol } O_2}{2 \text{ mol } H_2}\right) = 42.3 \text{ mol } O_2$$

2 Before attempting to do any calculations, you must balance the equation as shown in part (a) so that you can use the coefficients to do your conversions (see Chapter 8 for info on balancing equations).

a. The balanced equation is

$$2Al(s) + 3CuSO_4(aq) \rightarrow Al_2(SO_4)_3(aq) + 3Cu(s)$$

b. **6.920 mol Al**

$$\left(10.38 \text{ mol } CuSO_4\right)\left(\frac{2 \text{ mol } Al}{3 \text{ mol } CuSO_4}\right) = 6.920 \text{ mol } Al$$

c. **2.08 mol Cu**

$$\left(2.08 \text{ mol } CuSO_4\right)\left(\frac{3 \text{ mol } Cu}{3 \text{ mol } CuSO_4}\right) = 2.08 \text{ mol } Cu$$

d. **2.9 mol $CuSO_4$**

$$\left(0.96 \text{ mol } Al_2(SO_4)_3\right)\left(\frac{3 \text{ mol } CuSO_4}{1 \text{ mol } Al_2(SO_4)_3}\right) = 2.9 \text{ mol } CuSO_4$$

e. **13.34 mol Al**

$$\left(20.01 \text{ mol } Cu\right)\left(\frac{2 \text{ mol } Al}{3 \text{ mol } Cu}\right) = 13.34 \text{ mol } Al$$

③ Balance the equation first. Then go on to the calculations for the remaining questions.

a. The balanced equation is

$$2Fe(s) + 3S(s) \rightarrow Fe_2S_3(s)$$

b. **9.3 mol S**

$$(6.2 \text{ mol Fe})\left(\frac{3 \text{ mol S}}{2 \text{ mol Fe}}\right) = 9.3 \text{ mol S}$$

c. **5.30 mol Fe₂S₃**

$$(10.6 \text{ mol Fe})\left(\frac{1 \text{ mol Fe}_2S_3}{2 \text{ mol Fe}}\right) = 5.30 \text{ mol Fe}_2S_3$$

d. **1.2 mol Fe₂S₃**

$$(3.5 \text{ mol S})\left(\frac{1 \text{ mol Fe}_2S_3}{3 \text{ mol S}}\right) = 1.2 \text{ mol Fe}_2S_3$$

④ As always, begin by balancing the equation.

a. The balanced equation is

$$2C_2H_6(g) + 7O_2 \rightarrow 4CO_2(g) + 6H_2O(g)$$

b. **30.8 mol CO₂**

$$(15.4 \text{ mol C}_2H_6)\left(\frac{4 \text{ mol CO}_2}{2 \text{ mol C}_2H_6}\right) = 30.8 \text{ mol CO}_2$$

c. **97.7 mol C₂H₆**

$$(293 \text{ mol H}_2O)\left(\frac{2 \text{ mol C}_2H_6}{6 \text{ mol H}_2O}\right) = 97.7 \text{ mol C}_2H_6$$

d. **0.623 mol O₂**

$$(0.178 \text{ mol C}_2H_6)\left(\frac{7 \text{ mol O}_2}{2 \text{ mol C}_2H_6}\right) = 0.623 \text{ mol O}_2$$

⑤ Without a balanced equation (see Chapter 8), nothing else is possible.

a. The balanced equation is

$$2H_2O_2(l) \rightarrow 2H_2O(l) + O_2(g)$$

b. **87.9 g H₂O**

$$(2.94 \times 10^{24} \text{ molecules H}_2O_2)\left(\frac{1 \text{ mol H}_2O_2}{6.022 \times 10^{23} \text{ molecules H}_2O_2}\right)$$

$$\left(\frac{2 \text{ mol H}_2O}{2 \text{ mol H}_2O_2}\right)\left(\frac{18.02 \text{ g H}_2O}{1 \text{ mol H}_2O}\right) = 87.9 \text{ g H}_2O$$

c. **10.8 L O$_2$**

$$(32.9 \text{ g H}_2\text{O}_2)\left(\frac{1 \text{ mol H}_2\text{O}_2}{34.02 \text{ g H}_2\text{O}_2}\right)\left(\frac{1 \text{ mol O}_2}{2 \text{ mol H}_2\text{O}_2}\right)\left(\frac{22.4 \text{ L O}_2}{1 \text{ mol O}_2}\right) = 10.8 \text{ L O}_2$$

6 The conversion factors you need to do the calculations require a balanced equation, so do the balancing first.

a. The balanced equation is

$$N_2O_3(g) + H_2O(l) \rightarrow 2HNO_2(aq)$$

b. **8.81 L N$_2$O$_3$**

$$(36.98 \text{ g HNO}_2)\left(\frac{1 \text{ mol HNO}_2}{47.02 \text{ g HNO}_2}\right)\left(\frac{1 \text{ mol N}_2\text{O}_3}{2 \text{ mol HNO}_2}\right)\left(\frac{22.4 \text{ L N}_2\text{O}_3}{1 \text{ mol N}_2\text{O}_3}\right) = 8.81 \text{ L N}_2\text{O}_3$$

c. **4.65 × 10^{23} molecules H$_2$O**

$$(17.3 \text{ L N}_2\text{O}_3)\left(\frac{1 \text{ mol N}_2\text{O}_3}{22.4 \text{ L N}_2\text{O}_3}\right)\left(\frac{1 \text{ mol H}_2\text{O}}{1 \text{ mol N}_2\text{O}_3}\right)\left(\frac{6.022 \times 10^{23} \text{ molecules H}_2\text{O}}{1 \text{ mol H}_2\text{O}}\right)$$

$$= 4.65 \times 10^{23} \text{ molecules H}_2\text{O}$$

7 In this problem, the provided chemical equation is already balanced, so you can proceed directly to the calculations:

a. **4.03 L Cl$_2$**

$$(50.0 \text{ g PbCl}_2)\left(\frac{1 \text{ mol PbCl}_2}{278.10 \text{ g PbCl}_2}\right)\left(\frac{1 \text{ mol Cl}_2}{1 \text{ mol PbCl}_2}\right)\left(\frac{22.4 \text{ L Cl}_2}{1 \text{ mol Cl}_2}\right) = 4.03 \text{ L Cl}_2$$

b. **2.97 × 10^{22} formula units PbCl$_4$**

$$(13.71 \text{ g PbCl}_2)\left(\frac{1 \text{ mol PbCl}_2}{278.10 \text{ g PbCl}_2}\right)\left(\frac{1 \text{ mol PbCl}_4}{1 \text{ mol PbCl}_2}\right)\left(\frac{6.022 \times 10^{23} \text{ formula units PbCl}_4}{1 \text{ mol PbCl}_4}\right)$$

$$= 2.97 \times 10^{22} \text{ formula units PbCl}_4$$

c. **67.6 g PbCl$_2$**

$$(84.8 \text{ g PbCl}_4)\left(\frac{1 \text{ mol PbCl}_4}{349.00 \text{ g PbCl}_4}\right)\left(\frac{1 \text{ mol PbCl}_2}{1 \text{ mol PbCl}_4}\right)\left(\frac{278.10 \text{ g PbCl}_2}{1 \text{ mol PbCl}_2}\right) = 67.6 \text{ g PbCl}_2$$

8 Begin limiting-reagent problems by determining which reactant is the limiting reagent. The answers to other questions build on that foundation. To find the limiting reagent, simply pick one of the reactants as a candidate. How much of your candidate reactant do you have? Use that information to calculate how much of any other reactants you'd need for a complete reaction. You can deduce the limiting and excess reagents from the results of these calculations.

a. **Iron(III) oxide.** In this problem, we chose iron(III) oxide as the initial candidate limiting reagent:

$$(50 \text{ g Fe}_2\text{O}_3)\left(\frac{1 \text{ mol Fe}_2\text{O}_3}{159.70 \text{ mol Fe}_2\text{O}_3}\right)\left(\frac{3 \text{ mol CO}}{1 \text{ mol Fe}_2\text{O}_3}\right)\left(\frac{28.01 \text{ g CO}}{1 \text{ mol CO}}\right) = 26 \text{ g CO}$$

Iron(III) oxide turns out to be the correct choice, because more carbon monoxide is initially present than is required to react with all the iron(III) oxide. You need 26 g of CO for every 50 g of Fe_2O_3, but you're given 67 g of CO and only 50 g of Fe_2O_3. This means you'll run out of Fe_2O_3 long before you'll run out of CO.

b. **Carbon monoxide is the excess reagent, and 41 g of it will remain after the reaction is completed.** Because iron(III) oxide is the limiting reagent, the excess reagent is carbon monoxide. To find the amount of excess reagent that will remain after the reaction reaches completion, first calculate how much of the excess reagent will be consumed. This calculation is identical to the one performed in part (a), so 26 g of carbon monoxide will be consumed. Next, subtract the quantity consumed from the amount originally present to obtain the amount of carbon monoxide that will remain: $67 \text{ g} - 26 \text{ g} = 41 \text{ g}$.

c. **35 g of iron; 41 g of carbon dioxide.** To answer this question, do two calculations, each starting with the assumption that all of the limiting reagent is consumed:

$$\left(50 \text{ g Fe}_2\text{O}_3\right)\left(\frac{1 \text{ mol Fe}_2\text{O}_3}{159.70 \text{ mol Fe}_2\text{O}_3}\right)\left(\frac{2 \text{ mol Fe}}{1 \text{ mol Fe}_2\text{O}_3}\right)\left(\frac{55.85 \text{ g Fe}}{1 \text{ mol Fe}}\right) = 35 \text{ g Fe}$$

$$\left(50 \text{ g Fe}_2\text{O}_3\right)\left(\frac{1 \text{ mol Fe}_2\text{O}_3}{159.70 \text{ mol Fe}_2\text{O}_3}\right)\left(\frac{3 \text{ mol CO}_2}{1 \text{ mol Fe}_2\text{O}_3}\right)\left(\frac{44.01 \text{ g CO}_2}{1 \text{ mol CO}_2}\right) = 41 \text{ g CO}_2$$

9 To find the limiting reagent, simply pick one of the reactants as a candidate. Calculate how much of the other reagents you'd need to completely react with all of your available candidate reagent. Deduce the limiting and excess reagents from these calculations.

a. **Sodium.** In this example, we chose sodium as the initial candidate limiting reagent:

$$\left(25 \text{ g Na}\right)\left(\frac{1 \text{ mol Na}}{22.99 \text{ g Na}}\right)\left(\frac{2 \text{ mol H}_2\text{O}}{2 \text{ mol Na}}\right)\left(\frac{18.02 \text{ g H}_2\text{O}}{1 \text{ mol H}_2\text{O}}\right) = 20 \text{ g H}_2\text{O}$$

Sodium turns out to be the correct choice, because more water is initially present than is required to react with all the sodium. It takes 25 g of sodium to react with 20 g of H_2O. You're given only 25 g of sodium and 40.2 g of water, so the sodium will limit the reaction.

b. **Water is the excess reagent, and 20.2 g of it will remain after the reaction is completed.** Sodium is the limiting reagent, so the excess reagent is water. The calculation in part (a) reveals how much water is consumed in a complete reaction: 20 g. Because 40.2 g of water is initially present, 20.2 g of water will remain after the reaction: $40.2 \text{ g} - 20 \text{ g} = 20.2 \text{ g}$.

c. **43 g of sodium hydroxide; 12 L of hydrogen gas.** To answer this question, do two calculations, each starting with the assumption that all of the limiting reagent is consumed:

$$\left(25 \text{ g Na}\right)\left(\frac{1 \text{ mol Na}}{22.99 \text{ g Na}}\right)\left(\frac{2 \text{ mol NaOH}}{2 \text{ mol Na}}\right)\left(\frac{40.00 \text{ g NaOH}}{1 \text{ mol NaOH}}\right) = 43 \text{ g NaOH}$$

$$\left(25 \text{ g Na}\right)\left(\frac{1 \text{ mol Na}}{22.99 \text{ g Na}}\right)\left(\frac{1 \text{ mol H}_2}{2 \text{ mol Na}}\right)\left(\frac{22.4 \text{ L H}_2}{1 \text{ mol H}_2}\right) = 12 \text{ L H}_2$$

(10) The first step is to identify the limiting reagent. Simply pick one of the reactants as a trial candidate for the limiting reagent, and calculate how much of the other reagents are required to react completely with the candidate.

a. **Chlorine gas.** In this example, we chose aluminum as the candidate limiting reagent:

$$(29.3 \text{ g Al})\left(\frac{1 \text{ mol Al}}{26.98 \text{ g Al}}\right)\left(\frac{3 \text{ mol Cl}_2}{2 \text{ mol Al}}\right)\left(\frac{22.4 \text{ L Cl}_2}{1 \text{ mol Cl}_2}\right) = 36.5 \text{ L Cl}_2$$

For every 29.3 g of Al, you need 36.5 L of Cl_2 gas. More chlorine gas is required (36.5 L) to completely react with the available aluminum than is available (34.6 L), so chlorine gas is the limiting reagent.

b. **Aluminum is the excess reagent, and 1.5 g of it will remain after the reaction is completed.** To calculate how much excess reagent (aluminum, in this case) will remain after a complete reaction, first calculate how much will be consumed:

$$(34.6 \text{ L Cl}_2)\left(\frac{1 \text{ mol Cl}_2}{22.4 \text{ L Cl}_2}\right)\left(\frac{2 \text{ mol Al}}{3 \text{ mol Cl}_2}\right)\left(\frac{26.98 \text{ g Al}}{1 \text{ mol Al}}\right) = 27.8 \text{ g Al}$$

Subtract that quantity from the amount of aluminum originally present to calculate the remaining amount: $29.3 \text{ g} - 27.8 \text{ g} = 1.5 \text{ g}$.

c. **137 g of aluminum chloride.** This calculation starts with the assumption that all of the limiting reagent, chlorine gas, is consumed.

$$(34.6 \text{ L Cl}_2)\left(\frac{1 \text{ mol Cl}_2}{22.4 \text{ L Cl}_2}\right)\left(\frac{2 \text{ mol AlCl}_3}{3 \text{ mol Cl}_2}\right)\left(\frac{133.33 \text{ g AlCl}_3}{1 \text{ mol AlCl}_3}\right) = 137 \text{ g AlCl}_3$$

(11) The reaction equation is already balanced, so you can proceed with the calculations.

a. **88.46% is the percent yield.** You're given an actual yield (21.61 g of sulfurous acid) and asked to calculate the percent yield. To do so, you must calculate the theoretical yield. The question clearly shows that sulfur dioxide is the limiting reagent, so begin the calculation of theoretical yield by assuming that all the sulfur dioxide is consumed:

$$(19.07 \text{ g SO}_2)\left(\frac{1 \text{ mol SO}_2}{64.07 \text{ g SO}_2}\right)\left(\frac{1 \text{ mol H}_2\text{SO}_3}{1 \text{ mol SO}_2}\right)\left(\frac{82.07 \text{ g H}_2\text{SO}_3}{1 \text{ mol H}_2\text{SO}_3}\right) = 24.43 \text{ g H}_2\text{SO}_3$$

So the theoretical yield of sulfurous acid is 24.43 g. With this information, you can calculate the percent yield:

$$\text{Percent Yield} = \left(\frac{21.61 \text{ g}}{24.43 \text{ g}}\right) \times 100 = 88.46\%$$

b. **74.6% is the percent yield.** You're asked to calculate a percent yield, having been given an actual yield (27.59 g of sulfurous acid). In this case, the limiting reagent is water, so begin the calculation by assuming that all the water is consumed:

$$(8.11 \text{ g H}_2\text{O})\left(\frac{1 \text{ mol H}_2\text{O}}{18.02 \text{ g H}_2\text{O}}\right)\left(\frac{1 \text{ mol H}_2\text{SO}_3}{1 \text{ mol H}_2\text{O}}\right)\left(\frac{82.07 \text{ g H}_2\text{SO}_3}{1 \text{ mol H}_2\text{SO}_3}\right) = 37.0 \text{ g H}_2\text{SO}_3$$

So the theoretical yield of sulfurous acid is 37.0 g. Using that information, calculate the percent yield:

$$\text{Percent Yield} = \left(\frac{27.59\text{ g}}{37.0\text{ g}} \right) \times 100 = 74.6\%$$

(12) With a balanced equation at your disposal, you can calculate with impunity.

a. **16.1 L of nitrogen is produced.** In this question, you're given the percent yield and asked to calculate an actual yield. To do so, you must know the theoretical yield. The question clearly indicates that hydrazine is the limiting reagent, so calculate the theoretical yield of nitrogen by assuming that all the hydrazine is consumed:

$$\left(23.4\text{ g N}_2\text{H}_4 \right) \left(\frac{1\text{ mol N}_2\text{H}_4}{32.05\text{ g N}_2\text{H}_4} \right) \left(\frac{1\text{ mol N}_2}{1\text{ mol N}_2\text{H}_4} \right) \left(\frac{22.4\text{ L N}_2}{1\text{ mol N}_2} \right) = 16.4\text{ L N}_2$$

So the theoretical yield of nitrogen is 16.4 L. Algebraically rearrange the percent yield equation to solve for the actual yield:

$$\text{Actual Yield} = \frac{(\text{Percent Yield})(\text{Theoretical Yield})}{100}$$

Substitute for your known values and solve:

$$\text{Actual Yield} = \frac{(98)(16.4\text{ L})}{100} = 16.1\text{ L}$$

b. **83.3% is the percent yield.** In this question, you're given the initial amounts of the reactants and an actual yield. You're asked to calculate a percent yield. To do so, you need to know the theoretical yield. The added wrinkle in this problem is that you don't initially know which reagent is limiting. As with any limiting reagent problem, pick a candidate limiting reagent. In this example, we chose hydrazine as the initial candidate:

$$\left(84.8\text{ g N}_2\text{H}_4 \right) \left(\frac{1\text{ mol N}_2\text{H}_4}{32.05\text{ g N}_2\text{H}_4} \right) \left(\frac{1\text{ mol O}_2}{1\text{ mol N}_2\text{H}_4} \right) \left(\frac{32.00\text{g O}_2}{1\text{ mol O}_2} \right) = 84.7\text{ g O}_2$$

More oxygen is required to react with the available hydrazine than is initially present, so oxygen is the limiting reagent. Knowing this, calculate a theoretical yield of water by assuming that all of the oxygen is consumed:

$$\left(54.7\text{ g O}_2 \right) \left(\frac{1\text{ mol O}_2}{32.00\text{ g O}_2} \right) \left(\frac{2\text{ mol H}_2\text{O}}{1\text{ mol O}_2} \right) \left(\frac{18.02\text{ g H}_2\text{O}}{1\text{ mol H}_2\text{O}} \right) = 61.6\text{ g H}_2\text{O}$$

So the theoretical yield of water is 61.5 g. Use this information to calculate the percent yield:

$$\text{Percent Yield} = \left(\frac{51.33\text{ g}}{61.6\text{ g}} \right) \times 100 = 83.3\%$$

If you're ready to test your skills a bit more, take the following chapter quiz that incorporates all the chapter topics.

Whaddya Know? Chapter 14 Quiz

Ready for a quiz? The 15 questions in this section will test the skills you learned in this chapter. When you're done, check out the section that follows for answers and explanations.

Use the following equation for problems 1 and 2:

$$2KClO_3(s) \rightarrow 2KCl(s) + 3O_2(g)$$

1 How many moles of oxygen gas are produced when 34 moles of $KClO_3$ decomposes completely?

2 The above reaction produces 4.2 moles of potassium chloride when run to completion. How many moles of potassium chlorate did you begin with?

Use the following equation for problems 3–4:

$$C_3H_8(g) + 5O_2(g) \rightarrow 3CO_2(g) + 4H_2O(g)$$

3 78.2 grams of propane combusts with excess oxygen present. What mass, in grams, of carbon dioxide is produced?

4 If a combustion reaction involving propane produces 22 grams of carbon dioxide, how many grams of oxygen did the reaction use?

Use the following equation for problems 5–6:

$$Cu(s) + 2AgNO_3(aq) \rightarrow Cu(NO_3)_2(aq) + 2Ag(s)$$

5 In the above reaction, how many grams of copper are needed to produce 17 kilograms of silver?

6 15 grams of copper is put into a solution of silver nitrate. How many grams of silver are produced by this reaction?

Use the following equation for problems 7–9. Assume this reaction happens at STP:

$$2SO_2(g) + O_2(g) \rightarrow 2SO_3(g)$$

7 39 liters of sulfur dioxide gas reacts with excess oxygen gas. How many grams of sulfur trioxide gas are produced?

8 8.3×10^{23} molecules of oxygen react with excess sulfur dioxide. What is the volume of the resulting sulfur trioxide product?

9 If 56 grams of sulfur trioxide gas is produced in the above reaction, how many molecules of sulfur dioxide did you begin with?

Use the following equation for problems 7–9. Assume this reaction happens at STP:

$$Ca(OH)_2(aq) + CO_2(g) \rightarrow CaCO_3(s) + H_2O(l)$$

10 45 dm³ of carbon dioxide gas reacts with excess calcium hydroxide. How many grams of water are produced from this reaction?

11 34 grams of calcium carbonate are produced when how many moles of calcium hydroxide react with excess carbon dioxide?

Use the following reaction for problems 12–13:

$$2Be(s) + O_2 \rightarrow 2BeO(s)$$

12 You react 45 grams of solid beryllium with 30 grams of oxygen gas. Identify the limiting reactant in this reaction and determine how many grams of beryllium oxide are produced.

13 200 dm³ of oxygen gas at STP reacts with 6.5×10^{23} atoms of beryllium. Identify the limiting reactant in this reaction and determine how many grams of beryllium oxide are produced.

Use the following reaction for problems 14–15:

$$Ca + 2HCl \rightarrow CaCl_2 + H_2$$

14 You run the above reaction to completion in your chemistry lab. The amount of calcium chloride produced from the reaction is 5.6 grams. The theoretical maximum you could have produced from the reaction was 7.2 grams. What was your percentage yield?

15 A chemist conducts the reaction above with 70 g of HCl and excess calcium in a lab. Their reaction produces 101.5 g of $CaCl_2$. What is their percent yield?

Answers to Chapter 14 Quiz

(1) **51 mol O_2.** To solve this problem you are simply going to be using the mole ratio given in the equation. The numbers used for the mole value of oxygen and potassium chlorate are taken directly from the coefficients in the balanced chemical equation. The worked-out solutions is here:

$$\left(34 \text{ mol KClO}_3\right)\left(\frac{3 \text{ mol O}_2}{2 \text{ mol KClO}_3}\right) = 51 \text{ mol O}_2$$

(2) **4.2 mol KCl.** To solve this problem you will again use the mole ratio and the coefficients from the balanced chemical equation given in the problem. Make sure you pay attention to the names of the formulas, however. In this case you are only given the names and not the formulas. Do not mix up potassium chloride, KCl, with potassium chlorate, $KClO_3$, which is an easy and common thing to do. Make sure you pay attention to the small details like that. Solution here:

$$\left(4.2 \text{ mol KCl}\right)\left(\frac{2 \text{ mol KClO}_3}{2 \text{ mol KCl}}\right) = 4.2 \text{ mol KCl}$$

(3) **234.07 g CO_2.** This is a very straightforward mass to mass conversion problem. You begin with the initial mass of propane and convert that to a mass of carbon dioxide. You will use the molar mass of propane and carbon dioxide along with the mole ratio shown in the coefficients. Solution here:

$$\left(78.2 \text{ g C}_3\text{H}_8\right)\left(\frac{1 \text{ mol C}_3\text{H}_8}{44.1 \text{ g C}_3\text{H}_8}\right)\left(\frac{3 \text{ mol CO}_2}{1 \text{ mol C}_3\text{H}_8}\right)\left(\frac{44.01 \text{ g CO}_2}{1 \text{ mol CO}_2}\right) = 234.07 \text{ g CO}_2$$

(4) **26.66 g O_2.** This problem is another mass to mass conversion problem. In this case you are given a mass of one of the products, CO_2, and told to convert back to a starting quantity of a reactant, O_2. That doesn't change anything about the problem for you. Just convert from one mass to another using the molar masses and the mole ratio, as shown here:

$$\left(22 \text{ g CO}_2\right)\left(\frac{1 \text{ mol CO}_2}{44.01 \text{ g CO}_2}\right)\left(\frac{5 \text{ mol O}_2}{3 \text{ mol CO}_2}\right)\left(\frac{32 \text{ g O}_2}{1 \text{ mol O}_2}\right) = 26.66 \text{ g O}_2$$

(5) **5007.65 g Cu.** This is another mass to mass practice problem but it has a slight wrinkle. The initial starting quantity of matter is given to you kilograms instead of grams. Make sure that you realize you need to convert those 17 kilograms to grams before doing the mass to mass conversion. You can do the conversion on its own or you can add it into the mass to mass problem and solve it all on one line. For the purposes of this solution the conversion is separate so that it is easier to see:

$$\left(17 \text{ kg Ag}\right)\left(\frac{1000 \text{ g Ag}}{1 \text{ kg Ag}}\right) = 17000 \text{ g Ag}$$

You then take this gram value and convert it to grams of copper:

$$\left(17000 \text{ g Ag}\right)\left(\frac{1 \text{ mol Ag}}{107.87 \text{ g Ag}}\right)\left(\frac{1 \text{ mol Cu}}{2 \text{ mol Ag}}\right)\left(\frac{63.55 \text{ g Cu}}{1 \text{ mol Cu}}\right) = 5007.65 \text{ g Cu}$$

(6) **50.92 g Ag.** This is a straightforward mass to mass problem. Just make sure you ignore the mention of silver nitrate in the wording. When you see something like that included in a problem, it is there to simply have the reaction make sense in the description but it doesn't play a role in actually solving the question. The only way that chemical would matter is if a quantity is included for it. Then you'd need to find the limiting reactant, which you'll be doing later on in this quiz:

$$(15 \text{ g Cu})\left(\frac{1 \text{ mol Cu}}{63.55 \text{ g Cu}}\right)\left(\frac{2 \text{ mol Ag}}{1 \text{ mol Cu}}\right)\left(\frac{107.87 \text{ g Ag}}{1 \text{ mol Cu}}\right) = 50.92 \text{ g Ag}$$

(7) **139.41 g SO_3.** This problem begins with a volume of sulfur dioxide and asks you to convert that volume to the mass of sulfur trioxide. It also mentions that you have excess oxygen gas present in the reaction. Remember that if you see something mentioned as "excess" or "sufficient" or anything that sounds like that in a problem, you can safely ignore it. Your initial value is given to you in liters, so you will be using the molar volume constant since this reaction is happening at STP. You then use the mole ratio to convert, and finally you multiply by the molar mass of sulfur trioxide, as shown here:

$$(39 \text{ L SO}_2)\left(\frac{1 \text{ mol SO}_2}{22.4 \text{ L SO}_2}\right)\left(\frac{2 \text{ mol SO}_3}{2 \text{ mol SO}_2}\right)\left(\frac{80.07 \text{ g SO}_3}{1 \text{ mol SO}_3}\right) = 139.41 \text{ g SO}_3$$

(8) **61.77 L SO_3.** To solve this problem you will need to convert from the number of molecules of oxygen to the volume of the sulfur trioxide gas produced. You begin by dividing by Avogadro's number. You then use the molar ratio found in the equation to convert from moles of oxygen to moles of sulfur trioxide gas. Finally, you multiply by the molar volume constant to convert to volume of sulfur trioxide. Remember, you can ignore the mention of sulfur dioxide as it is in excess:

$$(8.3 \times 10^{23} \text{ molecules O}_2)\left(\frac{1 \text{ mol O}_2}{6.02 \times 10^{23} \text{ molecules O}_2}\right)\left(\frac{2 \text{ mol SO}_3}{1 \text{ mol O}_2}\right)\left(\frac{22.4 \text{ L SO}_3}{1 \text{ mol SO}_3}\right) = 61.77 \text{ L SO}_3$$

(9) **4.2×10^{23} molecules SO_2.** To solve this problem you are converting from an initial mass to a number of molecules. You will begin by converting the mass of sulfur trioxide gas given to moles by dividing by the molar mass. You then convert from sulfur trioxide to sulfur dioxide using the mole ratio from the equation. Finally, you multiply by Avogadro's number to determine the number of molecules:

$$(56 \text{ g SO}_3)\left(\frac{1 \text{ mol SO}_3}{80.07 \text{ g SO}_3}\right)\left(\frac{2 \text{ mol SO}_2}{2 \text{ mol SO}_3}\right)\left(\frac{6.02 \times 10^{23} \text{ molecules SO}_2}{1 \text{ mol SO}_2}\right) = 4.2 \times 10^{23} \text{ molecules SO}_2$$

(10) **36.2 g H_2O.** You begin with a volume of carbon dioxide at STP and are converting it to a mass of water. Calcium hydroxide can be ignored because it is excess. Divide your volume of CO_2 by the molar volume constant then multiply your answer by the mole ratio of carbon dioxide to water found in the equation. Finally, multiply by the molar mass of water:

$$(45 \text{ dm}^3 \text{ CO}_2)\left(\frac{1 \text{ mol CO}_2}{22.4 \text{ dm}^3 \text{ CO}_2}\right)\left(\frac{1 \text{ mol H}_2\text{O}}{1 \text{ mol CO}_2}\right)\left(\frac{18.02 \text{ g H}_2\text{O}}{1 \text{ mol H}_2\text{O}}\right) = 36.2 \text{ g H}_2\text{O}$$

(11) **0.34 mol Ca(OH)₂.** This is a tricky question and it is the kind of thing you are going to see every once in a while in chemistry. In most problems that you find about stoichiometry you are usually going to convert from a mass/volume/quantity to another mass/volume/quantity or a mol to mol only conversion. In this case, however, you are starting with a mass of calcium carbonate and are converting it to moles of calcium hydroxide. You are not converting it to mass or anything else; you just stop at moles. To solve this you take your initial mass and divide by the molar mass of calcium carbonate and then you multiply by the mole ratio and that is it:

$$\left(34 \text{ g CaCO}_3\right)\left(\frac{1 \text{ mol CaCO}_3}{100.09 \text{ g CaCO}_3}\right)\left(\frac{1 \text{ mol Ca(OH)}_2}{1 \text{ mol CaCO}_3}\right) = 0.34 \text{ g Ca(OH)}_2$$

(12) **124.91 g BeO.** Limiting reactant is oxygen. This question gives you a starting mass of both reactants and asks you to determine how much product is produced. Because of those 2 starting quantities, you should recognize this is a limiting reactant problem. To solve it correctly you must identify which reactant is limiting and calculate the amount of product produced based on the starting quantity of the limiting reactant. If you were to use the nonlimiting reactant (the excess reactant) your answer would be too large. The simplest way to go about solving problems like this is just to perform a calculation using both starting quantities to the desired product and see which answer is smaller. The smaller answer is correct and also means that reactant is the limiting reactant:

$$\left(56 \text{ g Be}\right)\left(\frac{1 \text{ mol Be}}{9.01 \text{ g Be}}\right)\left(\frac{2 \text{ mol BeO}}{2 \text{ mol Be}}\right)\left(\frac{25.01 \text{ g BeO}}{1 \text{ mol BeO}}\right) = 124.91 \text{ g BeO}$$

$$\left(30 \text{ g O}_2\right)\left(\frac{1 \text{ mol O}_2}{32 \text{ g O}_2}\right)\left(\frac{2 \text{ mol BeO}}{1 \text{ mol O}_2}\right)\left(\frac{25.01 \text{ g BeO}}{1 \text{ mol BeO}}\right) = 46.89 \text{ g BeO}$$

You can see that 46.98 g is smaller quantity so that is the correct answer, and it means that oxygen is the limiting reactant.

(13) **27.00 g BeO.** Limiting reactant is Be. This is another limiting reactant problem. In this case you are given two starting quantities of matter, a volume of oxygen and a number of atoms of beryllium and asked to determine the grams of beryllium oxide that can be produced. To solve this you need to convert both of these starting quantities to grams of beryllium oxide to see which produces less. The lesser amount is the correct number and indicates the limiting reactant in the problem:

$$\left(200 \text{ dm}^3 \text{ O}_2\right)\left(\frac{1 \text{ mol O}_2}{22.4 \text{ dm}^3 \text{ O}_2}\right)\left(\frac{2 \text{ mol BeO}}{1 \text{ mol O}_2}\right)\left(\frac{25.01 \text{ g BeO}}{1 \text{ mol BeO}}\right) = 446.61 \text{ g BeO}$$

$$\left(6.5 \times 10^{23} \text{ atoms Be}\right)\left(\frac{1 \text{ mol Be}}{6.5 \times 10^{23} \text{ atoms Be}}\right)\left(\frac{2 \text{ mol BeO}}{2 \text{ mol Be}}\right)\left(\frac{25.01 \text{ g BeO}}{1 \text{ mol BeO}}\right) = 27.00 \text{ g BeO}$$

27 g is the lesser of the two possible values so it is the correct answer. It also indicates that beryllium is the limiting reactant in this problem.

(14) **77.78% yield.** You are told that a reaction is run to completion and the amount of product produced is 5.6 grams. The problem doesn't explicitly say that is the actual amount of your substance produced but the fact that it came from a reaction in a lab can be used to help you understand it is your actual amount needed for the percent yield formula. It also identifies the theoretical maximum you can have from this reaction as 7.2 grams. Using these two pieces of information, you can determine the percent yield like this:

$$\frac{\text{actual value}}{\text{theoretical value}} = \left(\frac{5.6\ \text{g}}{7.2\ \text{g}}\right)(100) = 77.78\%$$

(15) **95.3%.** This problem gives you the initial starting quantity of one of the reactants and the amount of product produced by the reaction in the lab. The value given to you in the problem of 101.5 g $CaCl_2$ represents the actual yield of the reaction. To determine the percent yield of the reaction you will need to calculate the theoretical yield of $CaCl_2$ using the initial starting quantity of 70g HCl, as shown here:

$$(70\ \text{g HCl})\left(\frac{1\ \text{mol HCl}}{36.46\ \text{g HCl}}\right)\left(\frac{1\ \text{mol CaCl}_2}{2\ \text{mol HCl}}\right)\left(\frac{110.98\ \text{g CaCl}_2}{1\ \text{mol CaCl}_2}\right) = 106.5\ \text{g CaCl}_2$$

Based on the above problem the theoretical yield of the reaction given a starting quantity of 70 g of HCl is 106.5 $CaCl_2$. You can now compare this to the actual yield of 101.5 $CaCl_2$ given in the problem using the percent yield formula.

$$\frac{\text{actual yield}}{\text{theoretical yield}} = \frac{101.5\ \text{g}}{106.5\ \text{g}} \times 100 = 95.3\%$$

So the percent yield of this reaction was 95.3%.

5

Examining Changes in Energy

In This Unit . . .

IN THIS CHAPTER

» **Looking at states of matter**

» **Moving between phases**

» **Kinetic and potential energy**

Chapter **15**

Understanding States of Matter in Terms of Energy

Walk into a room and turn on the light. Look around — what do you see? You may see a table, some chairs, a lamp, a computer humming away. But really all you see is matter and energy. There are many kinds of matter and many kinds of energy, but when all is said and done, you're left with these two things — matter and energy. Scientists used to believe that these two were separate and distinct, but now they realize that they're linked.

When asked, children often report that solids, liquids, and gases are composed of different kinds of matter. This assumption is understandable, given the striking differences in the properties of these three states. Nevertheless, for a given type of matter at a given pressure, the fundamental difference between a solid, a liquid, and a gas actually is the *amount of energy* within the particles of matter. Understanding the states of matter (phases) in terms of energy and pressure helps to explain the different properties of those states and how matter moves between the states. We explain what you need to know in this chapter.

Changing States of Matter

First remember the three main states of matter that were discussed in Chapter 3: solid, liquid, and gas. There is a distinct difference in energy between those three states of matter for any given substance and when the internal energy of the substance changes, the state of matter the substance is currently in can change as well. When a substance goes from one state of matter to another, the process is a *change of state*. Some rather interesting things, which we describe in the following sections, occur during this process.

Melting point

Imagine taking a big chunk of ice out of your freezer and putting it into a large pot on your stove. If you measure the temperature of that chunk of ice, you may find it to be −5°C or so. If you take temperature readings while heating the ice, you find that the temperature of the ice begins to rise as the heat from the stove causes the ice particles to begin vibrating faster and faster in their crystal lattice. After a while, some of the particles move so fast that they break free of the lattice, and the crystal lattice (which keeps a solid solid) eventually breaks apart. The solid begins to go from a solid state to a liquid state — a process called *melting*. The temperature at which melting occurs is called the *melting point (mp)* of the substance. The melting point for ice is 32°F, or 0°C.

If you watch the temperature of ice as it melts, you see that the temperature remains steady at 0°C until all the ice has melted. During this change of state *(phase change)*, the temperature remains constant; the energy that you're adding goes to work breaking the lattice, freeing the particles as a liquid. The liquid contains more energy than the ice (because the particles in liquids move faster than the particles in solids), even if both the ice and liquid water are at the same temperature. This is true at all phase changes — during the phase change itself, the temperature remains constant.

Boiling point

If you heat a pot of cool water (or if you continue to heat the pot of now-melted ice mentioned in the preceding section), the temperature of the water rises and the particles move faster and faster as they absorb the heat. The temperature rises until the water reaches the next change of state — boiling. As the particles move faster and faster as they heat up, they begin to break the attractive forces between each other and move freely as steam — a gas. The process by which a substance moves from the liquid state to the gaseous state is called *boiling*. The temperature at which a liquid begins to boil is called the *boiling point (bp)*. The bp is dependent on atmospheric pressure, but for water at sea level, it's 212°F, or 100°C. The temperature of the boiling water remains constant until all the water has been converted to steam.

You can have both water and steam at 100°C. They have the same temperature, but the steam has a lot more energy (because the particles move independently and pretty quickly). Because steam has more energy, steam burns are normally a lot more serious than boiling water burns — loads more energy is transferred to your skin.

You can summarize the process of water changing from a solid to a liquid in this way:

ice → water → steam

Because the basic particle in ice, water, and steam is the water molecule (written as H_2O), the same process can also be shown as

$$H_2O(s) \rightarrow H_2O(l) \rightarrow H_2O(g)$$

Here the (s) stands for solid, the (l) stands for liquid, and the (g) stands for gas. This second depiction is much better, because unlike H_2O, most chemical substances don't have different names for the solid, liquid, and gas forms.

Freezing point

If you cool a gaseous substance, you can watch the phase changes that occur. The phase changes are

» **Condensation:** Going from a gas to a liquid

» **Freezing:** Going from a liquid to a solid

The gas particles have a high amount of energy, but as they're cooled, that energy is reduced. The attractive forces then have a chance to draw the particles closer together, forming a liquid. This process is called *condensation*. The particles are now in clumps (as is characteristic of particles in a liquid state), but as more energy is removed by cooling, the particles start to align themselves, and a solid is formed. This is known as *freezing*. The temperature at which this occurs is called the *freezing point (fp)* of the substance.

REMEMBER

The freezing point is the same as the melting point — it's the temperature at which the liquid is able to become a solid or the solid becomes a liquid. In the same fashion, the condensation temperature is the same as the boiling point.

You can represent water changing states from a gas to a solid like this:

$$H_2O(g) \rightarrow H_2O(l) \rightarrow H_2O(s)$$

Sublimate this!

Most substances go through the logical progression from solid to liquid to gas as they're heated — or vice versa as they're cooled. But a few substances go directly from the solid to the gaseous state without ever becoming a liquid. Scientists call this process *sublimation*. Dry ice — solid carbon dioxide, written as $CO_2(s)$ — is the classic example of sublimation. (Mothballs and certain solid air fresheners also go through the process of sublimation.) You can see dry ice particles becoming smaller as the solid begins to turn into a gas, but no liquid is formed during this phase change. (If you've seen dry ice, then you remember that a white cloud usually accompanies it — magicians and theater productions often use dry ice for a cloudy or foggy effect. The white cloud you normally see isn't the carbon dioxide gas — the gas itself is colorless. The white cloud is the condensation of the water vapor in the air due to the cold of the dry ice.)

The process of sublimation is represented as

$$CO_2(s) \rightarrow CO_2(g)$$

The reverse of sublimation is *deposition* — going directly from a gaseous state to a solid state. One method of purifying solid iodine is to heat it gently. It sublimates and then it will undergo deposition on a cooler surface. Many times that cooler surface is a piece of laboratory apparatus that resembles a large test tube full of ice. It's commonly called a *cold finger*.

EXAMPLE

Q. Describe how the motion of atoms changes as a substance changes phase from a solid to a liquid.

A. As a phase change occurs, particle motion increases considerably. In a solid, the motion of particles is minimal, though there is still particle motion. That motion is vibrational only and does not result in the particles moving from their place in the structure in any significant way. As the solid changes to a liquid, particles begin to have more freedom of movement. Particles in a liquid can move by each other and slide over one another.

YOUR TURN

 Describe how the spacing of atoms changes as a substance changes from solid to liquid to gas.

Figuring Out Phase Diagrams

A *phase diagram* is a graph representing the relationship among all the states of matter of a specific substance. The most common type of phase diagram relates the states to temperature and pressure. The pressure is the vertical axis, and the temperature is the horizontal axis. Different substances have different ranges of temperature and/or pressure. The lines may be longer or shorter, and the angles may vary, but the solid (*s*), liquid (*l*), and gas (*g*) regions always have the same basic relationship to each other. The phase diagram allows you to predict which state of matter exists at a certain temperature and pressure combination. Figure 15-1 shows a general form of a phase diagram.

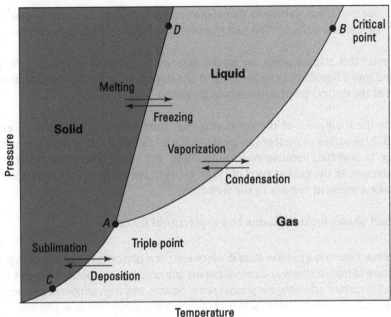

FIGURE 15-1:
A general
phase
diagram.

Knowing how to read the phase diagram for a particular substance is really useful. With a phase diagram, you can predict the type of phase change a substance will undergo if the temperature is changed at constant pressure or the pressure is changed at constant temperature.

Notice that the phase diagram has the following three general areas corresponding to the three states of matter (solid, liquid, and gas). The lines separating each area correspond to phase changes:

>> **A to C:** This line represents the relationship between the vapor pressure of the solid and the temperature for *sublimation* (going directly from a solid to a gas without first becoming a liquid). Crossing the solid/gas line from left to right is sublimation, and the reverse process is *deposition*. The heat of sublimation or deposition is applicable for changes in this region.

>> **A to B:** This line represents the relationship between the vapor pressure of a liquid and the temperature. Crossing the liquid/gas line in the diagram from left to right represents vaporization, and crossing from right to left represents condensation. Crossing this line involves the heat of vaporization or the heat of condensation.

>> **A to D:** This line represents the relationship between the melting point and the pressure. Crossing the solid/liquid line in the diagram from left to right is melting (fusion), and crossing from right to left is freezing (solidification). The heat of fusion or solidification is related to these types of changes.

The B point shown in this phase diagram is the *critical point* of the substance, the point beyond which the liquid and gas phases are indistinguishable from each other. Both liquids and gases contain randomly arranged particles; the real difference between liquids and gases is that in a liquid, the particles are in contact, while in a gas, the particles are widely separated. Increasing the pressure on a gas pushes its particles closer together. As the pressure continues to increase, the particles eventually come into contact, and there is no longer any difference between the gas and a liquid. The two phases become identical in appearance.

At or beyond this critical point, no matter how much pressure is applied, the gas cannot be condensed into a liquid. The temperature at the critical point is the *critical temperature*, and the pressure at the critical point is the *critical pressure*.

Point A is the *triple point* of the substance, the combination of temperature and pressure at which all three states of matter can exist. At the triple point, all phases are present. Any type of change is possible, because melting, boiling, and sublimation (and their reverses) occur simultaneously. In the case of good old H_2O, going to the triple point would produce boiling ice water. Take a moment to bask in the weirdness.

Other weird phases include plasma and supercritical fluids:

>> **Plasma:** Plasma is a gas-like state in which electrons pop off gaseous atoms to produce a mixture of free electrons and *cations* (atoms or molecules with positive charge). For most types of matter, achieving the plasma state requires very high temperatures, very low pressures, or both. Matter at the surface of the sun, for example, exists as plasma.

>> **Supercritical fluids:** Supercritical fluids exist under high-temperature, high-pressure conditions. At temperatures and pressures higher than the critical point, the phase boundary between liquid and gas disappears, and the matter exists as a kind of liquidy gas or gassy liquid. Supercritical fluids can diffuse through solids like gases do but can also dissolve things like liquids do.

TIP

Phase diagrams are useful tools for describing the states of a given type of matter across different temperatures and pressures. In Figure 15-2, you see phase diagrams for H_2O and CO_2. As described earlier, you can see how this diagram shows you each compound's phase of matter in any temperature and pressure conditions.

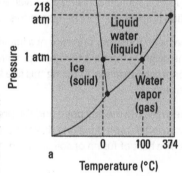
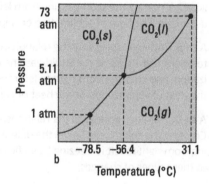

FIGURE 15-2: The phase diagrams for water, H_2O, and carbon dioxide, CO_2.

Q. Ethanol (C_2H_6O) has a freezing point of −114°C. 1-propanol (C_3H_8O) has a melting point of −88°C. At 25°C (where both compounds are liquids), which one is likely to have the higher vapor pressure, and why?

EXAMPLE

A. Ethanol has the higher vapor pressure at 25°C. First, notice that a freezing point and a melting point are the same thing — that point is the temperature at which a substance undergoes the liquid-to-solid or solid-to-liquid phase transition. Next, compare the freezing/melting points of ethanol and propanol. Much colder temperatures must be achieved to freeze ethanol than to freeze propanol. This suggests that ethanol molecules have fewer attractive forces among themselves than propanol molecules do. At 25°C, both compounds are in liquid phase. In pure liquids whose particles have less intermolecular (between-molecule) attraction, the vapor pressure is higher because the molecules at the surface of the liquid can more easily escape into vapor (gas) phase.

YOUR TURN

2 A cup of water is put into a freezer and cools to the solid phase within an hour. The water remains at that temperature for six months. After six months, the cup is retrieved from the freezer. The cup is empty. What happened?

3 Look at the phase diagram for carbon dioxide (CO_2) in Figure 15-2b. If you put carbon dioxide under a pressure of 4.5 atm at a temperature of 23°C, in what phase of matter would the carbon dioxide be? What are the triple point and the critical point of carbon dioxide according to the phase diagram?

Taking a Look at Energy and Temperature

Matter is one of two components of the universe. Energy is the other. *Energy* is the ability to do work. Energy can take several forms, such as heat energy, light energy, electrical energy, and mechanical energy. But two general categories of energy are especially important to chemists: kinetic energy and potential energy.

Moving right along: Kinetic energy

Kinetic energy is energy of motion. A baseball flying through the air toward a batter has a large amount of kinetic energy (as we're sure anyone who's ever been hit with a baseball will agree). Collisions between moving particles and the subsequent transfer of kinetic energy cause chemical reactions to occur. Chemists sometimes study moving particles, especially gases, because understanding the kinetic energy of these particles helps determine whether a particular reaction may take place.

The kinetic energy of moving particles can be transferred from one particle to another. Have you ever shot pool? You transfer kinetic energy from your moving pool stick to the cue ball to (hopefully) the ball you're aiming at.

Kinetic energy can be converted into other types of energy. In a hydroelectric dam, the kinetic energy of the falling water is converted into electrical energy. In fact, a scientific law — *the law of conservation of energy* — states that in ordinary chemical reactions (or physical processes), energy is neither created nor destroyed but can be converted from one form to another. (This law doesn't hold in nuclear reactions, though.)

Sitting pretty: Potential energy

Suppose you throw a football up into a tree, where it gets stuck. You gave that ball kinetic energy — energy of motion — when you threw it. But where's that energy now? It's been converted into the other major category of energy: potential energy.

Potential energy is stored energy. Objects may have potential energy stored in terms of their position. That football up in the tree has potential energy due to its height. If the ball were to fall, that potential energy would be converted to kinetic energy. (Watch out!)

Potential energy due to position isn't the only type of potential energy. In fact, chemists really aren't all that interested in potential energy due to position. Chemists are far more interested in the energy stored (potential energy) in *chemical bonds*, which are the forces that hold atoms together in compounds.

It takes a lot of energy to run a human body. What if you couldn't store the energy you extract from food? You'd have to eat all the time just to keep your body going. But humans can store energy in terms of chemical bonds. And then later, when we need that energy, our bodies can break those bonds and release it.

The same is true of the fuels we commonly use to heat our homes and run our automobiles. Energy is stored in these fuels — gasoline, for example — and is released when chemical reactions take place.

Measuring Energy

Measuring kinetic energy is fairly easy. You can do it with a relatively simple instrument: a thermometer. Meanwhile, measuring potential energy can be a difficult task. The potential energy of a ball stuck up in a tree is related to the mass of the ball and its height above the ground. The potential energy contained in chemical bonds is related to the type of bond and the number of bonds that can potentially break. Because measuring potential energy is a bit advanced for someone taking a chemistry class, we devote the following sections to measuring kinetic energy.

Taking a look at temperature

When you measure, say, the air temperature in your backyard, you're really measuring the average kinetic energy (the energy of motion) of the gas particles there. The faster those particles are moving, the higher the temperature is.

All the particles aren't moving at the same speed. Some are going very fast, and some are going relatively slow, but most are moving at a speed between the two extremes. The temperature reading from your thermometer is related to the *average* kinetic energy of the particles.

You probably use the Fahrenheit scale to measure temperatures, but most chemists and other scientists use either the Celsius (°C) or Kelvin (K) temperature scale. (The degree symbol isn't used with K.) Figure 15-3 compares the three temperature scales using the freezing point and boiling point of water as reference points.

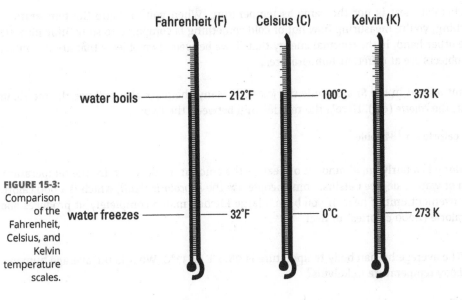

FIGURE 15-3: Comparison of the Fahrenheit, Celsius, and Kelvin temperature scales.

As you can see from Figure 15-3, water boils at 100°C (373 K) and freezes at 0°C (273 K). To get the Kelvin temperature, you take the Celsius temperature and add 273 (it's actually 273.15, but 273 works just fine in basic chemistry classes). Mathematically, the relationship looks like this:

$$K = °C + 273$$

You may want to know how to convert from Fahrenheit to Celsius and vice versa (because most people in the United States still think in degrees Fahrenheit). Here's the first equation you need:

$$°C = \frac{5}{9}(°F - 32)$$

When converting to degrees Celsius, be sure to subtract 32 from your Fahrenheit temperature before multiplying by $\frac{5}{9}$.

$$°F = \frac{9}{5}(°C) + 32$$

When converting to degrees Fahrenheit, be sure to multiply your Celsius temperature by $\frac{9}{5}$ and *then* add 32.

Most of the time, this book uses the Celsius scale. But when describing the behavior of gases, it uses the Kelvin scale.

Feeling the heat

To a chemist, heat is not the same as temperature. When you measure the temperature of something, you're measuring how hot or cold something is compared to something else. *Heat*, on the other hand, is the thermal energy that flows between two objects that are in contact if those objects are at different temperatures.

The unit of heat in the SI system is the *joule* (J). Many chemists also still use the metric unit of heat, the *calorie* (cal). Here's the relationship between the two:

1 calorie = 4.184 joules

The calorie is a fairly small amount of heat — the amount it takes to raise the temperature of 1 gram of water 1 degree Celsius. Some people use the *kilocalorie (kcal)*, which is 1,000 calories, as a convenient unit of heat. If you burn a large kitchen match completely, it produces about 1 kilocalorie (1,000 calories) of heat.

Q. The average human body temperature is 98.6°F or 37°C. What is the average human body temperature in kelvins?

EXAMPLE

A. Whenever you're presented with a temperature in degrees Celsius, you simply need to add 273 to it to get your answer in kelvins.

37° C + 273 = 310 K

So in the Kelvin scale, the human body has an average temperature of 310 K.

YOUR
TURN

④ Perform the following temperature conversions:

    a. 36°F to degrees Celsius

    b. 353 K to degrees Celsius

    c. 56°C to kelvins

    d. 76.00°C to degrees Fahrenheit

Practice Questions Answers and Explanations

(1) In a solid, particles are tightly packed together in a rigid and fixed shape. They have no real freedom of movement, and there is minimal free space between these particles. In a liquid, particles are spread out more and do not join in a rigid, fixed structure. Instead, liquids take the form of whatever container they're placed in. In a gas, particles are highly energetic and are moving very, very fast compared to the other phases. There's a large amount of space between particles. Gases have no definite shape or volume.

(2) **The frozen water sublimed, moving directly from a solid to a gaseous state.** Although this process occurs slowly at the temperatures and pressures found within normal household freezers, it does occur. Try it.

(3) **Carbon dioxide would be a gas. The triple point is 5.11 atm and −56.4°C. The critical point is at 73 atm and 31.1°C.** To determine the phase of the compound, plot the given temperature and pressure values on the phase diagram; if you look for where the two given measurements meet for carbon dioxide, you see they come together in the gas area of the phase diagram. To find the triple point, identify where the phase boundaries representing all three phases come together. Finally, to determine the critical point, identify where the phase boundary between liquid carbon dioxide and gaseous carbon dioxide goes away.

(4) When performing temperature conversions, make sure you know where you're starting from and where you're going. Identify which formula to use and get to work. One thing to keep in mind, though: You'll probably never convert directly between the Fahrenheit and Kelvin scales in a chemistry class but there is one problem in the practice quiz that asks you to do it just in case. When you approach this problem, try to think of it as two steps where you convert from Fahrenheit to Celsius and then from Celsius to Kelvin.

a. $\frac{5}{9}(36°F - 32) = \textbf{2.2°C}$

b. $353\ K - 273 = \textbf{80°C}$

c. $56°C + 273 = \textbf{329 K}$

d. $\frac{9}{5}(76.00°C) + 32 = \textbf{168.8°F}$

If you're ready to test your skills a bit more, take the following chapter quiz that incorporates all the chapter topics.

Whaddya Know? Chapter 15 Quiz

Ready for a quiz? The 10 questions in this section will test the skills you learned in this chapter. When you're done, check out the section that follows for answers and explanations.

1 Describe how the spacing of particles change as matter undergoes phase change from a gas to a liquid to a solid.

2 Name the phase change associated with the change of state listed here:
a. liquid to solid
b. solid to gas
c. gas to solid
d. gas to liquid

3 For questions 3–7 refer to this figure.

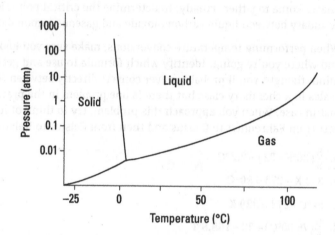

4 What phase of matter is Substance X in when it is at a pressure of 15 atm and a temperature of 15°C?

5 How would you describe the general pressure at which substance X is a gas?

6 What phase change occurs when you begin with substance X at a pressure of 1 atm and a temperature of –15°C and then raise the temperature to 50°C?

7 What phase change occurs if you have substance X at a pressure of 0.01 atm and a temperature of 63°C and then raise the pressure to 3 atm?

8 What phase of matter is substance X in at 0.01 atm and 75°C?

9 What does the critical point represent on a phase diagram?

10 What does the triple point represent on a phase diagram?

11 Perform the following temperature conversions:

a. 32°C to Kelvin

b. 75°F to Celsius

c. 315 K to °C

d. –15°C to Kelvin

e. 100°F to Kelvin

f. 15°C to °F

Answers to Chapter 15 Quiz

(1) As matter changes from a gas to a liquid to a solid it is losing energy. This loss of energy is what causes the phase change. In practical terms this usually means the temperature of the matter is being reduced. As the temperature of the matter goes down the motion of its particles slows. The particles become closer and closer together and far more organized as they go from a gas to a liquid to a solid. Gas particles are very spaced apart and not organized in anyway. Liquid particles are spaced somewhat apart and have a volume but they are able to flow over one another and are not rigidly organized. Solid particles have a definite order to them and the lowest amount of particle motion.

(2) Phase changes occur as matter moves from one state to another:

   a. freezing

   b. sublimation

   c. deposition

   d. condensation

(3) **Liquid.** To determine this, locate where pressure and temperature are on the phase diagram. Pressure is on the Y axis and temperature is on the X axis. On the y axis for pressure there is no specific notation for 15 atm but you can make a general estimate of where it is. Then once you know, this trace over to where you estimate a temperature of 15°C would be on the X axis. This will give you phase of matter that exists at these conditions, a liquid.

(4) In general, substance X appears to be a gas at **mostly lower pressure** until you begin to move up in temperature somewhat significantly. Most substances are a gas at low pressure just as substance X is shown to be on the phase diagram.

(5) **Melting.** To determine, this locate 1 atm on the phase diagram, which thankfully is labeled in this case, and then trace over from there to find the initial temperature of -15°C. At these conditions you can see the substance is a solid. Once you have done this, trace along the x axis from your initial point until you reach the new temperature of 50°C. You will see this new point is in the liquid phase. That means the phase change that occurred here was from a solid to a liquid, which is called melting.

(6) **Condensation.** To determine this, locate 63°C on the X axis. This is not specifically labeled but you should be able to make a good enough estimation based on the marking of the figure. Then trace up the Y axis until you hit 0.01 atm, your initial pressure. At these conditions substance X is a gas. Finally, increase the pressure to 3 atm by tracing up the y until you reach the general area of 3 atm. The new phase of matter here is a liquid. The phase change that occurred here was from a gas to a liquid, condensation.

(7) Substance X is a **gas** at 0.01 atm and 75°C.

(8) The critical point on a phase diagram represents the temperature at which the substance can only be a gas no matter how much you increase the pressure.

(9) The triple point on a phase diagram represents the conditions (temp and pressure) at which all 3 states of matter are in equilibrium with one another for the substance. This is the point where all of the phase change lines come together on the phase diagram. At this point you have a substance that is constantly undergoing phase change between solid, liquid, and gas.

(10) Pay careful attention to the units you are converting between. As a hint/reminder, remember that you can never have negative Kelvin, so if you get that as one of your answers you need to go and redo that problem. Finally, a small but minor point that is often overlooked by people is that Kelvin is just Kelvin; there is no degree sign (°) after it so make sure you aren't adding one in there!

a. $32°C + 273 = 305 K$

b. $\frac{5}{9}(75°F - 32) = 23.9°C$

c. $315 K - 273 = 42°C$

d. $-15°C + 273 = 258 K$

e. This requires you to convert from °F to Kelvin temperature. In this book you were not given a specific conversion for that, so instead of looking it up somewhere else (you can't do that on a test anyway), just break this into two steps. First convert from °F to °C, then convert from °C to Kelvin.

$$\frac{5}{9}(100°F - 32) = 37.78°C$$

$$37.78°C + 273 = 310.78 K$$

f. $\frac{9}{5}(15.00°C) + 32 = 59°F$

Chapter 16

Warming Up to Thermochemistry

Reactants form products, but something else goes on during a chemical reaction that you really should be interested in: energy changes. In this chapter, we focus on one specific type of energy change that takes place: the change in heat.

Many chemical reactions that you are most familiar with are *exothermic reactions*, reactions that release energy. The burning of natural gas to run the stove when you fry chicken, the burning of wood in a campfire to keep warm, and the combustion of gasoline in your automobile's engine are all example of exothermic reactions. In many cases, the energy produced is the reaction product that's desired.

In other reactions, energy is absorbed as the reaction takes place. These reactions are *endothermic*, and energy appears on the reactants' side at the start. Cooking is a great example of endothermic reactions. Crack a couple of eggs into a hot frying pan, and energy from the pan is absorbed into the eggs during the myriad chemical reactions called cooking.

Thermochemistry is the study of the energy changes that take place during a chemical reaction. In this chapter, we investigate the concepts associated with thermochemistry. We'll discover how to calculate the amount of heat produced during a specific chemical reaction and how much heat is needed to cause a desired reaction to take place. So turn on the air conditioning and grab a glass of cold tea — this chapter may heat things up a bit.

Understanding the Basics of Thermodynamics

To get a grasp on thermochemistry and the broader science of *thermodynamics*, we first need to explore some of the basics that go into the study of *heat* and *energy*.

Heat

Heat (normally represented as q) is the energy that flows when a system and surroundings are at different temperatures. The energy always flows from the area of higher temperature to the area of lower temperature. If q has a positive value ($+q$), the system has absorbed energy via heat transfer from the surroundings (an *endothermic* process). If q has a negative value ($-q$), the system has transferred heat to the surroundings (an *exothermic* process). So an exothermic process occurs when heat energy is given off from the system to its surroundings (this usually feels hot), and an endothermic process occurs when heat energy is taken in by the system (this usually feels cold).

The transferred heat is an *extensive property*, which means its amount depends on the amount of matter involved. For example, more fire means more heat, so you get more heat from a large fire than from a small fire, even if the large fire isn't hotter. *Temperature* (the average kinetic energy of matter) is an *intensive property*, a property that's not affected by the amount of matter involved. The temperature of your latte may be the same as the temperature of your bathwater, but much more energy (heat) was required to warm up the bathwater.

Energy

To study energy, it helps to divide the universe into two parts: the *system* and the *surroundings*. For chemists, the system may consist of the contents of a beaker or tube. This is an example of a *closed system*, one that allows exchange of energy with the surroundings but doesn't allow exchange of matter. Closed systems are common in chemistry. Though energy may move between system and surroundings, the total energy of the universe is constant.

Energy itself is divided into potential energy and kinetic energy:

- **Potential energy (*PE*):** Potential energy is energy due to position. *Chemical energy* is a kind of potential energy arising from the positions of particles within systems and from the energy stored within bonds.

- **Kinetic energy (*KE*):** Kinetic energy is the energy of motion. *Thermal energy* is a kind of kinetic energy arising from the movement of particles within systems.

The total *internal energy* of a system (E) is the sum of its potential and kinetic energies. When a system moves between two states (as it does in a chemical reaction), the internal energy may change as the system exchanges energy with the surroundings. The difference in energy (ΔE) between the initial and final states derives from heat (q) added to or lost from the system and from work (w) done by the system or on the system.

REMEMBER

We can summarize these energy explanations with the help of a couple of handy formulas:

$$E_{total} = KE + PE$$

$$\Delta E = E_{final} - E_{initial} = q + w$$

What kind of work can atoms and molecules do in a chemical reaction? One kind that's easy to understand is *pressure-volume work*. Consider the following reaction:

$$CaCO_3(s) \rightarrow CaO(s) + CO_2(g)$$

Solid calcium carbonate decomposes into solid calcium oxide and carbon dioxide gas. At constant pressure (P), this reaction proceeds with a change in volume (V). The added volume comes from the production of carbon dioxide gas. As gas is made, it expands, pushing against the surroundings. The carbon dioxide gas molecules do work as they push into a greater volume:

$$w = -P\Delta V$$

The negative sign in this equation means that the system loses internal energy because of the work it does on the surroundings. If the surroundings did work on the system, thereby decreasing the system's volume, then the system would gain internal energy.

So pressure-volume work can partly account for changes in internal energy during a reaction. When pressure-volume work is the only kind of work involved, any remaining changes come from heat flow. *Enthalpy (H)* corresponds to the heat content of a closed system at constant pressure. An *enthalpy change (ΔH)* in such a system corresponds to heat flow. The enthalpy change equals the change in internal energy minus the energy used to perform pressure-volume work:

$$\Delta H = \Delta E - (-P\Delta V) = \Delta E + P\Delta V$$

Like E, P, and V, H is a *state function*, meaning that its value has only to do with the state of the system and nothing to do with how the system got to that state. Heat (q) is *not* a state function; it's simply a form of energy that flows from warmer objects to cooler objects.

REMEMBER

Now breathe. The practical consequences of all this theory are the following:

>> Chemical reactions usually involve the flow of energy in the form of heat, *q*.

>> Chemists monitor changes in heat by measuring changes in temperature.

>> At constant pressure, the change in heat content equals the change in enthalpy, *ΔH*.

>> Knowing *ΔH* values helps you explain and predict chemical behavior.

Units of energy

The two units that you use in thermochemistry calculations are the *joule* and the *calorie*:

>> **Joule (J):** The joule is the SI unit of energy. It has the units of $kg \cdot m^2/s^2$. Remember that $1\ J = 1\ kg \cdot m^2/s^2$.

>> **Calorie (cal):** The calorie is the amount of energy needed to raise the temperature of 1 gram of water 1 degree Celsius. By definition, a calorie is exactly 4.184 J.

Although both units may be used, most older textbooks (and older teachers) tend to use calories, while newer texts tend to use joules. We tend to use both, just so you're familiar with both.

The *liter-atmosphere* (L·atm) is also used. Here's how the joule, the calorie, and the liter-atmosphere are related:

$$1 \text{ J} = 0.2390 \text{ cal}$$

$$101.3 \text{ J} = 1 \text{ L·atm}$$

We're sure you've heard the term *calories* before. Who hasn't talked about cutting back on calories or wondered how many calories that triple bacon cheeseburger contains? But those calories are different; they're nutritional Calories (notice the capital C). Nutritional Calories are really kilocalories, 1,000 calories. So that 300-Calorie candy bar really contains 300,000 calories of energy that must either be used up or, more likely, stored as fat. Now that's sticker shock! Food sellers count on 300 Calories sounding better than 300,000 calories.

TIP

You can usually tell from the context which type of calorie is being discussed. In general chemistry, calories are used much more often than nutritional Calories.

Engineers often use the British thermal unit (BTU). A BTU was originally defined as the amount of heat required to raise the temperature of 1 pound (0.454 kg) of liquid water by 1 degree Fahrenheit (0.556°C) at a constant pressure of 1 atmosphere. It's equivalent to 1,054 joules and is commonly used when referring to the power output of steam engines and air-conditioning units.

EXAMPLE

Q. Is the melting of ice an exothermic or endothermic process?

A. Endothermic process. To melt something, heat/energy must be absorbed. In this case, ice is the system. The system is absorbing energy from the surroundings, so the process is endothermic.

Q. Is the process of water vapor condensing on a cold can of soda pop exothermic or endothermic?

A. Exothermic. For something to condense from a gas to a liquid requires that substance to lose energy. The water vapor is the system and in this case the system is losing energy to the surroundings to undergo the phase change from gas to liquid. Losing energy makes the process exothermic.

Q. Gas is heated within a sealed cylinder. The heat causes the gas to expand, pushing a movable piston outward to increase the volume of the cylinder to 4.63 L. The initial and final pressures of the system are both 1.15 atm. The gas does 304 J of work on the piston. What was the initial volume of the cylinder? (*Note:* 101.3 J = 1 L·atm)

A. 2.02 L. You're given an amount of work, a constant pressure, and the knowledge that the volume of the system changes. So the equation to use here is $w = -P\Delta V = -P\left(V_{final} - V_{initial}\right)$. Because the system does work on the piston (and not the other way around), the sign of w is negative. Substituting your known values into the equation gives you

$$-304 \text{ J} = -(1.15 \text{ atm})(4.63 \text{ L} - V_{initial})$$

The units don't match, so convert joules to liter–atmospheres:

$$(304 \text{ J})\left(\frac{1 \text{ L} \cdot \text{atm}}{101.3 \text{ J}}\right) = 3.00 \text{ L} \cdot \text{atm}$$

Now you can solve for the initial volume:

$$-3.00 \text{ L} \cdot \text{atm} = (-1.15 \text{ atm})(4.63 \text{ L} - V_{initial})$$

$$\frac{-3.00 \text{ L} \cdot \text{atm}}{-1.15 \text{ atm}} = \frac{(-1.15 \text{ atm})(4.63 \text{ L} - V_{initial})}{-1.15 \text{ atm}}$$

$$2.61 \text{ L} = 4.63 \text{ L} - V_{initial}$$

$$V_{initial} = 2.02 \text{ L}$$

YOUR TURN

1 A fuel combusts at 3.00 atm constant pressure. The reaction releases 75.0 kJ of heat and causes the system to expand from 7.50 L to 20.0 L. What is the change in internal energy? (*Note:* 101.3 J = 1 L · atm)

2 Identify each process as an exothermic or endothermic process:

a. Evaporation

b. Sublimation

c. Freezing

Working with Specific Heat Capacity and Calorimetry

Heat is a form of energy that flows from warmer objects to cooler objects. But how much heat can an object hold? If objects have the same heat content, does that mean they're the same temperature? You can measure different temperatures, but how do these temperatures relate to heat flow? These kinds of questions revolve around the concept of heat capacity.

Specific heat capacity

REMEMBER

You'll encounter heat capacity in different forms, each of which is useful in different scenarios. Any system has a heat capacity. But how can you best compare heat capacities between chemical systems? You use *molar heat capacity* or *specific heat capacity* (or just *specific heat*). Molar heat capacity is simply the heat capacity of 1 mole of a substance. Specific heat capacity is simply the heat capacity of 1 gram of a substance. How do you know whether you're dealing with heat capacity, molar heat capacity, or specific heat capacity? Look at the units.

>> Heat capacity: $\dfrac{\text{Energy}}{\text{K}}$

>> Molar heat capacity: $\dfrac{\text{Energy}}{\text{mol} \cdot \text{K}}$

>> Specific heat capacity: $\dfrac{\text{Energy}}{\text{g} \cdot \text{K}}$

Specific heat capacity is the one that you will likely be dealing with the most in your chemistry class, so pay careful attention to it. For any given substance there is a number that represents the amount of energy, usually measured in joules, required to raise 1 g of that substance by 1°C, or 1 K. If you have a larger amount of a substance, raising the temperature will take longer. It is something you likely have some understanding of based on your everyday experience. To give you a practical and familiar example, imagine two pots of water, one big and one small, coming to a boil on a stove. The flame underneath each pot is the same, meaning the amount of energy being added to both is the same. Which one will boil faster? You probably already know the small one will boil faster and the big one will take longer. Why? Since there is more mass it takes longer for the water to heat to heat up.

The other major concept of specific heat is that different substances, of the same mass, will heat up and cool off at different rates based on what they are composed of. Another example you might have familiarity with is the idea of going swimming on one of the first hot days of the year. Let's say there is this lake you like to swim in. You know it is going to be the first really hot day of the year. You live where there is an actual winter, so it hasn't been very warm for a while; say the temperature yesterday was 50°F — not very warm. Yet today, because this happens in some places, the temperature is a nice 83°F and sunny. The air will feel quite warm to you and likely the ground, depending on the material, will feel quite hot in some places. Yet it is likely that if you were to go jump into your favorite lake, you would probably not have a good time because it would likely still feel very, very cold. This is because water has a relatively high specific heat compared to other substances, meaning that it generally it takes more energy to heat water up, though it also works the other way as well. By the end of the summer, that lake will likely be feeling pretty warm, and it'll stay warmer relative to the air temperature for quite some time because of water's high specific heat. If it takes more energy and time to heat water up, it also will hold its heat for a correspondingly long time before cooling down.

Specific heat values are known for many different substances, and they are something you can usually look up if you know the substance. Table 16-1 lists the specific heat capacities and molar heat capacities of some common substances. You can have variations on this where the unit of mass is a kilogram or the unit of energy is a kilojoule, but the concept remains the same.

TABLE 16-1 Specific and Molar Heat Capacities of Selected Substances at 25°C

Substance	Specific Heat Capacity (J/g° · C)	Molar Heat Capacity (J/mol° · C)
Al (solid)	0.90	24.3
C (diamond)	0.50	6.0
C (graphite)	0.72	8.6
Cu (solid)	0.39	24.5
Fe (solid)	0.44	24.8
H_2O (solid)	2.09	37.7
H_2O (liquid)	4.18	75.3
H_2O (gas)	2.03	36.4

You can see in the table that liquid water has a relatively high heat capacity. This capacity allows the oceans and other bodies of water to absorb large amounts of heat, helping to keep the Earth's temperature moderate. But the table also shows that water's three states of matter have different heat capacities. Be careful that you use the heat capacity value for the correct state.

The different allotropes of an element also have different heat capacities. (*Allotropes* are different forms of an element where the atoms of the element are bonded together in different ways.) Table 16-1 includes the heat capacities of graphite and diamond, two allotropic forms of carbon.

Calorimetry

In order to measure the energy change that takes place during a chemical process (the study of *calorimetry*), chemists use an instrument called a *calorimeter.* Chemists use two general types of calorimeters: constant-pressure and constant-volume:

>> **Constant-pressure calorimeter:** Constant-pressure calorimetry directly measures an enthalpy change (ΔH) for a reaction because it monitors heat flow at constant pressure: $\Delta H = q_p$.

Typically, heat flow is observed through changes in the temperature of a reaction solution. If a reaction warms a solution, then that reaction must have released heat into the solution. In other words, the change in heat content of the reaction ($q_{reaction}$) has the same magnitude as the change in heat content for the solution ($q_{solution}$) but has the opposite sign:

$q_{solution} = -q_{reaction}$.

Measuring $q_{solution}$ allows you to calculate $q_{reaction}$, but how can you measure $q_{solution}$? You do so by measuring the difference in temperature (ΔT) before and after the reaction:

REMEMBER

$q_{solution} = (\text{mass of solution})(\text{specific heat of solution})(\Delta T)$

In other words, $q = mC_p\Delta T$. Here, m is the mass of the solution and C_p is the specific heat capacity of the solution at constant pressure. ΔT is equal to $T_{final} - T_{initial}$.

When you use this equation, be sure that all your units match. For example, if your C_p has units of J/g·K, don't expect to calculate heat flow in kilocalories. A common source of error in solving specific-heat problems is the need to use the correct temperature units. Be sure to pay attention to whether your temperature is in kelvins or degrees Celsius.

You can make a constant-pressure calorimeter easily by stacking two Styrofoam coffee cups together. Using two stacked cups really insulates the system from the surroundings. In addition, you need to provide some way to stir the solution, and you need a thermometer to measure the temperature change. You can run the thermometer and stirrer through a lid, which helps provide additional insulation. Figure 16-1 illustrates how a coffee cup calorimeter may be constructed.

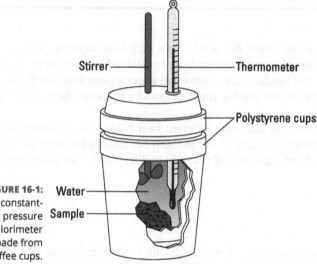

FIGURE 16-1:
A constant-pressure calorimeter made from coffee cups.

Stirrer — Thermometer

Polystyrene cups

Water

Sample

Styrofoam has a very low thermal conductivity, so hardly any of the heat of the reaction is absorbed by the cups. Therefore, the heat changes that occur are contained within the solution and recorded by the thermometer. Obviously, Styrofoam isn't a perfect insulator, but it is the one you'll most likely encounter in your chemistry class.

>> **Constant-volume (bomb) calorimeter:** Constant-volume calorimeters (commonly called *bomb calorimeters*) are used to measure energy changes that occur during combustion reactions. Figure 16-2 is a diagram of a typical bomb calorimeter. Though you probably won't encounter one of these during your lab sessions in school, knowing how they function can be important.

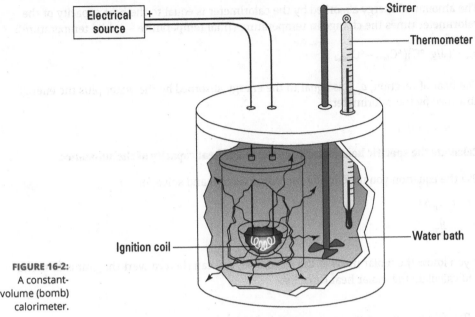

Stirrer

Thermometer

Water bath

Ignition coil

FIGURE 16-2:
A constant-volume (bomb) calorimeter.

To use a bomb calorimeter, follow these steps:

1. **Place a weighed sample of the substance to be tested into a cup and place it into a heavy-walled reaction vessel (called the bomb).**

    An electrical ignition apparatus is in contact with the sample.

2. **Evacuate the air in the bomb and replace it with oxygen.**

3. **Place the bomb in an insulated container filled with a known amount of water.**

4. **Insert a thermometer in the water to measure temperature changes while a stirrer circulates the water around the bomb.**

5. **After giving everything an opportunity to stabilize, ignite the sample electrically.**

    The sample burns in the oxygen gas. The energy given off by the combustion of the sample is absorbed by the water inside the calorimeter and by the calorimeter itself.

6. **Compare the temperature of the water before the combustion occurs to the temperature after combustion.**

    The heat absorbed by the water (q_{water}) is the specific heat capacity of the water (in J/g°C) times the mass of water (in grams) times the change in temperature (final temperature [°C$_{final}$] minus initial temperature [°C$_{initial}$]):

$$q_{water} = (J/g \cdot °C)(g)(°C_{final} - °C_{initial})$$

The amount of energy absorbed by the calorimeter is equal to the heat capacity of the calorimeter times the change in temperature (final temperature – initial temperature):

$$q_{cal} = (J/g \cdot °C)(°C_{final} - °C_{initial})$$

The heat of reaction, q_{rxn}, is equal to the energy absorbed by the water plus the energy absorbed by the calorimeter:

$$q_{rxn} = q_{water} + q_{cal}$$

7. **Calculate the specific heat capacity or the molar heat capacity of the substance.**

   Use the equation you used in the preceding section and solve for C_p:

   $$q = mC_p \Delta T$$

   $$C_p = \frac{q}{m\Delta T}$$

   If you know the molar mass of the substance, you can then convert the grams to moles and calculate the molar heat capacity.

EXAMPLE

Q. Paraffin wax is sometimes incorporated in to sheetrock to act as an insulator. During the day, the wax absorbs heat and melts. During the cool nights, the wax releases heat and solidifies. At sunrise, a small hunk of solid paraffin within a wall has a temperature of 298 K. The rising sun warms the wax, which has a melting temperature of 354 K. If the hunk of wax has a mass of 0.257 g and a specific heat capacity of 2.50 J/(g·K), how much heat must the wax absorb to bring it to its melting point?

A. The wax must absorb 36.0 J. You're given two temperatures, a mass, and a specific heat capacity. You're asked to find an amount of heat energy. You have all you need to proceed with $q = mC_p\Delta T$:

$$q = (0.257\,g)\left(2.50\frac{J}{g \cdot K}\right)(354\,K - 298\,K) = 36.0\,J$$

Q. How much heat does it take to raise the temperature of 50.0 g of solid iron from 20.0°C to 35.0°C?

A. It takes 330 J. Looking at Table 11–1, you see that the specific heat capacity of iron (Fe) is 0.44 J/g°·C. You also know the mass of the compound is 50.0 g, and you can determine the temperature change by subtracting the initial temperature from the final temperature. The question asks how much heat is produced, so you're solving for q. Substitute in the given values to get this answer:

$$q = mC_p\Delta T$$
$$= (50\,g)(0.44\,J/g \cdot °C)(35°C - 20°C)$$
$$= 330\,J$$

3 A 375 g plug of lead is heated and placed in an insulated container filled with 0.500 L of water. Prior to the immersion of the lead, the water is at 293 K. After a time, the lead and the water assume the same temperature, 297 K. The specific heat capacity of lead is 0.127 J/g·K, and the specific heat capacity of water is 4.18 J/g·K. How hot was the lead before it entered the water? (*Hint:* You'll need to use the density of water.)

4 At some point, all laboratory chemists learn the same hard lesson: Hot glass looks just like cold glass. Heath discovered this when he picked up a hot beaker someone left on his bench. At the moment Heath grasped the 413 K glass beaker, 567 J of heat flowed out of the beaker and into his hand. The glass of the beaker has a heat capacity of 0.84 J/g·K. If the beaker was 410 K the instant Heath dropped it, then what is the collective mass of the shards of beaker now littering the lab floor?

5 25.4 g of sodium hydroxide (NaOH) is dissolved in water within an insulated calorimeter. The heat capacity of the resulting solution is 4.18×10^3 J/K. The temperature of the water prior to the addition of NaOH was 296 K. If NaOH releases 44.2 kJ/mol as it dissolves, what is the final temperature of the solution?

Absorbing and Releasing Heat: Endothermic and Exothermic Reactions

You can monitor heat flow by measuring changes in temperature, but what does any of this have to do with chemistry? Chemical reactions transform both matter and energy. Though reaction equations usually list only the matter components of a reaction, you can also consider heat energy as a reactant or product. When chemists are interested in heat flow during a reaction (and when the reaction is run at constant pressure), they may list an enthalpy change (ΔH) to the right of the reaction equation. As we explain in the preceding section, at constant pressure, heat flow equals ΔH:

$$q_p = \Delta H = H_{final} - H_{initial}$$

REMEMBER

If the ΔH listed for a reaction is negative, then that reaction releases heat as it proceeds — the reaction is *exothermic* (*exo-* = out). If the ΔH listed for the reaction is positive, then that reaction absorbs heat as it proceeds — the reaction is *endothermic* (*endo-* = in). In other words, exothermic reactions release heat as a product, and endothermic reactions consume heat as a reactant.

TIP

The sign of the ΔH tells you the direction of heat flow, but what about the magnitude? The coefficients of a chemical reaction represent molar equivalents (see Chapter 8 for details), so the value listed for the ΔH refers to the enthalpy change for one molar equivalent of the reaction. Here's an example:

$$CH_4(g) + 2O_2(g) \rightarrow CO_2(g) + 2H_2O(g) \quad \Delta H = -802 \text{ kJ}$$

This reaction equation describes the combustion of methane, a reaction you might expect to release heat. The enthalpy change listed for the reaction confirms this expectation: For each mole of methane that combusts, 802 kJ of heat is released. The reaction is highly exothermic. Based on the stoichiometry of the equation, you can also say that 802 kJ of heat is released for every 2 mol of water produced. (Flip to Chapter 9 for the scoop on stoichiometry.)

So reaction enthalpy changes (or reaction "heats") are a useful way to measure or predict chemical change. But they're just as useful in dealing with physical changes, like freezing and melting, evaporating and condensing, and others. For example, water (like most substances) absorbs heat as it melts (or *fuses*) and as it evaporates. Here are the molar enthalpies for such changes:

>> **Molar enthalpy of fusion:** $\Delta H_{fus} = 6.01 \text{ kJ}$

>> **Molar enthalpy of vaporization:** $\Delta H_{vap} = 40.68 \text{ kJ}$

Be sure to pay attention to the states of the reactants and products. You also want to remember two other conventions that apply to thermochemical equations:

>> **Because the enthalpy change is an extensive property, if you use a multiplier on the equation, you use that same multiplier on the ΔH.** For example, if you want to write the

thermochemical equation for the production of 2 mol of water, you have to multiply the entire preceding equation, including the ΔH, by 2.

$$2\left[H_2(g)+\frac{1}{2}O_2(g)\rightarrow H_2O(g)\quad \Delta H=-241.8\text{ kJ}\right]$$
$$2H_2(g)+O_2(g)\rightarrow 2H_2O(g)\quad \Delta H=-483.6\text{ kJ}$$

» **If you reverse a thermochemical equation, you reverse the sign on the enthalpy change.** For example, if you want to write the thermochemical equation for the decomposition of a mole of water, simply reverse the initial equation for the formation of a mole of water and change the sign of the ΔH.

$$H_2O(g)\rightarrow H_2(g)+\frac{1}{2}O_2(g)\quad \Delta H=+241.8\text{ kJ}$$

This equation tells you that if you want to decompose a mole of water (about 18 grams), you must supply 241.8 kJ of energy.

Q. Here's a balanced chemical equation for the oxidation of hydrogen gas to form liquid water, along with the corresponding enthalpy change:

EXAMPLE

$$2H_2(g)+O_2(g)\rightarrow 2H_2O(l)\quad \Delta H=-572\text{ kJ}$$

How much electrical energy must be expended to perform electrolysis of 3.76 mol of liquid water, converting that water into hydrogen gas and oxygen gas?

A. 1.08×10^3 kJ must be expended. First, recognize that the given enthalpy change is for the reverse of the electrolysis reaction, so you must reverse its sign from -572 to 572. Second, recall that heats of reaction are proportional to the amount of substance reacting (2 mol of H_2O in this case), so the calculation is

$$\left(3.76\text{ mol }H_2O\right)\left(\frac{572\text{ kJ}}{2\text{ mol }H_2O}\right)=1.08\times10^3\text{ kJ}$$

---

6 Carbon dioxide gas can be decomposed into oxygen gas and carbon monoxide:

YOUR TURN

$$2CO_2(g)\rightarrow O_2(g)+2CO(g)\quad \Delta H=486\text{ kJ}$$

How much heat is released or absorbed when 9.67 g of carbon monoxide combines with oxygen to form carbon dioxide?

7 How much heat must be added to convert 4.77 mol of 268 K ice into steam? The specific heat capacity of ice is 38.1 J/mol·K. The specific heat capacity of water is 75.3 J/mol·K. The molar enthalpies of fusion and vaporization are 6.01 kJ and 40.68 kJ, respectively.

Summing Heats with Hess's Law

If you can find some related reactions and their heat changes (ΔH), you have the option of a method called *Hess's law calculations*. Enthalpy is a *state function*, which means the path taken during the reaction doesn't matter; all that matters are the initial and final states. The enthalpy change is the same whether the reaction occurs in one step or in a series of steps. The basis of Hess's law is that you can calculate the enthalpy change for a desired reaction by manipulating associated reactions until you reach the desired reaction. Hess's law states that if you can write a desired reaction as the sum of two or more known reactions, then the enthalpy change for the desired reaction is simply the sum of the enthalpy changes for the known reactions.

For the chemist, *Hess's law* is a valuable tool for dissecting heat flow in complicated, multistep reactions. For the confused or disgruntled chemistry student, Hess's law is a breath of fresh air. In essence, the law confirms that heat behaves the way we'd like it to behave: predictably.

Imagine that the product of one reaction serves as the reactant for another reaction. Now imagine that the product of the second reaction serves as the reactant for a third reaction. What you have is a set of coupled reactions, connected in series like the cars of a train:

$A \rightarrow B$ and $B \rightarrow C$ and $C \rightarrow D$

Therefore,

$A \rightarrow B \rightarrow C \rightarrow D$

You can think of these three reactions adding up to one big reaction, $A \rightarrow D$. What is the overall enthalpy change associated with this reaction ($\Delta H_{A \rightarrow D}$)? Here's the good news:

$$\Delta H_{A \rightarrow D} = \Delta H_{A \rightarrow B} + \Delta H_{B \rightarrow C} + \Delta H_{C \rightarrow D}$$

Enthalpy changes are additive. But the good news gets even better. Imagine that you're trying to figure out the total enthalpy change for the following multistep reaction:

$$X \rightarrow Y \rightarrow Z$$

Here's a wrinkle: For technical reasons, you can't measure this enthalpy change ($\Delta H_{X \rightarrow Z}$) directly but must calculate it from tabulated values for ($\Delta H_{X \rightarrow Y}$) and ($\Delta H_{Y \rightarrow Z}$). No problem, right? You simply look up the tabulated values and add them. But here's another wrinkle: When you look up the tabulated values, you find the following:

$$\Delta H_{X \rightarrow Y} = -37.5 \frac{kJ}{mol}$$

$$\Delta H_{Z \rightarrow Y} = -10.2 \frac{kJ}{mol}$$

Gasp! You need $\Delta H_{Y \rightarrow Z}$, but you're provided only $\Delta H_{Z \rightarrow Y}$! Relax. The enthalpy change for a reaction has the same magnitude and opposite sign as the reverse reaction. If you reverse a reaction when you're writing it out, you simply need to change a positive sign to a negative or vice versa. So if $\Delta H_{Z \rightarrow Y} = -10.2$ kJ/mol, then $\Delta H_{Y \rightarrow Z} = 10.2$ kJ/mol. It really is that simple:

$$\Delta H_{X \rightarrow Z} = \Delta H_{X \rightarrow Y} + \left(-\Delta H_{Z \rightarrow Y} \right)$$
$$= -37.5 \frac{kJ}{mol} + 10.2 \frac{kJ}{mol} = -27.3 \frac{kJ}{mol}$$

Q. Calculate the reaction enthalpy for the following reaction:

$$PCl_5(g) \rightarrow PCl_3(g) + Cl_2(g)$$

Use the following data:

Reaction1: $4PCl_3(g) \rightarrow P_4(s) + 6Cl_2(g)$ $\Delta H = 821$ kJ
Reaction2: $P_4(s) + 10Cl_2(g) \rightarrow 4PCl_5(g)$ $\Delta H = -1,156$ kJ

A. The reaction enthalpy is 83.8 kJ. Reaction enthalpies are given for two reactions. Your task is to manipulate and add Reactions 1 and 2 so the sum is equivalent to the target reaction. First, reverse Reactions 1 and 2 to obtain Reactions 1′ and 2′, and add the two reactions. Identical species that appear on opposite sides of the equations cancel out (as occurs with species P_4 and Cl_2):

Reaction1′: $P_4(s) + 6Cl_2(g) \rightarrow 4PCl_3(g)$ $\Delta H = -821$ kJ
Reaction2′: $4PCl_5(g) \rightarrow P_4(s) + 10Cl_2(g)$ $\Delta H = 1,156$ kJ
        Sum: $4PCl_5(g) \rightarrow 4PCl_3(g) + 4Cl_2(g)$ $\Delta H = 335$ kJ

Finally, divide the sum by 4 to yield the target reaction equation:

$$PCl_5(g) \rightarrow PCl_3(g) + Cl_2(g) \qquad \Delta H = 83.8 \text{ kJ}$$

YOUR
TURN

8 Calculate the reaction enthalpy for the following reaction:

$$N_2O_4(g) \rightarrow N_2(g) + 2O_2(g)$$

Use the following data:

Reaction 1: $N_2(g) + 2O_2(g) \rightarrow 2NO_2(g)$ $\Delta H = 1{,}032\,kJ$

Reaction 2: $2NO_2(g) \rightarrow N_2O_4(g)$ $\Delta H = -886\,kJ$

Practice Questions Answers and Explanations

(1) **78.8 kJ decrease.** You're given a constant pressure, an amount of heat, a change in volume, and the knowledge that there's been a change in internal energy. Changes in heat content at constant pressure are equivalent to a change in enthalpy, so the equation to use here is

$$\Delta H = \Delta E + P\Delta V$$

To do the math properly, you must make sure that all your units match, so convert the given heat energy from kJ to L·atm:

$$75.0 \text{ kJ} = 7.50 \times 10^4 \text{ J} = \left(7.50 \times 10^4 \text{ J}\right)\left(\frac{1 \text{ L·atm}}{101.3 \text{ J}}\right) = 7.40 \times 10^2 \text{ L·atm}$$

Because heat is released from the system, the change in enthalpy (ΔH) is negative. Substitute your known values into the equation:

$$-7.40 \times 10^2 \text{ L·atm} = \Delta E + (3.00 \text{ atm})(20.0 \text{L} - 7.50 \text{ L})$$
$$-7.40 \times 10^2 \text{ L·atm} = \Delta E + (3.00 \text{ atm})(12.5 \text{ L})$$
$$-7.40 \times 10^2 \text{ L·atm} = \Delta E + (37.5 \text{ L·atm})$$
$$-778 \text{ L·atm} = \Delta E$$

Solving for ΔE gives you -778 L·atm, which is equivalent to -78.8 kJ. Because the sign is negative, the internal energy of the system decreases by 78.8 kJ.

(2) Consider whether the system absorbs or loses heat.

    a. **Endothermic process.** Energy is being absorbed by the system, which in this case is the liquid evaporating.

    b. **Endothermic process.** Sublimation is one of the less well known phase changes; it involves going from a solid directly to a gas. This involves the system, the solid, absorbing a good bit of energy from the surroundings very quickly, making this an endothermic process.

    c. **Exothermic process.** Anything freezing loses heat to its surroundings, which makes it an exothermic process.

(3) **473 K.** The key to setting up this problem is to realize that whatever heat flows out of the lead flows into the water, so $q_{lead} = q_{water}$. Calculate each quantity of heat by using $q = mC_p\Delta T$. Recall that $\Delta T = T_{final} - T_{initial}$. The unknown in the problem is the initial temperature of lead:

$$q_{lead} = (375 \text{ g})\left(0.127\frac{J}{g \cdot K}\right)(297 \text{ K} - T_{initial})$$

To calculate q_{water}, you must first calculate the mass of 0.500 L water by using the density of water:

$$(0.500 \text{ L})\left(1.00\frac{kg}{L}\right) = 0.500 \text{ kg} = 500 \text{ g}$$

So q_{water} is

$$q_{water} = (500 \text{ g})\left(4.18\frac{J}{g \cdot K}\right)(297 \text{ K} - 293 \text{ K}) = 8.36 \times 10^3 \text{ J}$$

Setting q_{lead} equal to $-q_{water}$ and solving for $T_{initial}$ yields 473 K:

$$(375g)\left(0.127\frac{J}{g\cdot K}\right)(297K - T_{initial}) = -8.36 \times 10^3 \, J$$

$$297K - T_{initial} = -176K$$

$$T_{initial} = 473K$$

(4) $\mathbf{2.3 \times 10^2 \, g.}$ To solve this problem, apply $q = mC_p\Delta T$ to the beaker. The unknown is the mass of the beaker, m. Because heat flowed out of the beaker, the sign of q must be negative.

$$-576J = m\left(0.84\frac{J}{g\cdot K}\right)(410 \, K - 413 \, K)$$

$$\frac{-576J}{\left(0.84\frac{J}{g\cdot K}\right)(410 \, K - 413 \, K)} = m$$

$$2.3 \times 10^2 g = m$$

(5) $\mathbf{303 \, K.}$ Solve this problem in two parts. First, calculate the amount of heat released during dissolution of the NaOH by determining the number of moles of NaOH; then multiply that by the enthalpy change given for NaOH in the problem (44.2 kJ):

$$(25.4 \text{ g NaOH})\left(\frac{1 \text{ mol NaOH}}{40.0 \text{ g NaOH}}\right)\left(\frac{44.2 \text{ kJ}}{1 \text{ mol NaOH}}\right) = 28.1 \text{ kJ}$$

Because you're given the solution's heat capacity — not its molar heat capacity or its specific heat capacity — you can simply substitute the released heat, q, into the following equation:

$$q = (\text{heat capacity})(T_{final} - T_{initial})$$

Because the given heat capacity uses units of joules, you must convert the heat from kilojoules to joules before plugging in the value:

$$2.81 \times 10^4 J = \left(4.18 \times 10^3 \frac{J}{K}\right)(T_{final} - 296 \, K)$$

$$\frac{2.81 \times 10^4 J}{\left(4.18 \times 10^3 \frac{J}{K}\right)} = \frac{\left(4.18 \times 10^3 \frac{J}{K}\right)}{\left(4.18 \times 10^3 \frac{J}{K}\right)}(T_{final} - 296 \, K)$$

$$6.72 \, K = (T_{final} - 296 \, K)$$

$$6.72 \, K - (-296) = T_{final}$$

$$303 \, K = T_{final}$$

(6) $\mathbf{-83.9 \, kJ \text{ is released.}}$ Solve this problem with a chain of conversion factors. Convert from grams of CO to moles and then from moles to kilojoules. Be sure to adjust the sign of the enthalpy and incorporate the stoichiometry of the given reaction equation. To do so, divide the given mass of CO by the molar mass of CO (28.01 g). You then multiply by the given ΔH and divide by the coefficient of 2 for the CO in the equation.

$$(9.67 \text{ g CO})\left(\frac{1 \text{ mol CO}}{28.01 \text{ g CO}}\right)\left(\frac{-486 \text{ kJ}}{2 \text{ mol CO}}\right) = -83.9 \text{ kJ}$$

Because the sign is negative, 83.9 kJ is released, not absorbed.

(7) 2.60×10^5 J. The total heat required is the sum of several individual heats: heat to warm the ice to the melting point (273 K), heat to convert the ice to liquid water, heat to warm the liquid water to the boiling point, and heat to convert the liquid water to steam. Be careful to match your units (joules versus kilojoules). To perform each of the conversions, you need to multiply the given mole value by the corresponding molar enthalpies for each phase change. You then use the specific heat formula, $q = mC_p\Delta T$, to calculate the amount of energy needed to raise the temperature of the H_2O between the phase changes.

Warm ice to melting point: $q = (4.77\,\text{mol})\left(38.1\dfrac{\text{J}}{\text{mol}\cdot\text{K}}\right)(273\,\text{K} - 268\,\text{K}) = 909\,\text{J}$

Convert ice to liquid water: $q = (4.77\,\text{mol})\left(6.01\times10^3\dfrac{\text{J}}{\text{mol}}\right) = 2.87\times10^4\,\text{J}$

Warm water to boiling point: $q = (4.77\,\text{mol})\left(75.3\dfrac{\text{J}}{\text{mol}\cdot\text{K}}\right)(373\,\text{K} - 273\,\text{K}) = 3.59\times10^4\,\text{J}$

Convert liquid water to steam: $q = (4.77\,\text{mol})\left(4.068\times10^4\dfrac{\text{J}}{\text{mol}}\right) = 1.94\times10^5\,\text{J}$

$q_{total} = 909\,\text{J} + 2.87\times10^4\,\text{J} + 3.59\times10^4\,\text{J} + 1.94\times10^5\,\text{J} = 2.60\times10^5\,\text{J}$

The sum of all heats is 2.60×10^5 J (or 2.60×10^2 kJ).

(8) **–146 kJ.** Reverse Reaction 1 to get Reaction 1′ (–1,032 kJ). Reverse Reaction 2 to get Reaction 2′ (886 kJ). Add Reactions 1′ and 2′, yielding $\Delta H = -146\,\text{kJ}$.

Reaction 1: $2NO_2(g) \rightarrow N_2(g) + 2O_2(g)$	$\Delta H = -1{,}032\,\text{kJ}$
Reaction 2′: $N_2O_4(g) \rightarrow 2NO_2(g)$	$\Delta H = 886\,\text{kJ}$
Sum: $N_2O_4(g) \rightarrow N_2(g) + 2O_2(g)$	$\Delta H = -146\,\text{kJ}$

If you're ready to test your skills a bit more, take the following chapter quiz that incorporates all the chapter topics.

Whaddya Know? Chapter 16 Quiz

Ready for a quiz? The 10 questions in this section will test the skills you learned in this chapter. When you're done, check out the section that follows for answers and explanations.

1 You have a gas being heated in a closed and sealed cylinder. The heating of the gas causes the molecular motion to increase and the gas to expand. This cylinder has a piston in it that can move if pressure is applied by the expanding gas. This movement increases the volume of the cylinder to 18.2 L. The initial pressure of the system was 5.7 atm. After adding heat to the gas and raising the temperature, the final pressure of the system, after the piston moved, was the same. What was the initial volume of the cylinder? The amount of work done by the piston was 804 J.

2 A gas in an expandable container has a volume of 2.50 L undergoes a reaction that gives off energy. This heats the gas and causes the container to expand to a volume of 5.00L. The gas in the container is at a constant pressure of 10.0 atm. What is the amount of work done on the container in Joules?

3 What phase changes are endothermic?

4 What phase changes are exothermic?

5 How much energy, in J, is released when 650.6 g of iron metal is cooled from 456 °C to 22 °C? C_{Fe} = 0.444 J/(g*°C)

6 A 25.2 g sample of a material with a specific heat of 1.50 cal/(g*°C) experienced a 45°C increase in temperature. What was the sample's energy change in calories?

7 When 1,500 joules of energy are lost from a 0.12–kilogram object, the temperature decreases from 45°C to 40°C. What is the specific heat of this object, in J/(g*°C)?

8 A piece of aluminum absorbs 6700 joules of energy. This causes the temperature of the aluminum to rise from 32°C to 60°C. What is the mass of the aluminum in grams? The specific heat of aluminum is 0.89 J/(g*°C).

9 Using the following equation:

$$CH_4(g) + 2O_2(g) \rightarrow CO_2(g) + 2H_2O(g) \quad \Delta H = -802 \text{ kJ}$$

How much energy is released or absorbed when 10 moles of oxygen gas combine with methane (CH_4) gas to produce carbon dioxide and water?

10 Using the following equation:

$$H_2(g) + Cl_2(g) \rightarrow 2HCl(g) \quad \Delta H = -92 \text{ kJ}$$

How much energy is released or absorbed when 5 moles of gaseous hydrogen chloride decompose into hydrogen and chlorine?

11 Using the following equation:

$$C(g) + 2H_2(g) \rightarrow CH_4(g) \quad \Delta H = -75 \text{ kJ}$$

How much energy is released or absorbed when 4.2 moles of gaseous hydrogen combine with carbon to produce methane gas?

12 Calculate the reaction enthalpy for the following reaction:

$$N_2(g) + 2O_2(g) \rightarrow N_2O_4(g)$$

Use the following data:

Reaction 1: $N_2(g) + 2O_2(g) \rightarrow 2NO_2(g) \quad \Delta H = 1{,}032 \text{ kJ}$
Reaction 2: $N_2O_4(g) \rightarrow 2NO_2(g) \quad\quad\quad \Delta H = 886 \text{ kJ}$

13 Calculate the reaction enthalpy for the following reaction:

$$2H_2O(l) + 2F_2(g) \rightarrow O_2(g) + 4HF(g)$$

Use the following data:

Reaction 1: $2HF(g) \rightarrow H_2(g) + F_2(g) \quad\quad \Delta H = 542 \text{ kJ}$
Reaction 2: $2H_2O(l) \rightarrow 2H_2(g) + O_2(g) \quad \Delta H = 572 \text{ kJ}$

Use the chart to answer questions 14 and 15:

Element	Specific Heat (J/ g*°C)
Potassium	0.75
Sulphur	0.73
Calcium	0.650
Iron	0.444
Nickel	0.440

14 Given the above specific heats and that each sample had the same mass and all other conditions were equal, which of the materials would cool the slowest?

15 Given the above specific heats and that each sample had the same mass and all other conditions were equal, which of the materials would require the most energy to raise the temperature by 10°C?

Answers to Chapter 16 Quiz

(1) This problem begins by describing a system and a process that happens. After the first few sentences you are told an amount of work, an initial and final pressure that does not change (making it constant), and the knowledge that the volume of the system changes. So the equation to use here is $w = -P\Delta V = -P(V_{final} - V_{initial})$ and you will end up solving for the initial volume. Because the system does work on the piston (and not the other way around), the sign of w is negative. Substituting your known values into the equation gives you

$$-804 \text{ J} = -(5.7 \text{ atm})(18.2 \text{L} - V_{initial})$$

The units don't match, so convert joules to liter–atmospheres:

$$(804 \text{ J})\left(\frac{1 \text{ L} \cdot \text{atm}}{101.3 \text{ J}}\right) = 7.94 \text{ L} \cdot \text{atm}$$

Now you can solve for the initial volume:

$$-7.94 \text{ L} \cdot \text{atm} = (-5.7 \text{ atm})(18.2 \text{L} - V_{initial})$$

$$\frac{-7.94 \text{ L} \cdot \text{atm}}{-5.7 \text{ atm}} = \frac{(-5.7 \text{ atm})(18.2 \text{L} - V_{initial})}{-5.7 \text{ atm}}$$

$$1.39 \text{ L} = 18.2 \text{L} - V_{initial}$$

$$V_{initial} = 16.81 \text{ L}$$

(2) **-2532.5 J.** In this problem you are given an initial and final volume of a gas in a container along with the constant pressure of that system. Using these pieces of information, you can solve for the amount of work done on the container. This will be in L*atm, which you will then have to convert to Joules.

Here is the solution worked out:

$$w = -P\Delta V = -P(V_{final} - V_{initial})$$

$$w = -10 \text{ atm}(5\text{L} - 2.5 \text{ L})$$

$$w = -10 \text{ atm}(2.5 \text{ L})$$

$$w = -25 \text{ atm} \cdot \text{L}$$

$$w = (-25 \text{ atm} \cdot \text{L})\left(\frac{101.3 \text{ J}}{1\text{L} \cdot \text{atm}}\right)$$

$$w = -2532.5 \text{ J}$$

(3) The three phase changes that are endothermic are:

Sublimation (solid → gas)

Melting (solid→liquid)

Boiling/Evaporation (liquid → gas)

All three of these phase changes involve the system gaining energy. That energy gain results in the matter undergoing a change of state. This change of state is a result of the atoms having more kinetic energy and thus more molecular motion. This increase in motion causes an increase in the spacing between molecules.

(4) The three phase changes that are exothermic are:

Deposition (gas → solid)

Freezing (liquid → solid)

Condensation (gas→liquid)

All three of these phase changes involve the system losing energy to their surroundings. That energy loss results in the matter undergoing a change of state. This loss of energy causes the particles of matter to have less kinetic energy and thus move slower. This slowing of the motion causes the gas particles to become closer together.

(5) **–125,368 J.** This is a very straightforward specific heat problem. You will need to use the specific heat formula to solve it. You are given a mass (650.6 g) of your metal; you are given the initial and final temperature of the metal, which you can use to calculate the change in temperature (456°C – 22°C = 434°C); and you are told the specific heat of iron (C_{Fe} = 0.444 J/(g*°C)). Plug these values into your specific heat formula and solve for q:

$$q = mC_p \Delta T$$

$$q = (650.6 \text{ g}) \left(0.444 \frac{J}{g \cdot ^\circ C} \right) (434 ^\circ C)$$

$$q = 125,368 \text{ J}$$

$$q = -125,368 \text{ J}$$

Since the piece of iron lost heat, you would need to make the energy change have a negative sign, as is shown at the end of the solution.

(6) **1701 cal.** This problem gives you a mass, a specific heat value, and a temperature change. It asks for the energy change in calories of the material. Since the specific heat included in the problem has calories in it, you will not need to do any conversions. The units will work out as you solve the problem.

$$q = mC_p \Delta T$$

$$q = (25.2 \text{ g}) \left(1.50 \frac{cal}{g \cdot ^\circ C} \right) (45 ^\circ C)$$

$$q = 1701 \text{ cal}$$

(7) **2.5 $\frac{J}{g \cdot ^\circ C}$.** This is another specific heat problem. You are given an amount of energy lost (q), a mass in kilograms, and a temperature change. You are asked for the specific heat of the object in J/(g*°C). Since the specific heat is asked for with grams as the mass unit, you will need to convert your kilograms to grams before you use it in the equation, then you solve for C_p. In this solution, you can see how the equation has been manipulated algebraically to solve for C_p.

$$(0.12 \text{ kg})\left(\frac{1000 \text{ g}}{1 \text{ kg}}\right) = 120 \text{ g}$$

$$q = mC_p \Delta T$$

$$\frac{q}{m\Delta T} = \frac{mC_p \Delta T}{m \Delta T}$$

$$\frac{q}{m\Delta T} = C_p$$

$$\frac{(1500 \text{ J})}{(120 \text{ g})(5 \text{ °C})} = C_p$$

$$2.5 \frac{\text{J}}{\text{g} \cdot \text{°C}} = C_p$$

(8) **268.86 g.** In this problem, you are given an amount of joules (6700 J), a specific heat (0.89 J/(g*°C)), and a temperature change (60°C–32°C = 28°C). You are asked to solve for the mass. To do so, just rearrange the equation and plug your numbers in to solve:

$$q = mC_p \Delta T$$

$$\frac{q}{C_p \Delta T} = \frac{mC_p \Delta T}{C_p \Delta T}$$

$$\frac{q}{C_p \Delta T} = m$$

$$\frac{6700 \text{ J}}{\left(0.89 \frac{\text{J}}{\text{g} \cdot \text{°C}}\right)(28 \text{ °C})} = m$$

$$268.86 \text{ g} = m$$

(9) **–4010 kJ is released.** The enthalpy change matches with the reaction you are shown in the problem, so you do not need to change the sign of the ΔH. You can tell this by reading the wording of the problem and seeing that oxygen and methane are the reactants, as shown in the reaction, and carbon dioxide and water are the products as described. This matches the reaction shown exactly. The next thing to notice is the coefficient in front of oxygen (2) as you will be using this in your calculation. Once you know this, you can solve:

$$(10 \text{ mol O}_2)\left(\frac{-802 \text{ kJ}}{2 \text{ mol O}_2}\right) = -4010 \text{ kJ}$$

The sign of the answer is negative, which means energy is released.

(10) **230 kJ is absorbed.** In this example you will need to switch the sign of the given enthalpy change because the reaction shown is the reverse of the one described in the problem. You then solve using the coefficient of 2 in front of hydrogen chloride:

$$(5 \text{ mol HCl})\left(\frac{92 \text{ kJ}}{2 \text{ mol HCl}}\right) = 230 \text{ kJ}$$

The sign of the answer is positive, which means energy is absorbed.

(11) **-157.5 kJ of energy is released.** This sign of the enthalpy given for this reaction does not need to be adjusted because the equation matches the reaction as described in the problem. Solve as follows:

$$\left(4.2\,\text{mol H}_2\right)\left(\frac{-75\,\text{kJ}}{2\,\text{mol H}_2}\right) = -157.5\,\text{kJ}$$

The sign of the answer is negative, which means energy is released.

(12) **146 kJ.** To solve this you need to manipulate the two equations to match the equation given when you add them up and then adjust their reaction enthalpies accordingly and add the enthalpies together to find the correct answer. The reactions guide you to the right answer by showing you how to alter the enthalpy numbers. In this problem you need to reverse reaction 2, which makes the sign negative for the enthalpy as shown here:

Reaction 1: $N_2(g) + 2O_2(g) \rightarrow 2NO_2(g)$ $\Delta H = 1{,}032\,\text{kJ}$

Reaction 2: $2NO_2(g) \rightarrow N_2O_4(g)$ $\Delta H = -886\,\text{kJ}$

      Sum: $N_2(g) + 2O_2(g) \rightarrow N_2O_4(g)$ $\Delta H = 146\,\text{kJ}$

(13) **-512 kJ.** To solve this equation you need to flip the first equation and reverse the sign of the enthalpy change for it. You also need to double it to make it match the required coefficient for hydrogen fluoride.

Reaction1′: $2H_2(g) + 2F_2(g) \rightarrow 4HF(g)$ $\Delta H = 2(-542\,\text{kJ})$

Reaction2′: $2H_2O(l) \rightarrow 2H_2(g) + O_2(g)$ $\Delta H = 572\,\text{kJ}$

      Sum: $2H_2O(l) + 2F_2(g) \rightarrow O_2(g) + 4HF(g)$ $\Delta H = -512\,\text{kJ}$

(14) **Potassium.** The material that would cool the slowest is the one that has the highest specific heat. Something with a high specific heat retains its temperature better and longer than comparable materials with lower specific heats. If everything else is equal, the highest specific heat would cool the slowest, making potassium the answer.

(15) **Potassium.** The material that would require the most energy to raise the temperature by 10°C is the material that has the highest specific heat, if all other conditions are equal. The actual amount of the temperature raise doesn't matter at all. This answer would be the same regardless of how many degrees you are trying to raise the temperature since the mass of the objects are the same.

Chapter **17**

Obeying Gas Laws

At first pass, gases may seem to be the most mysterious of the states of matter. Nebulous and wispy, gases easily slip through your grip. For all their diffuse fluidity, however, gases are actually the best understood of the states. The key thing to understand about gases is that they tend to behave in the same ways physically, if not chemically. For example, gases expand to fill the entire volume of any container in which you put them. Also, gases are easily compressed into smaller volumes. Even more so than liquids, gases easily form homogeneous mixtures. Because so much open space occurs between individual gas particles, these particles are pretty laid back about the idiosyncrasies of their neighbors. In this chapter, we examine the kinetic molecular theory (how gas particles behave) along with the application of that theory (the gas laws).

Working with the Kinetic Molecular Theory

Kinetic molecular theory first made a name for itself when scientists attempted to explain and predict the properties of gases and, in particular, how those properties changed with varying temperature and pressure. The idea emerged that the particles of matter within a gas (atoms or molecules) undergo a serious amount of motion as a result of the kinetic energy within them.

Kinetic energy is the energy of motion. Gas particles have a lot of kinetic energy and constantly zip about, colliding with one another or with other objects. The picture is complicated, but scientists simplified things by making several assumptions about the behavior of gas particles.

A theory is useful to scientists if it describes the physical system they're examining and allows them to predict what will happen if they change some variable. The *kinetic molecular theory of gases* does just that. It has limitations — all theories do — but it's one of the most useful theories in chemistry. Here are the theory's basic postulates (assumptions, hypotheses, axioms — pick your favorite word), which you can accept as being true:

>> **Gases are composed of tiny particles, either atoms or molecules.** Unless you're discussing matter at greatly elevated temperatures, the particles referred to as gases tend to be relatively small. The more massive particles clump together to form liquids or even solids. So gases are normally small with relatively low atomic and molecular weights.

>> **The gas particles are so small when compared to the distances between them that the volume the gas particles themselves take up is negligible and is assumed to be zero.** These gas particles do take up some volume — that's one of the properties of matter. But the gas particles are small, so if a container doesn't hold many of them, you say that their volume is negligible when compared to the volume of the container or the space between the gas particles. Because of all that space between the gas particles, they can be squeezed together to compress the gas. Solids and liquids can't be squeezed, because their particles are *much* closer together. (Chapter 16 covers the various states of matter, if you want to have a look-see at the differences among solids, liquids, and gases.)

The concept of a *negligible* quantity is used a lot in chemistry. In the real world, you can compare this negligible concept to finding a dollar in the street. If you have no money at all, then that dollar represents a sizable quantity of cash (perhaps your next meal). But if you're a multimillionaire, then that dollar doesn't represent much at all. You may not even pick it up. (Though let's be realistic — a dollar is a dollar.) Its value is negligible when compared to the rest of your wealth. And sure, the gas particles have a volume, but it's so small that it's insignificant when compared with the distance between the gas particles and the volume of the container.

>> **The gas particles are in constant random motion, moving in straight lines and colliding with the inside walls of the container.** The first key phrase you find in this statement is *random motion*. When we discuss the behavior of gases in this chapter and use the kinetic molecular theory, all the motion of gas particles is assumed to be completely random. The particles move in straight lines, without being influenced by any outside attractions or repulsions.

Now, these gas particles are always moving. (Gases have a higher kinetic energy — energy of motion — associated with them than solids or liquids do; see Chapter 16.) The particles continue to move in straight lines until they collide with something — either with each other or with the inside walls of the container. The particles also all move in different directions, so the collisions with the inside walls of the container tend to be uniform over the entire inside surface. You can observe this uniformity by blowing up a balloon. The balloon is almost spherical because the gas particles hit all points of the inside walls the same. The collision of the gas particles with the inside walls of the container is called *pressure*.

The idea that the gas particles are in constant, random, straight-line motion explains why gases uniformly mix if put in the same container. It also explains why, when you drop a bottle of cheap perfume at one end of the room, the people at the other end of the room are able to smell it right away.

>> **The gas particles are assumed to have negligible attractive or repulsive forces between each other.** In other words, the gas particles are assumed to be totally

independent, neither attracting nor repelling each other. That said, it's hair-splitting time: This assumption is actually false. If it were true, chemists would never be able to liquefy a gas, which they can do. But the reason you can accept this assumption as true (or at least useful) is that the attractive and repulsive forces are generally so small that they can safely be ignored. The assumption is most valid for nonpolar gases, such as hydrogen and nitrogen, because the attractive forces involved are London forces. However, if the gas molecules are polar, as in water and HCl, this assumption can become a problem.

>> **The gas particles may collide with each other. These collisions are assumed to be elastic, with the total amount of kinetic energy of the two gas particles remaining the same.** If gas particles hit each other, no kinetic energy is lost, but kinetic energy may be transferred from one gas particle to the other. For example, imagine two gas particles — one moving fast and the other moving slow — colliding. Kinetic energy is transferred from the faster particle to the slower particle. The slow-moving particle bounces off the faster particle and moves away at a greater speed than before, and the faster particle bounces off the slower particle and moves away at a slower speed. The total amount of kinetic energy remains the same, but one gas particle loses energy and the other gains energy. This transfer of energy is the principle behind pool — you transfer kinetic energy from your moving pool stick to the cue ball to the ball you're aiming at.

>> **The Kelvin temperature is directly proportional to the *average* kinetic energy of the gas particles.** The gas particles aren't all moving with the same amount of kinetic energy. A few are moving relatively slow, and a few are moving very fast, but most are somewhere in between these two extremes. *Temperature*, particularly as measured using the Kelvin temperature scale, is defined as the average kinetic energy of particles (in this case, gas particles) in a system. If you heat the gas so that the Kelvin temperature (K) increases, the average kinetic energy of the gas also increases. (*Remember:* To calculate the Kelvin temperature, add 273 to the Celsius temperature: $K = °C + 273$.)

A gas that obeys all the postulates of the kinetic molecular theory is called an *ideal gas*. Obviously, no real gas obeys the assumptions made in the second and fourth postulates *exactly* (all gas particles actually do have small measures of volume and attractive or repulsive force), but a nonpolar gas at high temperatures and low pressure (concentration) approaches ideal gas behavior.

EXAMPLE

Q. Why don't real-life gases behave like ideal gases at very high pressure?

A. Under very high-pressure conditions, gas particles are much closer to one another and undergo more collisions. Because of their proximity, real-life gas particles are more likely to experience mutual attractions or repulsions. Because of their more frequent collisions, gases at very high pressure are more likely to reveal the effects of any inelasticity in their collisions (losing energy upon colliding). Real-life gases are more likely to behave like ideal gases when they're at very low pressure.

Q. Describe what happens to the motion of particles at the molecular level as they increase in temperature.

A. According to the kinetic molecular theory of gases, as particles increase in temperature, they are gaining energy. As these particles gain energy, their motion increases, and they begin to move faster and faster. This leads to an increase in spacing between the particles and an increase in the average speed of the particles in the system.

1 Why does it make sense that the noble gases (which we introduce in Chapter 5) exist as gases at normal temperature and pressure?

2 Ice floats in water. Based on the usual assumptions of kinetic molecular theory, why is this weird?

3 At the same temperature, how would the pressure of an ideal gas differ from that of a gas with mutually attractive particles? How would it differ from that of a gas with mutually repulsive particles?

Measuring and Converting Pressure

When working with gases, you need to consider the ways in which people use and measure a gas's pressure. Different methods allow you to measure the pressure of a confined gas as well as atmospheric pressure.

If you get a complete weather report, the atmospheric pressure is normally included. You can get an idea about changes in the weather by observing whether the atmospheric pressure is rising or falling. The atmospheric pressure is measured using a barometer, and Figure 17-1 shows the components of one.

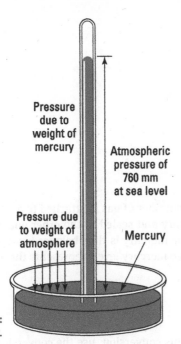

FIGURE 17-1:
A barometer.

A barometer is composed of a long glass tube that's closed at one end and totally filled with a liquid. You can use water, but the tube would have to be *very* long (about 35 feet long), making for a rather inconvenient barometer. Barometers typically use mercury because it's a very dense liquid. The tube filled with mercury is inverted into an open container of mercury so that the open end of the tube is under the surface of the mercury in the container. The force of gravity pulls the mercury in the tube *down*, causing it to drain out into the container, while the weight of the gases in the atmosphere exerts pressure downward on the mercury in the open container and forces it *up* into the tube. Sooner or later, these forces balance, and the mercury in the tube comes to rest at a certain height in the tube. The greater the pressure of the atmosphere, the higher the mercury column that can be measured; the lower the pressure of the atmosphere (for example, at the top of a tall mountain), the shorter the column. At sea level, the column is 760 millimeters high, the so-called normal atmospheric pressure.

Atmospheric pressure can be expressed in a number of ways. It can be expressed in millimeters of mercury (mm Hg); atmospheres (atm), a unit of pressure where 1 atmosphere is the pressure at sea level; torr, a unit of pressure where 1 torr equals 1 millimeter of mercury; pounds per square inch (psi); pascals (Pa), a unit of pressure where 1 pascal equals 1 newton per square meter (don't worry about what a newton is; it's just a way to express force); or kilopascals (kPa), where 1 kilopascal equals 1,000 pascals.

So you can express the atmospheric pressure at sea level as

760 mm Hg = 1 atm = 760 torr = 14.69 psi = 101,325 Pa = 101.325 kPa

Note that sometimes you also hear atmospheric pressure reported in inches of mercury (1 atm = 29.921 in. Hg). Variety is the spice of life.

You can measure the pressure of a gas confined in a container by using an apparatus called a *manometer* (pronounced man–*ah*–muh–ter). Figure 17-2 shows the components of a manometer.

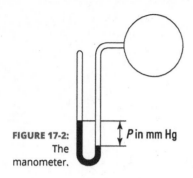

FIGURE 17-2:
The
manometer.

P in mm Hg

A manometer is kind of like a barometer. The container of gas is attached to a U-shaped piece of glass tubing that's partially filled with mercury and sealed at the other end. Gravity pulls down the mercury column at the closed end. The mercury is then balanced by the pressure of the gas in the container. The difference in the two mercury levels represents the amount of gas pressure.

EXAMPLE

Q. Convert 456.6 mm Hg to atmospheres.

A. The pressure is 0.6008 atm. To perform this conversion, use the conversion factor $\frac{1 \text{ atm}}{760 \text{ mm Hg}}$. Multiply your initial measurement in millimeters of mercury by that conversion factor to get the pressure in atmospheres:

$$\left(456.6 \text{ atm}\right)\left(\frac{1 \text{ atm}}{760 \text{ mm Hg}}\right) = 0.6008 \text{ atm}$$

YOUR
TURN

4 A student measures a barometer column and records that it has a height of 740 mm Hg. The student then needs to convert that pressure to kilopascals. What is the new value?

5 If you begin with a confined gas in a manometer with an internal pressure of 2.5 atm, what is the internal pressure of that gas when converted to torr?

Boyle's Law: Playing with Pressure and Volume

Each gas law that we discuss in this chapter deals with some of the four important variables you encounter when working with ideal gases:

>> Volume

>> Pressure

>> Temperature

>> Amount

Relationships among these four factors are the domain of the gas laws. Each variable is dependent upon the others, so altering one can change all the others as well. Each gas law shows a relationship between two properties while keeping the other two properties constant. (In other words, you take two properties, change one, and then see its effect on the second — while keeping the remaining properties constant.)

The first of these relationships that was formulated into a law concerns pressure and volume. Robert Boyle, an Irish gentleman regarded by some as the first chemist (or "chymist," as his friends might have said), is typically given credit for noticing that gas pressure and volume have an inverse relationship:

$$\text{Volume} = \frac{1}{\text{Pressure}} \times \text{Constant}$$

Boyle's law describes the pressure–volume relationship of gases if the temperature and amount are kept constant. Figure 17-3 illustrates the pressure–volume relationship using the kinetic molecular theory.

decrease
volume,
increase
pressure

FIGURE 17-3:
Pressure-
volume
relationship
of gases
(Boyle's law).

The left-hand cylinder in the figure contains a certain volume of gas at a certain pressure. When the volume is decreased, the same number of gas particles is now contained in a much smaller volume and the number of collisions increases significantly. Therefore, the pressure is greater.

REMEMBER

Boyle's law states that an inverse relationship exists between the volume and the pressure. As the volume decreases, the pressure increases, and vice versa. Boyle determined that the product of the pressure and the volume is a constant (*k*):

$$PV = k$$

Now consider a case where you have a gas at a certain pressure (P_1) and volume (V_1). If you change the volume to some new value (V_2), the pressure also changes to a new value (P_2). You can use Boyle's law to describe both sets of conditions:

$$P_1V_1 = k$$
$$P_2V_2 = k$$

The constant, *k*, is the same in both cases. So you can say

$$P_1V_1 = P_2V_2 \text{ (with temperature and amount constant)}$$

This equation is another statement of Boyle's law — and it's really a more useful one, because you'll normally deal with changes in pressure and volume. If you know three of the quantities in the preceding equation, you can calculate the fourth one with some very simple algebra.

The relationship between pressure and volume makes good sense in light of kinetic molecular theory. At a given temperature and number of particles, more collisions will occur at smaller volumes. These increased collisions produce greater pressure. And vice versa. Boyle had some dubious ideas about alchemy, among other things, but he really struck gold with the pressure-volume relationship in gases.

In short, pressure and volume have an indirect relationship. As one value increases, the other decreases. You can prove this idea numerically.

EXAMPLE

Q. You have 5.00 L of a gas at a pressure of 1.00 atm, and then you decrease the volume to 2.00 L. What's the new pressure?

A. The new pressure is 2.50 atm. To find the answer, follow these steps:

1. Use the following equation: $P_1V_1 = P_2V_2$.

2. Identify the variables from your problem and then substitute them in for each variable in the equation: 1.00 atm for P_1, 5.00 L for V_1, and 2.00 L for V_2. You get $(1.00 \text{ atm})(5.00 \text{ L}) = P_2(2.00 \text{ L})$

3. Solve for P_2:

$$\frac{(1.00 \text{ atm})(5.00 \text{ L})}{2.00 \text{ L}} = P_2 = 2.50 \text{ atm}$$

The answer makes sense, because you decreased the volume and the pressure increased, which is exactly what Boyle's law says.

Q. A sealed plastic bag is filled with 1 L of air at standard temperature and pressure (STP); the values of STP are 273 K and 101.325 kilopascals (kPa). You accidentally sit on the bag. The maximum pressure the bag can withhold before popping is 500 kPa. What is the internal volume of the bag at the instant of popping?

A. The volume is 0.2 L. The problem tells you that the bag has an initial volume, V_1, of 1 L and an initial pressure, P_1, of 101.325 kPa (the pressure at STP). The temperature doesn't matter because it remains constant. The pressure inside the bag reaches 500 kPa before popping, so that value represents P_2. You can also leave your pressure units in kilopascals, as opposed to converting them to atmospheres (atm), because both P_1 and P_2 are in the same unit (kPa). You need to convert units only when the units on the left and right side of the equation don't agree. So the only missing variable in this problem is the final volume. Solve for the final volume, V_2, by rearranging Boyle's law and then plugging in the known values:

$$V_2 = \frac{P_1 V_1}{P_2}$$

$$V_2 = \frac{(1\ \text{L})(101.325\ \text{kPa})}{(500\ \text{kPa})} = 0.2\ \text{L}$$

YOUR TURN

6 An amateur entomologist captures a particularly excellent ladybug specimen in a plastic jar. The internal volume of the jar is 0.5 L, and the air within the jar is initially at 1 atm. The bug-lover is so excited by the catch that he squeezes the jar fervently in his sweaty palm, compressing it such that the final pressure within the jar is 1.25 atm. What is the final volume of the ladybug's prison?

---

7 A container has an internal volume of 3 L. This volume is divided equally in two by a gas-tight seal. On one half of the seal, neon gas resides at 5 atm (think about what the volume of this portion of the gas is). The other half of the container is kept under vacuum. Suddenly the internal seal is broken and the gas from the first chamber expands to fill the entire container. What is the final pressure within the container?

Charles's Law and Absolute Zero: Looking at Volume and Temperature

History attributes the following law to French chemist Jacques Charles. Charles discovered a direct, linear relationship between the volume and the temperature of a gas:

$$\text{Volume} = \text{Temperature} \times \text{Constant}$$

Charles's law deals with volume and temperature, keeping the pressure and amount constant. You run across situations dealing with this relationship in everyday life, especially in terms of the heating and cooling of balloons.

Figure 17-4 shows the temperature–volume relationship.

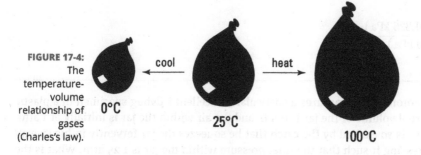

FIGURE 17-4:
The temperature-volume relationship of gases (Charles's law).

cool → heat

0°C   25°C   100°C

Look at the balloon in the middle of Figure 17-4. What do you think would happen to the balloon if you placed it in the freezer or took it outside in subzero weather? It'd get smaller. Inside the freezer or in arctic weather, the external pressure, or atmospheric pressure, is the same, but the gas particles inside the balloon aren't moving as fast, so the volume shrinks to keep the pressure constant. If you heat the balloon, the balloon expands and the volume increases. This correspondence is a *direct relationship* — as the temperature increases, the volume increases, and vice versa.

Jacques Charles developed the mathematical relationship between temperature and volume. He also discovered that you must use the Kelvin (K) temperature when working with gas law expressions and calculations.

Charles's law says that the volume is directly proportional to the Kelvin temperature. Mathematically, the law looks like this:

$$V = bT \quad \text{or} \quad \frac{V}{T} = b$$

(where b is a constant)

If the temperature of a gas with a certain volume (V_1) and Kelvin temperature (T_1) is changed to a new Kelvin temperature (T_2), the volume also changes (V_2).

$$\frac{V_1}{T_1} = b \qquad \frac{V_2}{T_2} = b$$

The constant, b, is the same, so

$$\frac{V_1}{T_1} = \frac{V_2}{T_2}$$

with the pressure and amount of gas held constant and temperature expressed in kelvins.

If you have any three of the quantities, you can calculate the fourth.

Not to be outdone by the French, another Irish scientist took Charles's observations and ran with them. William Thomson, eventually to be known as Lord Kelvin, took stock of all the data available in his mid-19th century heyday and noticed a few things:

>> Plotting the volume of a gas versus its temperature always produced a straight line.

>> Extending these various lines caused them all to converge at a single point, corresponding to a single temperature at zero volume. This temperature — though not directly accessible in experiments — was about –273°C. Kelvin took the opportunity to enshrine himself in the annals of scientific history by declaring that temperature to be *absolute zero*, the lowest temperature possible.

REMEMBER

This declaration had at least two immediate benefits. First, it happened to be correct. Second, it allowed Kelvin to create the Kelvin temperature scale, with absolute zero as the Official Zero. Why do we care so much about absolute zero? The most important reason is that it actually is the bottom of possible temperatures. This means that you can't just perpetually move to lower and lower temperatures; instead there is simply a bottom that you eventually reach as an object gets colder and colder. It is at this zero point, in theory, that all molecular motion would completely and utterly cease. Nothing would move. Since temperature is a measure of the kinetic energy of a substance, remember that the hotter something is, the faster its particles move. This applies in the opposite direction as well, as something becomes colder and colder the motion of the molecules slows down. At absolute zero this motion would theoretically completely stop. Nothing would move at all, not even electrons. It's a very interesting concept and one that you should read more about if it sounds interesting to you. We have never been able to simulate absolute zero in a lab (and we never will be able to) but there are groups of scientists, at this very minute, who are working hard away in their labs to try and reach colder and colder temperatures every day. Now, why does this matter to you? Using the Kelvin scale (where °C = K – 273), everything makes a whole lot more sense. For example, doubling the Kelvin temperature of a gas doubles the volume of that gas.

WARNING

When you work with Charles's law, converting Celsius temperatures to the Kelvin scale is crucial. If you don't make the conversion, your answer will be incorrect.

Q. Suppose you live in Alaska and are outside in the middle of winter, where the temperature is −23°C. You blow up a balloon so that it has a volume of 1.00 L. You then take it inside your home, where the temperature is a toasty 27°C. What's the new volume of the balloon?

EXAMPLE

A. The new volume is 1.20 L. To solve, follow these steps:

1. **Convert your temperatures to Kelvin by adding 273 to the Celsius temperature.**

   $-23°C + 273 = 250$ K (outside)
   $27°C + 273 = 300$ K (inside)

2. **Solve for V_2, using the following setup.**

   $$\frac{V_1}{T_1} = \frac{V_2}{T_2}$$

3. **Multiply both sides by T_2 so V_2 is on one side of the equation by itself.**

   $$\frac{V_1 T_2}{T_1} = V_2$$

4. **Substitute the values to calculate the following answer.**

   $$\frac{(1.00\ \text{L})(300.\ \text{K})}{250.\ \text{K}} = V_2 = 1.20\ \text{L}$$

This answer is reasonable, because Charles's law says that if you increase the Kelvin temperature, the volume increases.

Q. A red rubber dodge ball, which sits in a 20.0°C basement, is filled with 3.50 L of compressed air. Eager to begin practice for the impending dodge ball season, Vince reclaims the ball and takes it outside. After a few hours of practice, the well-sealed ball has a volume of 3.00 L. What's the temperature outside in degrees Celsius?

A. The temperature is −22°C. The question provides an initial temperature, an initial volume, and a final volume. You're asked to find the final temperature, T_2. Apply Charles's law, plugging in the known values and solving for the final temperature. But take care — Charles's law requires you to convert all temperatures to kelvins (where K = °C + 273). The correct temperature to use here is 293 K (20.0°C + 273). After identifying your variables, plug your numbers into the rearranged equation and solve:

$$\frac{V_1}{T_1} = \frac{V_2}{T_2}$$
$$T_2 = \frac{T_1 V_2}{V_1}$$
$$T_2 = \frac{(293\ \text{K})(3.00\ \text{L})}{3.50\ \text{L}} = 251\ \text{K}$$

After you determine your value in kelvins (251), you must convert back to degrees Celsius by subtracting 273 from your answer. Thus, you get −22°C.

8   A small hot-air balloon has an initial volume of 300 L. To move higher into the air, the pilot runs the burner for a long period of time. The long burn increases the temperature of the air within the balloon from 40.0°C to 50.0°C. What is the new volume of the balloon?

9   Always helpful, Danny persuades his little sister Suzie that her Very Special Birthday Balloon will last much longer if she puts it in the freezer in the basement for a while. The temperature in the house is 20.0°C. The balloon has an initial volume of 0.250 L. If the balloon has collapsed to 0.200 L by the time Suzie catches on to Danny's devious deed, what is the temperature inside the freezer in degrees Celsius?

Gay-Lussac's Law: Examining Pressure and Temperature

Gay-Lussac's law, named after the 19th-century French scientist Joseph-Louis Gay-Lussac, deals with the relationship between the pressure and temperature of a gas if its volume and amount are held constant. Imagine, for example, that you have a metal tank of gas. The tank has a certain volume, and the gas inside has a certain pressure. If you heat the tank, you increase the kinetic energy of the gas particles. So they're now moving much faster, and they're not only hitting the inside walls of the tank more often but also hitting with more force. The pressure has increased.

Gay-Lussac's law says that the pressure is directly proportional to the Kelvin temperature. Figure 17-5 shows this relationship.

Mathematically, Gay-Lussac's law is represented by the following equation, where P is the pressure, T is the temperature, and k is a constant (at constant volume and amount):

$$P = kT \left(\text{or } \frac{P}{T} = k \right)$$

FIGURE 17-5:
The
pressure-
temperature
relationship
of gases —
Gay-Lussac's
law.

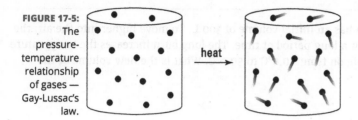

Consider a gas at a certain Kelvin temperature and pressure (T_1 and P_1), with the conditions being changed to a new temperature and pressure (T_2 and P_2):

$$\frac{P_1}{T_1} = \frac{P_2}{T_2}$$

Q. If you have a tank of gas at a pressure of 800 torr and a temperature of 250 K, and it's heated to 400 K, what's the new pressure?

EXAMPLE

A. 1,280 torr.

To find the new pressure as the problem asks, follow these steps:

1. **Starting with** $\frac{P_1}{T_1} = \frac{P_2}{T_2}$, **multiply both sides by** T_2 **so you can solve for** P_2.

$$\frac{P_1 T_2}{T_1} = P_2$$

2. **Substitute the values to calculate the following answer.**

$$\frac{(800 \text{ torr})(400 \text{ K})}{250 \text{ K}} = P_2 = 1,280 \text{ torr}$$

This answer is reasonable because if you heat the tank, the pressure should increase.

YOUR TURN

10 If you have a tank of gas with a pressure of 1.3 atm and a temperature of 360 K, what is the new pressure of the tank if you lower the temperature to a chilly 40 K?

11 Knowing the relationship between pressure and temperature, why should you never throw a pressurized and sealed can of gas into a fire? A fire's temperature is approximately 673 K (that's really, really hot).

Combining Pressure, Temperature, and Volume into One Law

All the examples in the preceding sections assume that two properties are held constant and one property is changed to see its effect on a fourth property. But life is rarely that simple. How do you handle situations in which two or even three properties change? You can treat each one separately, but it sure would be nice if you had a way to combine things so that wouldn't be necessary.

Actually, there is a way. You can combine Boyle's law, Charles's law, and Gay–Lussac's law into one equation. Trust us, you don't want us to show you exactly how it's done because it involves a lot of boring algebra, but the end result is called the *combined gas law*, and it looks like this:

$$\frac{P_1 V_1}{T_1} = \frac{P_2 V_2}{T_2}$$

Just like in the preceding examples, P is the pressure of the gas (in atmospheres, millimeters of mercury, torr, and so on), V is the volume of the gas (in appropriate units), and T is the temperature (in kelvins). The 1 and 2 stand for the initial and final conditions, respectively. The amount is still held constant: No gas is added, and no gas escapes. Six quantities are involved in this combined gas law, so knowing five allows you to calculate the sixth.

As a final gift, you may like the following equations. Sometimes working with the combined gas law can be a serious pain in the butt due to all the algebra involved with rearranging the equations endlessly. So here you go: the combined gas law solved for every variable! Please use this amazing new information sparingly.

Pressure $\quad P_1 = \dfrac{P_2 V_2 T_1}{T_2 V_1} \quad P_2 = \dfrac{P_1 V_1 T_2}{T_1 V_2}$

Volume $\quad V_1 = \dfrac{T_1 P_2 V_2}{T_2 P_1} \quad V_2 = \dfrac{T_2 P_1 V_1}{T_1 P_2}$

Temperature $\quad T_1 = \dfrac{P_1 V_1 T_2}{P_2 V_2} \quad T_2 = \dfrac{P_2 V_2 T_1}{P_1 V_1}$

EXAMPLE

Q. A weather balloon with a volume of 25.0 L at a pressure of 1.00 atm and a temperature of 27°C is allowed to rise to an altitude where the pressure is 0.500 atm and the temperature is –33°C. What's the new volume of the balloon?

A. The new volume is 40.0 L. Before you work out this problem, do a little reasoning. The temperature is decreasing, so that change should cause the volume to decrease (Charles's law). However, the pressure is also decreasing, which should cause the balloon to expand (Boyle's law). These two factors are competing, so at this point, you don't know which will win out. That's why doing numerical calculations with the combined gas law is so valuable.

You're looking for the new volume (V_2), so solve as follows:

1. **Rearrange the combined gas law to obtain the following equation (by multiplying each side by T_2 and dividing each side by P_2, which puts V_2 by itself on one side).**

$$\frac{P_1 V_1 T_2}{P_2 T_1} = V_2$$

2. **Identify your quantities.**

$P_1 = 1.00 \text{ atm} \qquad V_1 = 25.0 \text{ L} \qquad T_1 = 27°C + 273 = 300. \text{ K}$
$P_2 = 0.500 \text{ atm} \qquad V_2 = ? \qquad T_2 = -33°C + 273 = 240. \text{ K}$

3. **Substitute the values to calculate the following answer.**

$$V_2 = \frac{(1.00 \text{ atm})(25.0 \text{ L})(240. \text{ K})}{(0.500 \text{ atm})(300. \text{ K})}$$

$V_2 = 40.0 \text{ L}$

Because the volume increased overall in this case, Boyle's law had a greater effect than Charles's law.

---

12 A sealed box containing an inert poisonous gas is placed inside a larger vacuum–sealed box. The smaller box containing the poisonous gas has an internal volume of 0.800 L, a pressure of 290 atm, and a temperature of 283 K. After a set amount of time, the seal on the smaller box breaks and releases the gas into the larger box. After the poisonous gas expands to fill the newly available volume, the gas reaches STP (273 K and 1 atm). What is the total volume of the secondary box?

YOUR TURN

---

13 The volume of a whoopee cushion is 0.450 L at 27.0°C and 105 kPa. Danny has placed one such practical joke device on the chair of his unsuspecting Aunt Bertha. Unbeknownst to Danny, this particular whoopee cushion suffers from a construction defect that sometimes blocks normal outgassing and ruins the flatulence effect. So even when the cushion receives the full force of Aunt Bertha's ample behind, the blockage prevents deflation. The cushion sustains the pressure exerted by Bertha so that the internal pressure becomes 200. kPa. As she sits on the cushion, Bertha warms its contents a full 10.0°C. At last, and to Danny's profound satisfaction, the cushion explodes. What volume of air does it expel?

Dealing with Amounts: Avogadro's Law and the Ideal Gas Law

The combined gas equation gives you a way to calculate changes involving pressure, volume, and temperature. But you still have the problem of amount to deal with. In order to account for amount, you need to know another law.

REMEMBER

From his study of gases, Amedeo Avogadro determined that equal volumes of gases at the same temperature and pressure contain equal numbers of gas particles (that's the same Avogadro who lends his name to the number of particles per mole; see Chapter 12). So *Avogadro's law* says that the volume of a gas is directly proportional to the number of moles of gas (number of gas particles) at a constant temperature and pressure. Mathematically, Avogadro's law looks like this:

$$V = kn \quad \text{(at constant temperature and pressure)}$$

In this equation, k is a constant and n is the number of moles of gas. If you have a number of moles of gas (n_1) at one volume (V_1) and the number of moles changes due to a reaction (n_2), the volume also changes (V_2), giving you the equation

$$\frac{V_1}{n_1} = \frac{V_2}{n_2}$$

A very useful consequence of Avogadro's law is that the volume of a mole of gas can be calculated at any temperature and pressure. An extremely useful form to know when calculating the volume of a mole of gas is "1 mole of any gas at STP occupies 22.4 liters." STP in this case is not an oil or gas additive. It stands for *standard temperature and pressure:*

>> **Standard pressure:** 1.00 atm (760. torr or 760. mm Hg)

>> **Standard temperature:** 273 K

This relationship between moles of gas and liters gives you a way to convert the gas from a mass to a volume.

Now, if you take Boyle's law, Charles's law, Gay-Lussac's law, and Avogadro's law and throw them into a blender on high for a minute, you get the *ideal gas equation* — a way of working in volume, temperature, pressure, *and* amount. The ideal gas equation has the following form:

$$PV = nRT$$

The P represents pressure in atmospheres (atm), the V represents volume in liters (L), the n represents moles of gas, the T represents the temperature in kelvins (K), and the R represents the ideal gas constant, which is 0.08206 L·atm/K·mol.

REMEMBER

Using this value of the ideal gas constant, the pressure must be expressed in atmospheres, and the volume must be expressed in liters. You can calculate other ideal gas constants if you really want to use torr and milliliters, for example, but why bother? It's easier to memorize one value for R and then remember to express the pressure and volume in the appropriate units. Naturally, you'll *always* express the temperature in kelvins when working any kind of gas law problem.

EXAMPLE

Q. A 0.80 L container holds 10. mol of helium. The temperature of the container is 10.0°C. What's the internal pressure of the container?

A. The pressure is 290 atm. Consider your known and unknown variables. You're given volume, moles (number of particles), and temperature. You're asked to calculate pressure. The equation that fills the bill is the ideal gas law, $PV = nRT$. Rearrange the equation to solve for P so that $P = (nRT)/V$.

One key to using R in the ideal gas law is ensuring that all your units in the problem match with the units for R. In this problem, you have values in degrees Celsius, liters, and moles. You must convert the 10.°C to the Kelvin scale by adding 273 to the temperature, giving you 283 K. In addition, you can determine that your answer for pressure will have the unit atmospheres (atm), the pressure unit given in R.

Next, plug in your known values and solve:

$$P = \frac{(10.\ \text{mol})\left(0.08206\ \frac{\text{L} \cdot \text{atm}}{\text{mol} \cdot \text{K}}\right)(283\ \text{K})}{0.80\ \text{L}} = 290\ \text{atm}$$

That's nearly 300 times normal atmospheric pressure. Stay away from that container!

YOUR TURN

14 A container with a volume of 15.0 L contains oxygen. The gas is at a temperature of 29.0°C and a pressure of 98.69 atm. How many moles of gas occupy that container?

Mixing It Up with Dalton's Law of Partial Pressures

Gases mix. Gases mix better than liquids do and infinitely better than solids. So what's the relationship between the total pressure of a gaseous mixture and the pressure contributions of the individual gases? Here's a satisfyingly simple answer: Each individual gas within the mixture contributes a partial pressure, and adding the partial pressures yields the total pressure. This relationship is summarized by *Dalton's law of partial pressures* for a mixture of individual gases:

$$P_{total} = P_1 + P_2 + P_3 + \ldots + P_n$$

This relationship makes sense if you think about pressure in terms of kinetic molecular theory. Adding a gaseous sample into a particular volume that already contains other gases increases the number of particles in that space. Because pressure depends on the number of particles colliding with the container walls, increasing the number of particles increases the pressure proportionally.

REMEMBER
There's no one specific pressure unit you have to use when doing problems with Dalton's partial-pressures equation. As long as the pressure units for all the gases are the same, you're good to go. However, if all the pressures given aren't in the same units, then some conversion must take place!

EXAMPLE
Q. A chemist designs an experiment to study the chemistry of the atmosphere of the early Earth. She constructs an apparatus to combine pure samples of the primary volcanic gases that made up the atmosphere billions of years ago: carbon dioxide, ammonia, and water vapor. If the partial pressures of these gases are 50. kPa, 80. kPa, and 120. kPa, respectively, what's the pressure of the resulting mixture?

A. The pressure is 250 kPa. However difficult early-Earth atmospheric chemistry may prove to be, this particular problem is a simple one. Dalton's law states that the total pressure is simply the sum of the partial pressures of the component gases:

$$P_{total} = P_{CO_2} + P_{NH_3} + P_{H_2O}$$
$$= 50.\ kPa + 80.\ kPa + 120.\ kPa$$
$$= 250\ kPa$$

YOUR TURN
15 You have a system of gases in a closed cylinder. The pressure of each gas is as follows: Gas 1= 1.2 atm, Gas 2 = 7.8 atm, and Gas 3 = 0.35 atm. What is the total pressure inside the cylinder?

16 A chemist adds solid zinc powder to a solution of hydrochloric acid to initiate the following reaction:

$$Zn(s) + 2HCl(aq) \rightarrow ZnCl_2(aq) + H_2(g)$$

The chemist inverts a test tube and immerses the open mouth into the reaction beaker to collect the hydrogen gas that bubbles up from the solution. The reaction proceeds to equilibrium. At the end of the experiment, the water levels within the tube and outside the tube are equal. The pressure in the lab is 101.325 kPa, and the temperature of all components is 298 K. The vapor pressure of water at 298 K is 3.17 kPa. What is the partial pressure of hydrogen gas trapped in the tube?

Diffusing and Effusing with Graham's Law

"Wake up and smell the coffee." This command is usually issued in a scornful tone, but most people who've awakened to the smell of coffee remember the event fondly. The morning gift of coffee aroma is made possible by a phenomenon called *diffusion*. Diffusion is the movement of a substance from an area of higher concentration to an area of lower concentration. Diffusion occurs spontaneously, on its own. Diffusion leads to mixing, eventually producing a homogeneous mixture in which the concentration of any gaseous component is equal throughout an entire volume. Of course, that state of complete diffusion is an equilibrium state, and achieving equilibrium can take time.

REMEMBER

Different gases diffuse at different rates, depending on their molar masses (see Chapter 12 for details on molar masses). You can compare the rates at which two gases diffuse using *Graham's law*. Graham's law also applies to *effusion*, the process in which gas molecules flow through a small hole in a container. Whether gases diffuse or effuse, they do so at a rate inversely proportional to the square root of their molar mass. In other words, more massive gas molecules diffuse and effuse more slowly than less massive gas molecules. So for Gases A and B, the following applies:

$$\frac{\text{Rate A}}{\text{Rate B}} = \sqrt{\left(\frac{\text{Molar mass B}}{\text{Molar mass A}}\right)}$$

Note: For the problems in this book, round your molar masses to two decimal places before plugging them into the formula.

TIP

By far, the most important part of solving effusion problems is identifying which gas you'll identify as Gas A and which gas you'll identify as Gas B when you plug your values into the equation. Don't switch them up!

EXAMPLE

Q. How much faster does hydrogen gas effuse than neon gas?

A. 3.16 times faster. *Hydrogen gas* refers to H_2 because hydrogen is a diatomic element. Consult your periodic table (or your memory, if you're that good) to obtain the molar masses of hydrogen gas (2.02 g/mol) and neon gas (20.18 g/mol). Finally, plug those values into the appropriate places within Graham's law, and you can see the ratio of effusion speed. In this example, we chose hydrogen as Gas A and neon as Gas B.

$$\frac{\text{Rate } H_2}{\text{Rate Ne}} = \sqrt{\left(\frac{20.18 \text{ g/mol}}{2.02 \text{ g/mol}}\right)} = \sqrt{9.99}$$

The answer you get to this problem is 3.16. Putting this number over 1 can help you understand your answer. The ratio 3.16 / 1 means is that for every 3.16 mol of hydrogen gas that effuses, 1.00 mol of neon gas will effuse. This ratio is designed to compare rates. So hydrogen gas effuses 3.16 times faster than neon.

YOUR TURN

17 Mystery Gas A effuses 4.0 times faster than oxygen. What is the likely identity of the mystery gas?

Practice Questions Answers and Explanations

(1) The noble gases are described as *noble* because due to their full valence shells, they have very low reactivity. Only very weak attractive forces occur between the atoms of a noble gas. The particles don't significantly attract one another under any but the most extreme conditions (high pressure, low temperature). Therefore, only a small amount of heat is required to push these elements into the gas phase, so they're found as gases under most conditions.

(2) Kinetic molecular theory describes matter as moving from solid to liquid phase (melting) as you add energy to the sample. The added energy causes the particles to undergo greater motion and to collide with other particles more energetically. Usually, this means that a liquid is less dense than a solid sample of the same material, because the greater motion of the liquid particles prevents close packing. Solid water (ice), on the other hand, is less dense than liquid water because of the unique geometry of water crystals. Because solid water is less dense than liquid water, ice floats in water.

(3) At a given temperature, an ideal gas exerts greater pressure than a gas with mutually attractive particles, and it exerts lower pressure than a gas with mutually repulsive particles. On average, mutual attraction allows particles to occupy a smaller volume at a given kinetic energy because the molecules are held more tightly together. Mutual repulsion causes particles to attempt to occupy a larger volume because they want to stay as far apart from each other as possible.

(4) **98.7 kPa.** Multiply your initial value of millimeters of mercury by the conversion factor $\frac{101.325 \text{ kPa}}{760 \text{ mm Hg}}$ to get the pressure in kilopascals:

$$\left(740 \text{ mm Hg}\right)\left(\frac{101.325 \text{ kPa}}{760 \text{ mm Hg}}\right) = 98.7 \text{ kPa}$$

(5) **1,900 torr.** Multiply your initial value of atmospheres by the conversion factor $\frac{760 \text{ torr}}{1 \text{ atm}}$ to get the pressure in torr:

$$\left(2.5 \text{ atm}\right)\left(\frac{760 \text{ torr}}{1 \text{ atm}}\right) = 1,900 \text{ torr}$$

(6) **0.4 L.** You're given an initial pressure, an initial volume, and a final pressure. Boyle's law leaves you with one unknown: final volume. Solve for the final volume by plugging in the known values:

$$P_1 V_1 = P_2 V_2$$
$$V_2 = \frac{P_1 V_1}{P_2}$$
$$V_2 = \frac{\left(1 \text{ atm}\right)\left(0.5 \text{ L}\right)}{1.25 \text{ atm}}$$
$$V_2 = 0.4 \text{ L}$$

(7) **2.5 atm.** Under the initial conditions, gas at 5 atm resides in a 1.5 L volume (half of the container's 3 L internal volume). When the seal is removed, the entire 3 L of the container becomes available to the gas, which expands to occupy the new volume. Predictably, its pressure decreases. To calculate the new pressure, P_2, plug in the known values and solve:

$$P_1 V_1 = P_2 V_2$$

$$P_2 = \frac{P_1 V_1}{V_2}$$

$$P_2 = \frac{(5\text{ atm})(1.5\text{ L})}{3\text{ L}}$$

$$P_2 = 2.5\text{ atm}$$

(8) **310 L.** You're given the initial volume and initial temperature of the balloon as well as the balloon's final temperature. Apply Charles's law, plugging in the known values and solving for the final volume. Be careful — you need to express all temperatures in units of kelvins by adding 273 to Celsius temperatures, so your initial temperature is 313 K $(40.0°C + 273)$, and your final temperature is 323 K $(50.0°C + 273)$:

$$\frac{V_1}{T_1} = \frac{V_2}{T_2}$$

$$V_2 = \frac{V_1 T_2}{T_1}$$

$$V_2 = \frac{(300\text{ L})(323\text{ K})}{313\text{ K}}$$

$$V_2 = 310\text{ L}$$

(9) **−38.6°C.** Charles's law is the method here. The unknown is the final temperature, T2. You're given an initial temperature as well as the initial and final volumes. After converting the initial temperature to units of kelvins (20.0°C + 273 = 293 K), plug in the known values and solve for final temperature:

$$\frac{V_1}{T_1} = \frac{V_2}{T_2}$$

$$T_2 = \frac{T_1 V_2}{V_1}$$

$$T_2 = \frac{(293\text{ K})(0.200\text{ L})}{0.250\text{ L}}$$

$$T_2 = 234.4\text{ K}$$

The question asks for the answer in degrees Celsius, so convert the temperature: 234.4 K − 273 = −38.6°C.

REMEMBER

Pay attention to little details like final units when you're answering chemistry questions. They're easy to overlook, but they make all the difference between getting the answer right and getting it wrong.

10 **0.14 atm.** Begin by rearranging your equation for Gay-Lussac's law to solve for the variable you need. In this case, you need to solve for P_2. Then plug the given values into your rearranged equation and do the math:

$$\frac{P_1}{T_1} = \frac{P_2}{T_2}$$

$$P_2 = \frac{P_1 T_2}{T_1}$$

$$P_2 = \frac{(1.3 \text{ atm})(40 \text{ K})}{(360 \text{ K})}$$

$$P_2 = 0.14 \text{ atm}$$

11 Don't put a pressurized and sealed can of gas in a fire if you value your safety. When the imaginary can is thrown into the fire, the temperature of the gas inside begins to increase. This causes the gas molecules inside the can to move faster and faster. As they gain speed, the collisions inside the can increase, so the pressure begins to increase considerably. This corresponds to Gay-Lussac's law, which shows that pressure and temperature are in a direct relationship. Eventually, the pressure will build up to such a point that the can will explode.

12 **224 L.** All three factors (pressure, temperature, and volume) change between initial and final states, so you need to use the combined gas law. The initial values are given in the problem as 290 atm, 283 K, and 0.800 L. The final temperature (273 K) and pressure (1 atm) are known because the question states that the gas ends up at STP. So the only unknown is the final volume (V_2). Rearrange the combined gas law to solve for this value:

$$\frac{P_1 V_1}{T_1} = \frac{P_2 V_2}{T_2}$$

$$V_2 = \frac{P_1 V_1 T_2}{T_1 P_2}$$

$$V_2 = \frac{(290 \text{ atm})(0.80 \text{ L})(273 \text{ K})}{(283 \text{ K})(1 \text{ atm})}$$

$$V_2 = 224 \text{ L}$$

13 **0.244 L.** You're given an initial volume, initial temperature, and initial pressure. You're also given a final pressure. The problem doesn't specifically tell you what the final temperature is, but you're told the temperature increases by 10.0°C, so you can determine the final temperature by adding 10 to the initial temperature. (Be sure to then convert both temperatures to kelvins. The initial temperature is 27.0°C + 273 = 300. K, and the final temperature is 37°C + 273 = 310. K.) The only unknown is final volume. Rearrange the combined gas law to solve for final volume, V_2:

$$\frac{P_1 V_1}{T_1} = \frac{P_2 V_2}{T_2}$$

$$V_2 = \frac{P_1 V_1 T_2}{T_1 P_2}$$

$$V_2 = \frac{(105 \text{ kPa})(0.450 \text{ L})(310. \text{ K})}{(300. \text{ K})(200. \text{ kPa})}$$

$$V_2 = 0.244 \text{ L}$$

Even though the final temperature is higher than the initial temperature, the final volume is much smaller than the initial volume. In effect, 10 trifling degrees is no match for the pressure exerted by Bertha's posterior.

(14) **59.7 mol.** This problem simply requires the ideal gas law, arranged to solve for number of moles, n. The ideal gas law involves the constant R, so your units must match the units of the constant. You have atmospheres, liters, and degrees Celsius, so you must convert the Celsius temperature to the Kelvin scale by adding 273 to it: $29.0°C + 273 = 302$ K. Plug in all the values and solve:

$$PV = nRT$$

$$n = \frac{PV}{RT}$$

$$n = \frac{(98.69 \text{ atm})(15.0 \text{ L})}{\left(0.08206 \ \frac{\text{L} \cdot \text{atm}}{\text{mol} \cdot \text{K}}\right)(302 \text{ K})}$$

$$n = 59.7 \text{ mol}$$

(15) **9.35 atm.** This problem is very straightforward. One key concept to keep in mind when working with partial pressures is that all your gas units must be the same. In this case, all three of the given pressures are already in atmospheres. Simply add your three pressures together following Dalton's law, and you'll have your answer:

$$P_{\text{total}} = P_1 + P_2 + P_3$$

$$= 1.2 \text{ atm} + 7.8 \text{ atm} + 0.35 \text{ atm}$$

$$= 9.35 \text{ atm}$$

(16) **98.1 kPa.** The system has come to equilibrium, so the interior of the tube contains a gaseous mixture of hydrogen gas and water vapor. Because the water levels inside and outside the tube are equal, you know that the total pressure inside the tube equals the ambient pressure of the lab, 101.325 kPa. The total pressure includes the partial pressure contributions from hydrogen gas and from water vapor. Set up an equation using Dalton's law, rearrange the equation to solve for the pressure of just the hydrogen gas, plug in your numbers, and solve:

$$P_{\text{total}} = P_{\text{H}_2} + P_{\text{H}_2\text{O}}$$

$$P_{\text{H}_2} = P_{\text{total}} - P_{\text{H}_2\text{O}}$$

$$P_{\text{H}_2} = 101.325 \text{ kPa} - 3.17 \text{ kPa}$$

$$P_{\text{H}_2} = 98.1 \text{ kPa}$$

(17) **Hydrogen, H_2.** The question states that the ratio of the rates is 4.0. Oxygen gas is a diatomic element, so it's written as O_2 and has a molar mass of 32.00 g/mol. Substitute these known values into Graham's law to determine the molar mass of the unknown gas. The problem states that the unknown gas effuses at a rate 4.0 times faster than oxygen, so put the unknown gas over oxygen for the ratio. (In short, the unknown gas is A, and the oxygen is B.)

$$\frac{\text{Rate A}}{\text{Rate O}_2} = \sqrt{\frac{\text{Molar mass O}_2}{\text{Molar mass A}}}$$

$$4.0 = \sqrt{\frac{32.00 \text{ g/mol}}{\text{Molar mass A}}}$$

Square both sides of this equation to cancel out the square root and then solve for the molar mass of unknown Gas A:

$$16.00 = \frac{32.00 \text{ g/mol}}{\text{Molar mass A}}$$

$$\text{Molar mass A} = \frac{32.00 \text{ g/mol}}{16.00}$$

$$= 2.00 \text{ g/mol}$$

After you have the molar mass, you can check the periodic table to determine the identity of the element by matching up the molar mass. No element specifically matches the 2.00 g/mol; however, if you remember that hydrogen is diatomic and is always written as H_2, you can identify the unknown gas as hydrogen!

If you're ready to test your skills a bit more, take the following chapter quiz that incorporates all the chapter topics.

Whaddya Know? Chapter 17 Quiz

Ready for a quiz? The 15 questions in this section will test the skills you learned in this chapter. When you're done, check out the section that follows for answers and explanations.

1. In what phase of matter (solid, liquid, or gas) do the particles have the highest amount of energy? In what phase of matter do the particles have the least amount of energy?

2. Convert 130.54 kPa to mmHg and atm.

3. If gas at an initial pressure of 6.3 atm is allowed to expand from a volume of 22.4 L to a volume of 75.2 L, what is the pressure at the new volume?

4. You begin with a gas that has a volume of 3 L and a pressure of 10 atm, what happens to the volume of that gas when you change the pressure to 190 kPa?

5. Your friend gives you a balloon for a present. They tell you this balloon has a volume of 1.5 L. The temperature of the room you are in is 28°C. What will the volume of the balloon be if you go outside with it, where the temperature is a chilly 9°C, and allow the balloon to reach the same temperature as the outside air?

6. According to Charles's law, if you had a gas and could somehow lower the volume until you reached a theoretical volume of zero (which is obviously impossible). What would the temperature be? What is this temperature called and what happens to molecules at this temperature?

7. What must the final temperature be of a gas that was at 150 kPa at 29°C if the final pressure is 89 kPa? The gas is in a rigid metal container that holds 10 L of gas.

8. The gas inside of a flexible container at 25°C has a pressure of 0.25 atm. The size of the container starts out holding 3 L of gas. When it is heated to 30°C, its pressure is also increased to 0.27 atm. After this change, what is the new volume of the gas?

9. If you have 21 liters of gas held at a pressure of 105 kPa and a temperature of 450 K, what will be the volume of the gas if you decrease the pressure to 80 kPa and decrease the temperature to 750 K?

10. You have 6 L of nitrogen gas (N_2) at a pressure of 150 kPa and a temperature of 50°C, how many grams of nitrogen are present?

11. 2.50 grams of XeF_4 is introduced into an evacuated 3.00 liter container at 80.0°C. What is the pressure in the container?

12. Find the volume of a gas at 800.0 mm Hg and 40.0C if its volume at 720.0 mm Hg and 15.0°C is 6.84 L.

13 What is the mass in grams of 18.9 L of NH_3 at 31.0°C and 97.97 kPa?

14 You have 3 gases mixed together in a rigid, closed sphere. The pressure of the individual gases are as follows: Gas 1 = 5.6 atm, Gas 2 = 10.3 atm, Gas 3 = 1.4 atm. What is the total pressure of the all the gases in the sphere?

15 Compare the rate of effusion between oxygen gas (O_2) and methane gas (CH_4)? Which effuses faster and by how much?

Answers to Chapter 17 Quiz

(1) Gas particles have the most energy and move the fastest relative to solid and liquid particles. Gas particles move randomly throughout the entire volume of their container, they are highly energetic. Solid particles have the least energy and move the slowest relative to liquid and gas particles. They generally are structured and not very energetic.

(2)
$$(130.54 \text{ kPa})\left(\frac{760 \text{ mm Hg}}{101.325 \text{ kPa}}\right) = 979.13 \text{ mm Hg}$$

$$(130.54 \text{ kPa})\left(\frac{1 \text{ atm}}{101.325 \text{ kPa}}\right) = 1.2883 \text{ atm}$$

(3) **1.88 atm.** To solve problems involving gas variables it is always helpful to first identify which variables you have and what is given to you. Once you've done this you can determine which equation you are going to use and then go from there to solve it. In this problem you are told you have:

$$P_1 = 6.3 \text{ atm} \quad V_1 = 22.4 \text{ L} \quad P_2 = ? \quad V_2 = 75.2 \text{ L}$$

These are determined from reading the problem. Make sure you pay attention to keywords like "initial" or similar wording to help you identify what numbers go with which set of variables. Pay attention also to what the problem is asking. In this case, it says "what is the pressure at the new volume?" This could seem a bit confusing as it mentions pressure and volume in the question statement but if you read it closer you can clearly see it is asking you to identify the pressure at the second volume given in the problem. Problems like these usually are not written to confuse you, so don't think there is some extra trick to solving them. Just pay attention to exactly what the words say and don't get overwhelmed beyond that. Once you've identified the variables you know you have pressure and volume so you are going to use Boyle's law to solve this:

$$P_1V_1 = P_2V_2$$

$$P_2 = \frac{P_1V_1}{V_2}$$

$$P_2 = \frac{(6.3 \text{ atm})(22.4 \text{ L})}{(75.2 \text{ L})}$$

$$P_2 = 1.88 \text{ atm}$$

(4) **15.96 atm.** This problem is much the same as the last one. The only major difference here is that you have two pressures given to you that are not in the same units. To solve this correctly you need to convert one of the units to the other so that both pressures have the same units. It doesn't matter what the units are as long as they are the same for pressure. If you are using R in an ideal gas law problem like you will later in this quiz, you need to have the pressure units in atm. This problem doesn't need R, though, so you can convert them to whatever you want. As a piece of advice, if you need to convert units you might as well convert them to atm if that is an option as that can help you avoid errors later. Once that is done identify your variables, determine the correct equation, and solve your problem:

$$(190 \text{ kPa})\left(\frac{1 \text{ atm}}{101.325 \text{ kPa}}\right) = 1.88 \text{ atm}$$

$$P_1 = 10 \text{ atm} \quad V_1 = 3 \text{ L} \quad P_2 = 1.88 \text{ atm} \quad V_2 = ?$$

$$V_2 = \frac{P_1 V_1}{P_2}$$

$$P_2 = \frac{(10 \text{ atm})(3 \text{ L})}{(1.88 \text{ atm})}$$

$$P_2 = 15.96 \text{ atm}$$

5) In this problem you encounter temperatures given to you in Celsius. When doing gas law calculations you need to convert your Celsius temperatures to Kelvin temperature before you can use those temperatures in the problem.

$$28°C + 273 = 301 \text{ K}$$
$$9°C + 273 = 282 \text{ K}$$
$$P_1 = 150 \text{ kPa} \quad T_1 = 301 \text{ K} \quad T_2 = 282 \text{ K} \quad V_2 = ?$$
$$\frac{V_1}{T_1} = \frac{V_2}{T_2}$$
$$V_2 = \frac{V_1 T_2}{T_1}$$
$$V_2 = \frac{(1.5 \text{ L})(282 \text{ K})}{(301 \text{ K})}$$
$$V_2 = 1.41 \text{ L}$$

6) Compressing a gas to a volume of zero is of course impossible. However, in theory, if you could compress a gas to a volume of zero that would lead to the temperature of that gas being a −273°C or 0 K which is called absolute zero. At absolute zero all molecular motion stops. Nothing moves at any level. It is the absolute bottom of temperature and would indicate that the atoms present at absolute zero had zero energy. To prove this you could use any two values of your initial temperature and then volume and then set the final volume to 0. When you look at the equation below you can see that no matter what your starting initial volume and temperature, if the final volume were to be 0 then the final temperature will always be 0 as well. That zero would correspond to 0 K, absolute zero.

$$\frac{V_1}{T_1} = \frac{V_2}{T_2}$$
$$\frac{V_2 T_1}{V_1} = T_2$$

7) **179.19 K.** The challenge with this problem is to make sure you realize you can ignore the 10 L of gas. Why? It is the only volume mentioned in the problem and it doesn't change. Since there is only one volume, and it is held constant, you can safely not worry about it when you are given multiple pressures and multiple temperatures. If you were to plug it into the combined gas law you would just be using 10 L for both V_1 and V_2 and it wouldn't make

any difference to your answer which is why you ignore it. Remember also to convert from Celsius to Kelvin.

$29°C + 273 = 302$ K

$P_1 = 150$ kPa $P_2 = 89$ kPa $T_1 = 302$ K $T_2 = ?$

$$\frac{P_1}{T_1} = \frac{P_2}{T_2}$$

$$T_2 = \frac{P_2 T_1}{P_1}$$

$$T_2 = \frac{(89 \text{ kPa})(302 \text{ K})}{(150 \text{ kPa})}$$

$$T_2 = 179.19 \text{ K}$$

8 **2.82 L.** This is a straightforward combined gas law problem. You can identify this because you have all 3 variables present and each changes in some way. Remember to convert the temperatures to Kelvin.

$T_1 = 25°C + 273 = 298$ K $P_1 = 0.25$ atm $V_1 = 3$ L
$T_2 = 30°C + 273 = 303$ K $P_2 = 0.27$ atm $V_2 = ?$

$$\frac{V_1 P_1}{T_1} = \frac{V_2 P_2}{T_2}$$

$$V_2 = \frac{T_2 V_1 P_1}{T_1 P_2}$$

$$V_2 = \frac{(303 \text{ K})(3 \text{ L})(0.25 \text{ atm})}{(298 \text{ K})(0.27 \text{ atm})}$$

$$V_2 = 2.82 \text{ L}$$

9 **46 L.** This is another combined gas law problem. In this case make sure not to automatically convert the temperatures to Kelvin temperature. They are already in Kelvin so you do not need to add 273 to them.

$V_1 = 21$ L $T_1 = 450$ K $P_1 = 105$ kPa
$V_2 = ?$ $T_2 = 750$ K $P_2 = 80$ kPa

$$\frac{V_1 P_1}{T_1} = \frac{V_2 P_2}{T_2}$$

$$V_2 = \frac{T_2 V_1 P_1}{T_1 P_2}$$

$$V_2 = \frac{(750 \text{ K})(21 \text{ L})(105 \text{ kPa})}{(450 \text{ K})(80 \text{ kPa})}$$

$$V_2 = 46 \text{ L}$$

10 **9.25 g N_2.** This problem gives you a volume (6 L), a pressure (150 kPa), and a temperature ($50°C + 273 = 323$ K) along with identifying the chemical formula of the gas. There is no change taking place so you are not using anything derived from the combined gas law. Instead you need to determine the grams of the gas present. To do this you need to solve for the number of moles and then convert that mole value to grams using the molar mass of the gas. You do this using the ideal gas law. The ideal gas law involves using the gas constant R.

One of the units that R uses is atm which means that any pressure units you have in your problem need to be converted to atm as shown here:

$$(150 \text{ kPa})\left(\frac{1 \text{ atm}}{101.325 \text{ kPa}}\right) = 1.48 \text{ atm} = P$$

Once done with that you use the ideal gas law to solve:

$$PV = nRT$$

$$n = \frac{PV}{RT}$$

$$n = \frac{(1.48 \text{ atm})(6 \text{ L})}{\left(0.0821 \frac{\text{L} \cdot \text{atm}}{\text{mol} \cdot \text{K}}\right)(323 \text{ K})}$$

$$n = 0.33 \text{ mol N}_2$$

$$\text{grams of N}_2 = (0.33 \text{ mol N}_2)\left(\frac{28.02 \text{ g N}_2}{1 \text{mol N}_2}\right) = 9.25 \text{ g N}_2$$

(11) **0.12 atm.** This is another ideal gas law question. There is only one set of PTV values given to you with no change taking place. To determine the mole value you need to convert from grams to moles of XeF_4 as shown here:

$$(2.5 \text{ g XeF}_4)\left(\frac{1 \text{ mol XeF}_4}{207.28 \text{ g XeF}_4}\right) = 0.012 \text{ mol XeF}_4 = n$$

$$V = 3 \text{ L} \qquad T = 80°C + 273 = 353 \text{ K} \qquad P = ? \qquad n = 0.012 \text{ mol XeF}_4$$

$$PV = nRT$$

$$P = \frac{nRT}{V}$$

$$P = \frac{(0.012 \text{ mol XeF}_4)\left(0.0821 \frac{\text{L} \cdot \text{atm}}{\text{mol} \cdot \text{K}}\right)(353 \text{ K})}{(3 \text{ L})}$$

$$P = 0.12 \text{ atm}$$

(12) **6.69 L.** This is a combined gas law problem because you are given temperature, pressure, and volume values in an initial and final state. Remember to convert your Celsius temperature to Kelvin.

$$V_1 = ? \qquad T_1 = 40.0°C + 273 = 313 \text{ K} \qquad P_1 = 800.0 \text{ mm Hg}$$

$$V_2 = 6.84 \text{ L} \qquad T_2 = 15.0°C + 273 = 288 \text{ K} \qquad P_2 = 720.0 \text{ mm Hg}$$

$$\frac{V_1 P_1}{T_1} = \frac{V_2 P_2}{T_2}$$

$$\frac{T_1 V_2 P_2}{T_2 P_1} = V_2$$

$$V_2 = \frac{(313 \text{ K})(6.84 \text{ L})(720.0 \text{ mm Hg})}{(288 \text{ K})(800.0 \text{ mm Hg})}$$

$$V_2 = 6.69 \text{ L}$$

(13) **12.44 g NH_3.** This problem is asking you for a mass of a gas. The only way to get to a mass is to calculate moles using the ideal gas law and then convert that mole value to grams. Remember to convert your Celsius temperature to Kelvin and your kPa to atm.

$$(97.97 \text{ kPa})\left(\frac{1 \text{ atm}}{101.325 \text{ kPa}}\right) = 0.97 \text{ atm} = P$$

$$T = 31°C + 273 = 304 \text{ K} \qquad V = 18.9 \text{ L} \qquad n = ? \qquad P = 0.97 \text{ atm}$$

$$PV = nRT$$

$$n = \frac{PV}{RT}$$

$$n = \frac{(0.97 \text{ atm})(18.9 \text{ L})}{\left(0.0821 \frac{\text{L} \cdot \text{atm}}{\text{mol} \cdot \text{K}}\right)(304 \text{ K})}$$

$$n = 0.73 \text{ mol } NH_3$$

$$\text{grams of } N_2 = (0.73 \text{ mol } NH_3)\left(\frac{17.04 \text{ g } NH_3}{1 \text{ mol } NH_3}\right) = 12.44 \text{ g } NH_3$$

(14) **17.3 atm.** This problem gives you 3 different pressure values. Since nothing else is given you simply can use Dalton's law of partial pressures and add the pressure values together to determine the total pressure:

$$5.6 \text{ atm} + 10.3 \text{ atm} + 1.4 \text{ atm} = 17.3 \text{ atm total pressure}$$

(15) CH_4 will effuse faster at a rate of 1 mole of CH_4 for every 0.71 mole of O_2. When working on a gaseous effusion problem make sure you clearly define which gas is 1 and which gas is 2 in the formula you will be using. This is absolutely key to make sure you understand what the ratio is showing. The simplest way to do this is to identify the first gas given in the problem as gas 1(O_2, oxygen, and the second gas given as gas 2 (CH_4, methane). Once you know this, plug their molar masses into the formula and calculate their ratio:

$$\frac{R_1}{R_2} = \sqrt{\left(\frac{M_2}{M_1}\right)}$$

$$\frac{R_{O_2}}{R_{CH_4}} = \sqrt{\left(\frac{16.04}{32}\right)}$$

$$\frac{R_{O_2}}{R_{CH_4}} = 0.71$$

The hardest part about these problems sometimes is making sure you understand what the answer is showing you. In this case it means that 0.71 moles of O_2 will effuse for every 1 mole of CH_4 gas. It is a ratio and should be viewed as such. This means that CH_4 effuses faster, which makes sense because the molar mass of CH_4 is smaller than diatomic oxygen.

6 Studying Solutions

In This Unit . . .

Chapter **18**

Dissolving into Solutions

C ompounds can form mixtures. When compounds mix completely, right down to the level of individual molecules, chemists call the mixture a *solution*. Although most people think "liquid" when they think of solutions, a solution can be a solid, liquid, or gas. The only criterion is that the components are completely intermixed.

You encounter solutions all the time in everyday life. The air you breathe is a solution. That sports drink you use to replenish your electrolytes is a solution. That soft drink and that hard drink are both solutions. Your tap water is most likely a solution, too. In this chapter, we show you some of the properties of solutions and introduce you to the ways chemists represent a solution's concentration.

Seeing Different Forces at Work in Solubility

A *solution* is a homogeneous mixture, meaning that its properties are the same throughout. If you dissolve sugar in water and mix it really well, for example, your mixture is basically the same no matter where you sample it.

A solution is composed of a solvent and one or more solutes. The *solvent* is the substance that doesn't change state and is present in the largest amount, and the *solute* is the substance that changes state and is present in the lesser amount. You can determine which is which based on the quantities most of the time, but in a few cases of extremely soluble salts, such as lithium chloride, more than 5 g of salt can be dissolved in 5 mL of water. However, water is still considered the solvent, because it's the species that has not changed state.

A solution can have more than one solute. If you dissolve salt in water to make a brine solution and then dissolve some sugar in the same solution, you have two solutes, salt and sugar. You still have only one solvent, though: water.

There can also be solutions of gases. Our atmosphere, for example, is a solution. Because air is almost 79 percent nitrogen, nitrogen is considered the solvent, and the oxygen, carbon dioxide, and other gases are considered the solutes. Solids can also make solutions. Alloys, for example, are solutions of one metal in another metal. Brass is a solution of zinc in copper.

Understanding solubility

Why do some things dissolve in one solvent and not in another? For example, oil and water don't mix to form a solution, but oil dissolves in gasoline. A general rule of solubility says that *like dissolves like* in regard to polarity of both the solvent and solutes. Water, for example, is a polar material; it's composed of polar covalent bonds with the positive and negative ends of the molecule. (For a rousing discussion of water and its polar covalent bonds, see Chapter 15.) Water dissolves polar solutes, such as salts and alcohols. Oil, however, is composed of largely nonpolar bonds. So water doesn't act as a suitable solvent for oil.

When one liquid is added to another, the extent to which they intermix is called *miscibility*. Typically, liquids that have similar properties mix well — they're *miscible*. Liquids with dissimilar properties often don't mix well — they're *immiscible*.

You may understand miscibility in terms of the Italian Salad Dressing Principle. Inspect a bottle of Italian salad dressing that has been sitting in your refrigerator. The dressing consists of two distinct layers, an oily layer and a watery layer. Before using, you must shake the bottle to temporarily mix the layers. Eventually, they'll separate again because water is polar and oil is nonpolar. Polar and nonpolar liquids mix poorly, though occasionally with delicious consequences.

TIP

Comparing polarity between components is often a good way to predict solubility, regardless of whether those components are liquid, solid, or gas. Why is polarity such a good predictor? Because polarity is central to the tournament of forces that underlies solubility. So solids held together by ionic bonds (the most polar type of bond) or polar covalent bonds tend to dissolve well in polar solvents, like water.

REMEMBER

You know from your own experiences, we're sure, that only so much solute can be dissolved in a given amount of solvent. You've probably been guilty of putting far too much sugar in iced tea. No matter how much you stir, there's some undissolved sugar at the bottom of the glass. The reason is that the sugar has reached its maximum solubility in water at that temperature. *Solubility* is the maximum amount of solute that will dissolve in a given amount of a solvent at a specified temperature. Solubility normally has the units of grams solute per 100 milliliters of solvent (g/100 mL).

The solubility is related to the temperature of the solvent. For solids dissolving in liquids, solubility normally increases with increasing temperature. So if you heat that iced tea, the sugar at the bottom readily dissolves. However, for gases dissolving in liquids, such as oxygen dissolving in lake water, the solubility goes down as the temperature increases. This is the basis of *thermal pollution,* the addition of heat to water that decreases the solubility of the oxygen and affects the aquatic life.

Looking at forces in solutions

Introducing a solute into a solvent initiates a tournament of forces. Attractive forces between solute and solvent compete with attractive solute–solute and solvent–solvent forces. A solution forms only to the extent that solute–solvent forces dominate over the others. The process in which solvent molecules compete and win in the tournament of forces is called *solvation* or, in the specific case where water is the solvent, *hydration.* Solvated solutes are surrounded by solvent molecules. When solute ions or molecules become separated from one another and surrounded in this way, we say they're *dissolved.*

Imagine that the members of a ridiculously popular boy band exit their hotel to be greeted by an assembled throng of fans and the media. The band members attempt to cling to each other but are soon overwhelmed by the crowd's ceaseless, repeated attempts to get closer. Soon, each member of the band is surrounded by his own attending shell of reporters and hyperventilating teenage girls. That boy band was just dissolved much like solutes are dissolved by solvents.

The tournament of forces plays out differently among different combinations of components. In mixtures where solute and solvent are strongly attracted to one another, more solute can be dissolved. One factor that always tends to favor dissolution is *entropy,* a kind of disorder or "randomness" within a system. Dissolved solutes are less ordered than undissolved solutes.

Classifying solutions by concentration

You'll see three major types of solutions in a general chemistry class:

>> **Saturated:** Beyond a certain point, adding more solute to a solution doesn't result in a greater amount of solvation. At this point, the solution is in dynamic equilibrium; the rate at which solute becomes solvated equals the rate at which dissolved solute *crystallizes,* or falls out of solution. A solution in this state is *saturated.*

>> **Unsaturated:** An *unsaturated* solution is one that can accommodate more solute.

>> **Supersaturated:** A *supersaturated* solution is one in which more solute is dissolved than is necessary to make a saturated solution. A supersaturated solution is unstable; solute molecules may crash out of solution given the slightest perturbation. The situation is like that of Wile E. Coyote, who runs off a cliff and remains suspended in the air until he looks down — at which point he inevitably falls. Supersaturated solutions are most easily created by heating the solution to the point where it can accommodate more solute than it could normally handle.

REMEMBER

The concentration of solute required to make a saturated solution is the *solubility* of that solute. Solubility varies with the conditions of the solution. The same solute may have different solubility in different solvents, at different temperatures, and so on.

If a solution is unsaturated, then the amount of solute that is dissolved may vary over a wide range. A couple of rather indefinite terms describe the relative amount of solute and solvent that you can use:

>> You can say that the solution is *dilute,* meaning that, relatively speaking, it contains very little solute per given amount of solvent. If you dissolve 0.01 g of sodium chloride in a liter of water, for example, the solution is dilute. When asked for an example of a dilute solution, one student replied, "A $1 margarita." She was right — a lot of solvent (water) and a very little solute (tequila) are used in her example.

>> A solution may be *concentrated,* containing a large amount of solute per the given amount of solvent. If you dissolve 200 g of sodium chloride in a liter of water, for example, the solution is concentrated.

EXAMPLE

Q. Sodium chloride dissolves more than 25 times better in water than in methanol. Explain this difference, referring to the structure and properties of water, methanol, and sodium chloride.

A. Sodium chloride (NaCl) is an ionic solid, a lattice composed of sodium cations (atoms with positive charge) alternating with chlorine anions (atoms with negative charge). A *lattice* has a highly regular, idealized geometry and is held together by ionic bonds, the most polar type of bond. To dissolve NaCl, a solvent must be able to engage in very polar interactions with these ions and do so with near-ideal geometry. The structure and properties of water, which is polar, are better suited to this task than are those of methanol (see the following figure). The two O–H bonds of water (on the left) partially sum to produce a strong dipole along the mirror-image plane of the molecule that runs between the two hydrogen atoms. Methanol (on the right) is also polar, due largely to its own O–H bond, but it's less polar than water. In solution, water molecules can orient their dipoles cleanly and in either of two directions to interact favorably with Na^+ or Cl^- ions. Methanol molecules can engage in favorable interactions with these ions, too, but not nearly as well as water.

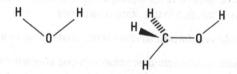

YOUR
TURN

1. *Lattice energy* is a measure of the strength of the interactions between ions in the lattice of an ionic solid, measured in kilojoules per mole (kJ/mol). The greater the lattice energy, the stronger the ion–ion interactions. Here's a table of ionic solids and their associated lattice energies. Rank these ionic solids in order of their solubility in water, starting with the most soluble.

Sodium Salt	Lattice Energy (kJ/mol)
NaBr	747
NaCl	787
NaF	923
NaI	704

2. Ethanol, CH_3CH_2OH, is miscible with water. Octanol, $CH_3(CH_2)_7OH$, isn't miscible in water. Is sucrose ($C_{12}H_{22}O_{11}$, table sugar) likely to be more soluble in ethanol or octanol? Why?

Concentrating on Molarity and Percent Solutions

Different solutes dissolve to different extents in different solvents under different conditions. How can anybody keep track of all these differences? Just referring to a solution as dilute or concentrated conveys a bit of information but nothing exact. Chemists need to know the actual concentration of a solution, a number. This section shows you two of the most important ways to describe concentration: *molarity* and *percent composition of solutions*.

Molarity

Molarity relates the amount of solute in moles to the volume of the solution:

$$\text{Molarity}\ (M) = \frac{\text{Moles of solute}}{\text{Liters of solution}}$$

Molarity is the concentration unit most often used by chemists, because it utilizes moles. The mole concept is central to chemistry, and molarity lets chemists easily work solutions into reaction stoichiometry.

To calculate molarity, you may have to use conversion factors to move between units. For example, if you're given the mass of a solute in grams, use the molar mass (usually rounded to two decimal places) of that solute to convert the given mass into moles (check out Chapter 12 for an introduction to molar mass). If you're given the volume of solution in milliliters or some other unit, you need to convert that volume into liters. Units are the very first thing to check if you get a problem wrong when using molarity. Make sure your units are correct!

The units of molarity are always moles per liter (mol/L or mol·L^{-1}). These units are often abbreviated as M and referred to as "molar." Thus, 0.25 M KOH(aq) is described as "point two-five molar potassium hydroxide," and it contains 0.25 mol of KOH per liter of solution. Note that this does *not* mean that there are 0.25 mol KOH per liter of *solvent* (water, in this case) — only the final volume of the solution (solute plus solvent) is important in molarity.

Like other units, the unit of molarity can be modified by standard prefixes, as in millimolar (mM, which equals 10^{-3} mol/L) and micromolar (μM, which equals 10^{-6} mol/L).

Molarity also proves very valuable when you need to create a solution of a certain concentration. For example, you can take 1 mol of KCl (formula mass of 74.55 g/mol) and dissolve and dilute the 74.55 grams into a volume of 1 L of solution in a volumetric flask. You then have a 1-molar solution of KCl. You can label that solution as 1 M KCl. You don't add the 74.55 g to 1 L of water. You want to end up with a final volume of 1 L. When preparing molar solutions, always dissolve and dilute to the required volume. This process is shown in Figure 18-1.

Percent composition of solutions

You've probably looked at a bottle of vinegar and seen "5% acetic acid," a bottle of hydrogen peroxide and seen "3% hydrogen peroxide," or a bottle of bleach and seen "5% sodium hypochlorite." Those percentages are expressing the concentration of that particular solute in each solution. *Percentage* is the amount per 100. Depending on the way you choose to express the percentage, the units of amount per 100 vary. Unfortunately, although the percentage of solute is often listed, the method (mass/mass, mass/volume, or volume/volume) is not. You can normally assume that the method is mass/mass, but we're sure you know about assumptions.

Three different percentages are commonly used. We give you the rundown in the following sections.

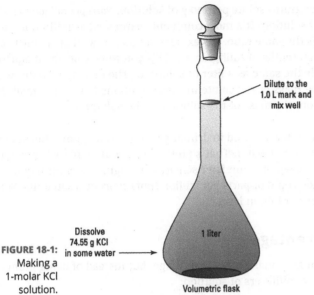

FIGURE 18-1:
Making a
1-molar KCl
solution.

Dilute to the
1.0 L mark and
mix well

Dissolve
74.55 g KCl
in some water

1 liter

Volumetric flask

Mass/mass (m/m) percentage

In *mass/mass percentage*, or *mass percentage*, the mass of the solute is divided by the mass of the solution and then multiplied by 100 to get the percentage. Another term that is sometimes used is *mass percentage (mass/mass percentage)*. Normally the mass unit is grams. Mathematically, it looks like this:

$$^m/_m\% = \frac{\text{grams solute}}{\text{grams solution}} \times 100\%$$

If, for example, you dissolve 5.0 g of sodium chloride in 50 g of water, the mass percent is

$$^m/_m\% = \frac{5.0 \text{ g NaCl}}{55 \text{ g water}} \times 100\% = 9.1\%$$

Therefore, the solution is a 9.1 percent (*m/m*) solution.

Mass percentage is the easiest percentage solution to make, but sometimes you may need to know the volume of the solution. In this case, you can use the mass/volume percentage.

Very dilute concentrations (as in the concentration of a contaminant in drinking water) are sometimes expressed as a special mass percent called *parts per million* (ppm) or *parts per billion* (ppb). Percentage and molarity, and even molality (see Chapter 19), are convenient units for the solutions that chemists routinely make in the lab or the solutions that are commonly found in nature. However, if you begin to examine the concentrations of certain pollutants in the environment, you find that those concentrations are very, very small.

Percentage is parts per hundred, or grams solute per 100 g of solution. *Parts per million (ppm)* is grams solute per million grams of solution. It's most commonly expressed as milligrams solute per kilogram solution, which is the same ratio. It's expressed this way so that chemists can easily weigh out milligrams or even tenths of milligrams; and if you're talking about aqueous solutions, a kilogram of solution is the same as a liter of solution. (The density of water is 1 g per milliliter, or 1 kg per liter. The mass of the solute in these solutions is so very small that it's negligible when converting from the mass of the solution to the volume.)

By law, the maximum contamination level of lead in drinking water is 0.05 ppm. This number corresponds to 0.05 mg of lead per liter of water. That's pretty dilute. But mercury is regulated at the 0.002 ppm level. Sometimes, even this unit isn't sensitive enough, so environmentalists have resorted to the parts per billion (ppb) or parts per trillion (ppt) concentration units. Some neurotoxins are deadly at the parts per billion level.

Mass/volume (m/V) percentage

Mass/volume percentage is very similar to mass/mass percentage, but instead of using grams of solution in the denominator, it uses milliliters of solution:

$$m/V\% = \frac{\text{grams solute}}{\text{mL solution}} \times 100\%$$

Using this concept, imagine you want to make 100 mL of a 15 percent (m/V) potassium nitrate solution. If you weren't familiar with how to use mass/volume percentage to describe concentration, you'd be out of luck.

Now because you're making 100 mL, you already know that you're going to weigh out 15 g of potassium nitrate (commonly called saltpeter — KNO_3). Now, here comes something that's a little different: You dissolve the 15 g of KNO_3 in a little bit of water and dilute it to exactly 100 mL in a volumetric flask. In other words, you dissolve and dilute 15 g of KNO_3 to 100 mL. You won't know exactly how much water you put in, but it's not important as long as the final volume is 100 mL.

Volume/volume (*V/V*) percentage

With *volume/volume percentages*, both the solute and solution are expressed in milliliters:

$$V/V\% = \frac{\text{mL solute}}{\text{mL solution}} \times 100\%$$

Ethyl alcohol (drinking alcohol) solutions are commonly made using volume/volume percentages. If you want to make 100 mL of a 50 percent ethyl alcohol solution, you take 50 mL of ethyl alcohol and dilute it to 100 mL with water. Again, it's a case of dissolving and diluting to the required volume. You can't simply add 50 mL of alcohol to 50 mL of water — you'd get less than 100 mL of solution. The polar water molecules attract the polar alcohol molecules, which tends to fill in the open framework of water molecules and prevents the volumes from simply being added together.

REMEMBER Clearly, paying attention to units is important when working with concentration. Only by observing which units are attached to a measurement can you determine whether you're working with molarity or with a mass-mass, mass-volume, or volume-volume percent solution.

EXAMPLE

Q. How would you make 350.0 g of a 5 percent (*m/m*) sucrose, or table sugar, solution?

A. You know that 5 percent of the mass of the solution is sugar, so you can multiply the 350.0 g by 0.05 to get the mass of the sugar:

$350.0 \text{ g} \times 0.05 = 17.5 \text{ g of sugar}$

The rest of the solution (350.0 g − 17.5 g = 332.5 g) is water. You can simply weigh out 17.5 g of sugar and add it to 332.5 g of water to get your 5 percent (*m/m*) solution.

Q. How would you prepare 2.00 L of a 0.550 M KCl solution?

A. First, calculate how much KCl you need to weigh out. The solution must have 0.550 mol of potassium chloride per liter. If you convert from moles of KCl to grams of KCl, you get the number of grams of KCl per liter. You're preparing 2.00 L, so multiplying by 2.00 L gives you the number of grams of KCl.

$$\frac{0.550 \text{ mol KCl}}{1 \text{ L}} \times \frac{74.55 \text{ g KCl}}{1 \text{ mol}} \times \frac{2.00 \text{ L}}{1} = 82.0 \text{ g KCl}$$

After doing the calculations, take that 82.0 g of KCl and dissolve and dilute it to 2.00 L.

Q. Calculate the molarity and the mass/volume percent solution obtained by dissolving 102.9 g H_3PO_4 into 642 mL final volume of solution. Be sure to use proper units. (*Hint:* 642 mL = 0.642 L.)

A. First, calculate the molarity. Before you can use the molarity formula, though, you must convert grams of H_3PO_4 to moles (see Chapter 12 for details):

$$\left(102.9 \text{ g } H_3PO_4\right)\left(\frac{1 \text{ mol } H_3PO_4}{98.0 \text{ g } H_3PO_4}\right) = 1.05 \text{ mol } H_3PO_4$$

$$\text{Molarity} = \left(\frac{\text{mol solute}}{\text{L solution}}\right) = \left(\frac{1.05 \text{ mol } H_3PO_4}{0.642 \text{ L}}\right) = 1.64 \text{ M } H_3PO_4$$

Next, calculate the mass/volume percent solution:

$$\left(\frac{102.9 \text{ g } H_3PO_4}{642 \text{ mL}}\right) \times 100 = 16.0\% \text{ or } \left(\frac{16.0 \text{ g } H_3PO_4}{100 \text{ mL}}\right)$$

Note that the convention in molarity is to divide moles by *liters*, but the convention in mass percent is to divide grams by *milliliters*. If you prefer to think only in terms of liters (not milliliters), then simply consider mass percent as kilograms divided by liters.

3 Calculate the molarity of these solutions:

    a. 2.0 mol NaCl in 0.872 L of solution

    b. 93 g $CuSO_4$ in 390 mL of solution

    c. 22 g $NaNO_3$ in 777 mL of solution

4 How many grams of solute are in each of these solutions?

    a. 671 mL of 2.0 M NaOH

    b. 299 mL of 0.85 M HCl

    c. 2.74 L of 0.258 M $Ca(NO_3)_2$

5 A 15.0 M solution of ammonia, NH_3, has a density of 0.90 g/mL. What is the mass percent of this solution?

Changing Concentrations by Making Dilutions

Real–life chemists in real–life labs don't make every solution from scratch. Instead, they make concentrated *stock solutions* and then make *dilutions* of those stocks as necessary for a given experiment.

To make a dilution, you simply add a small quantity of a concentrated stock solution to an amount of pure solvent. The resulting solution contains the amount of solute originally taken from the stock solution but disperses that solute throughout a greater volume. Therefore, the final concentration is lower; the final solution is less concentrated and more dilute.

How do you know how much of the stock solution to use and how much of the pure solvent to use? It depends on the concentration of the stock and on the concentration and volume of the final solution you want. You can answer these kinds of pressing questions by using the dilution equation, which relates concentration (C) and volume (V) between initial and final states:

$$C_1V_1 = C_2V_2$$

You can use the dilution equation with any units of concentration, provided you use the same units throughout the calculation. Because molarity is such a common way to express concentration, the dilution equation is sometimes expressed in the following way, where M_1 and M_2 refer to the initial and final molarity, respectively:

$$M_1V_1 = M_2V_2$$

Q. Suppose you want to prepare 500 mL of 2.0 M HCl from a stock solution of 12 M HCl. How much of the stock 12 M HCl is needed for this dilution?

EXAMPLE

A. You need 83.3 mL of the 12 M HCl.

To figure out the volume you need, follow these simple steps:

1. **Use the following formula.**

$$M_1V_1 = M_2V_2$$

In the preceding equation, V_1 is the old volume, or the volume of the original solution, M_1 is the molarity of the original solution, V_2 is the volume of the new solution, and M_2 is the molarity of the new solution.

2. **Solve for V_1 and substitute the values into the equation.**

$$V_1 = \frac{M_2V_2}{M_1}$$

$$V_1 = \frac{(2.0\,\text{M})(500.0\,\text{mL})}{12.0\,\text{M}}$$

$$V_1 = 83.3\,\text{mL}$$

3. **Take 83.3 mL of the 12.0 M HCl solution and dilute it to exactly 500.0 mL.**

Put about 400 mL of water into a 500.0 mL volumetric flask. Slowly add the 83.3 mL of the concentrated HCl as you stir, and then dilute to the final 500 mL with water.

Note: For safety, always add acid to water and never the reverse! Dissolving acid in water is an exothermic reaction that is less violent when acid is added to water than when water is added to acid.

Q. How would you prepare 500 mL of 0.2 M NaOH(*aq*) from a stock solution of 1.5 M NaOH?

A. Add 67 mL 1.5 M NaOH stock solution to 433 mL water.

Use the dilution equation, $M_1V_1 = M_2V_2$. The initial molarity, M_1, comes from the stock solution and is therefore 1.5 M. The final molarity is the one you want in your final solution, which is 0.200 M. The final volume is the one you want for your final solution, 500 mL, which is equivalent to 0.500 L. Using these known values, you can calculate the initial volume, V_1:

$$M_1V_1 = M_2V_2$$

$$V_1 = \frac{M_2V_2}{M_1}$$

$$V_1 = \frac{(0.200 \text{ M})(0.500 \text{ L})}{1.5 \text{ M}} = 0.067 \text{ L}$$

The calculated volume is equivalent to 67 mL. The final volume of the aqueous solution is to be 500 mL, and 67 mL of this volume comes from the stock solution. The remainder, $500 \text{ mL} - 67 \text{ mL} = 433 \text{ mL}$, comes from pure solvent (water, in this case). So to prepare the solution, add 67 mL of 1.5 M stock solution to 433 mL water. Mix and enjoy!

YOUR TURN

6 What is the final concentration in molarity of a solution prepared by diluting 2.50 mL of 3.00 M KCl(aq) up to 0.175 L final volume?

7 A certain mass of ammonium sulfate, $(NH_4)_2SO_4$, is dissolved in water to produce 1.65 L of solution. Then 80.0 mL of this solution is diluted with 120 mL of water to produce 200 mL of 0.200 M $(NH_4)_2SO_4$. What mass of ammonium sulfate was originally dissolved?

Altering Solubility with Temperature

REMEMBER Increasing temperature magnifies the effects of entropy on a system. Because the entropy of a solute is usually increased when it dissolves, increasing temperature usually increases solubility — for solid and liquid solutes, anyway. Another way to understand the effect of temperature on solubility is to think about heat as a reactant in the dissolution reaction:

$$\text{Solid solute} + \text{Water} + \text{Heat} \rightarrow \text{Dissolved solute}$$

Heat is usually absorbed into solute particles when a solute dissolves. Increasing temperature corresponds to added heat. So by increasing temperature, you supply a needed reactant in the dissolution reaction. (In those rare cases where dissolution releases heat, increasing temperature can decrease solubility.)

Gaseous solutes behave differently from solid or liquid solutes with respect to temperature. Increasing the temperature tends to decrease the solubility of gas in liquid. To understand this pattern, check out the concept of vapor pressure. Increasing temperature increases vapor pressure because added heat increases the kinetic energy of the particles in solution. With added energy, these particles stand a greater chance of breaking free from the intermolecular forces that hold them in solution. A classic, real-life example of temperature's effect on gas solubility is carbonated soda. Which goes flat (loses its dissolved carbon dioxide gas) more quickly: warm soda or cold soda?

REMEMBER

Comparing gas solubility in liquids with the concept of vapor pressure highlights another important pattern: Increasing pressure increases the solubility of a gas in liquid. Just as high pressures make it more difficult for surface-dwelling liquid molecules to escape into vapor phase, high pressures inhibit the escape of gases dissolved in solvent. *Henry's law* summarizes this relationship between pressure and gas solubility:

$$\text{Solubility} = \text{Constant} \times \text{Pressure}$$

TIP

The "constant" in Henry's law is *Henry's constant,* and its value depends on the gas, solvent, and temperature. A particularly useful form of Henry's law relates the change in solubility (*S*) that accompanies a change in pressure (*P*) between two different states:

$$\frac{S_1}{P_1} = \frac{S_2}{P_2}$$

According to this relationship, tripling the pressure triples the gas solubility, for example.

EXAMPLE

Q. Henry's constant for nitrogen gas (N_2) in water at 293 K is $0.69 \times 10^{-3}\,\text{mol}/(\text{L} \cdot \text{atm})$. The partial pressure of nitrogen in air at sea level is 0.78 atm. What is the solubility of N_2 in a glass of water at 20°C sitting on a coffee table within a beach house?

A. The solubility is $5.4 \times 10^{-4}\,\text{mol}/\text{L}$. This problem requires the direct application of Henry's law. The glass of water is at 20°C, which is equivalent to 293 K (just add 273 to any Celsius temperature to get the Kelvin equivalent). Because the glass sits within a beach house, you can assume the glass is at sea level. So you can use the provided values for Henry's constant and the partial pressure of N_2:

$$\text{Solubility} = \left(0.69 \times 10^{-3}\,\frac{\text{mol}}{\text{L} \cdot \text{atm}}\right)(0.78\,\text{atm}) = 5.4 \times 10^{-4}\,\text{mol}/\text{L}$$

YOUR TURN

8 A chemist prepares an aqueous solution of cesium sulfate, $Cs_2(SO_4)_3$, swirling the beaker in her gloved hand to promote dissolution. She notices something, momentarily furrows her brow, and then smiles knowingly. She nestles the beaker into a bed of crushed ice within a bucket. What did the chemist notice, why was she briefly confused, and why did she place the dissolving cesium sulfate on ice?

9   Deep-sea divers routinely operate under pressures of multiple atmospheres. One malady these divers must be concerned with is "the bends," a dangerous condition that occurs when divers rise too quickly from the depths, resulting in the rapid release of gas from blood and tissues. Why does the bends occur?

10 Reefus readies himself for a highly productive Sunday afternoon of football-watching by arranging bags of cheesy poofs and a six-pack of grape soda around his beanbag chair. At kickoff, Reefus cracks open his first grape soda and settles in for the long haul. Three hours later, covered in cheesy crumbs, Reefus marks the end of the fourth quarter by cracking open the last of the six-pack. The soda fizzes violently all over Reefus and the beanbag chair. What happened?

11 The grape soda preferred by Reefus is bottled under 3.5 atm of pressure. Reefus lives on a bayou at sea level (*Hint*: 1 atm). The temperature at which the soda is bottled is the same as the temperature in Reefus's living room. Assuming that the concentration of carbon dioxide in an unopened grape soda is 0.15 mol/L, what is the concentration of carbon dioxide in an opened soda that went flat while Reefus napped after the game?

Practice Questions Answers and Explanations

(1) **The order from most to least soluble is NaI, NaBr, NaCl, NaF.** As the question indicates, the greater the lattice energy is, the stronger the forces holding the ions together. Dissolving those ions means outcompeting those forces; a solution forms when attractive solute-solvent forces dominate over others, such as solute-solute bonds. Therefore, salts with lower lattice energy are typically more soluble than those with higher lattice energy.

(2) **Sugar should be more soluble in ethanol than in octanol.** Like dissolves like. Chemists know from experience that sugar dissolves well in water. Therefore, you expect sugar to dissolve best in solvents that are most similar to water. Because ethanol is more miscible with water than is octanol, you expect that ethanol has solvent properties (especially polarity) more like water than octanol does.

(3) Solve these kinds of problems by using the definition of molarity and conversion factors. In parts (b) and (c), you must first convert your mass in grams to moles. To do so, you divide by the molar mass from the periodic table (flip to Chapter 12 for details on molar mass and to Chapter 5 for the periodic table). In addition, be sure you convert milliliters to liters.

a. **2.3 M NaCl**

$$\text{Molarity} = \frac{\text{mol solute}}{\text{L solution}} = \frac{2.0 \text{ mol NaCl}}{0.872 \text{ L}} = 2.3 \text{ M NaCl}$$

b. **1.5 M CuSO}_4$**

$$\text{Moles} = \left(93 \text{ g CuSO}_4\right)\left(\frac{1 \text{ mol CuSO}_4}{159.61 \text{ g CuSO}_4}\right) = 0.58 \text{ mol CuSO}_4$$

$$\text{Volume} = \left(390 \text{ mL}\right)\left(\frac{1 \text{ L}}{1,000 \text{ mL}}\right) = 0.390 \text{ L}$$

$$\text{Molarity} = \left(\frac{0.58 \text{ mol CuSO}_4}{0.390 \text{ L}}\right) = 1.5 \text{ M CuSO}_4$$

c. **0.33 M NaNO}_3$**

$$\text{Moles} = \left(22 \text{ g NaNO}_3\right)\left(\frac{1 \text{ mol NaNO}_3}{85.00 \text{ g NaNO}_3}\right) = 0.26 \text{ mol NaNO}_3$$

$$\text{Volume} = 777 \text{ mL}\left(\frac{1 \text{ L}}{1,000 \text{ mL}}\right) = 0.777 \text{ L}$$

$$\text{Molarity} = \left(\frac{0.26 \text{ mol NaNO}_3}{0.777 \text{ L}}\right) = 0.33 \text{ M NaNO}_3$$

(4) Again, conversion factors are the way to approach these kinds of problems. Each problem features a certain volume of solution that contains a certain solute at a certain concentration. To begin each problem, convert your volume into liters — part (c) has already done this for you. Then rearrange the molarity formula to solve for moles:

$$Molarity = \frac{Moles\ of\ solute}{Liters\ of\ solution}$$

$$Moles\ of\ solute = (Molarity)(Liters\ of\ solution)$$

Plug your volume and given molarity (M) values into the formula and solve. Finally, take your mole value and convert it into grams by multiplying it by the molar mass of your compound from the periodic table (as we explain in Chapter 12).

a. **54 g NaOH**

$$Volume = 671\ mL\left(\frac{1\ L}{1,000\ mL}\right) = 0.671\ L$$

$$Moles\ NaOH = MV = \left(2.0\ \frac{mol}{L}\right)(0.671\ L) = 1.34\ mol\ NaOH$$

$$Mass = (1.34\ mol\ NaOH)\left(\frac{40.0\ g\ NaOH}{1\ mol\ NaOH}\right) = 54\ g\ NaOH$$

b. **9.1 g HCl**

$$Volume = (299\ mL)\left(\frac{1\ L}{1,000\ mL}\right) = 0.299\ L$$

$$Moles\ HCl = MV = \left(0.85\ \frac{mol}{L}\right)(0.299\ L) = 0.25\ mol\ HCl$$

$$Mass = (0.25\ mol\ HCl)\left(\frac{36.46\ g\ HCl}{1\ mol\ HCl}\right) = 9.1\ g\ NaOH$$

c. **116 g Ca(NO$_3$)$_2$**

$$Moles\ Ca(NO_3)_2 = MV = \left(0.258\frac{mol}{L}\right)(2.74\ L) = 0.707\ mol\ Ca(NO_3)_2$$

$$Mass = (0.707\ mol\ Ca(NO_3)_2)\left(\frac{164g\ Ca(NO_3)_2}{1\ mol\ Ca(NO_3)_2}\right) = 116g\ Ca(NO_3)_2$$

(5) **28%.** To calculate mass percent, you must know the mass of solute and the mass of solution. The molarity of the solution tells you the moles of solute per volume of solution. Starting with this information, you can convert to mass of solute by means of the gram formula mass (see Chapter 12 for details on calculating the gram formula mass):

$$Moles\ NH_3 = \left(15.0\ \frac{mol\ NH_3}{L}\right)(1\ L) = 15\ mol\ NH_3$$

$$Mass = (15.0\ mol\ NH_3)\left(\frac{17.04\ g\ NH_3}{1\ mol\ NH_3}\right) = 256\ g\ NH_3$$

So each liter of 15.0 M NH_3 contains 256 g of NH_3 solute. But how much mass does each liter of solution possess? Calculate the mass of the solution by using the density. Note that the problem lists the density in units of milliliters, so be sure to convert to the proper units. To keep the calculations easy, we converted the liters of solution to milliliters to match the density units:

$$\text{Volume} = (1\ \text{L})\left(\frac{1,000\ \text{mL}}{1\ \text{L}}\right) = 1,000\ \text{mL}$$

$$\text{Density} = \frac{\text{Mass}}{\text{Volume}}$$

$$\text{Mass} = (\text{Density})(\text{Volume})$$

$$= \left(0.90\frac{\text{g}}{\text{mL}}\right)(1,000\ \text{mL}) = 900\ \text{g of solution}$$

So 255 g of NH_3 is in every 900 g of 15.0 M NH_3. Now you can calculate the mass percent:

$$\text{Mass\%} = 100 \times \left(\frac{256\ \text{g}}{900\ \text{g}}\right) = 28\%$$

6 4.29×10^{-2} M Use the dilution equation, $M_1V_1 = M_2V_2$. In this problem, the initial molarity is 3.00 M, the initial volume is 2.50 mL (or 2.50×10^{-3} L), and the final volume is 0.175 L. Use these known values to calculate the final molarity, M_2:

$$M_1V_1 = M_2V_2$$

$$M_2 = \frac{M_1V_1}{V_2} = \frac{(3.00\ \text{M})(0.0025\ \text{L})}{(0.175\ \text{L})} = 4.29 \times 10^{-2}\ \text{M}$$

7 $109\text{g}(NH_4)_2SO_4$. First, use the dilution equation to find the concentration of the original solution. The initial volume of the solution is given to you as 80.0 mL and must be converted to liters to become 0.0800 L. The final volume is 2,000 mL (0.200 L when converted), and the final concentration is 0.200 M $(NH_4)_2SO_4$. You then solve for the initial concentration:

$$M_1V_1 = M_2V_2$$

$$M_1 = \frac{M_2V_2}{V_1} = \frac{(0.200\ \text{M})(0.200\ \text{L})}{(0.0800\ \text{L})} = 0.500\ \text{M}$$

This calculation means that the original solution contained 0.500 mol of $(NH_4)_2SO_4$ per liter of solution. The question indicates that 1.65 L of this original solution was prepared, so you then take the initial concentration and use the molarity formula to solve for the mole value of $(NH_4)_2SO_4$. When you have the mole value, you multiply it by the molar mass of $(NH_4)_2SO_4$ to determine the initial mass of $(NH_4)_2SO_4$ dissolved.

$$(1.65\ \text{L})\left(\frac{0.500\ \text{mol}\ (NH_4)_2SO_4}{1\ \text{L}}\right) = 0.825\ \text{mol}\ (NH_4)_2SO_4$$

$$\left(0.825\ \text{mol}\ (NH_4)_2SO_4\right)\left(\frac{132.16\text{g}\ (NH_4)_2SO_4}{1\ \text{mol}\ (NH_4)_2SO_4}\right) = 109\text{g}\ (NH_4)_2SO_4$$

8 As the chemist swirled the beaker of dissolving cesium sulfate, the beaker was becoming noticeably warmer. This observation momentarily confused her, because it suggested that the dissolution of cesium sulfate released heat, a state of affairs opposite to that usually observed with dissolving salts. Having diagnosed the situation, she cleverly turned it to her advantage. With typical salts, increasing temperature increases solubility in water, so heating a dissolving mixture can promote dissolution. In the case of cesium sulfate, however, the reverse is true: By cooling the dissolving mixture, the chemist promoted solubility of the cesium sulfate. (Again, though, it's far more common for solubility to increase as temperature increases.)

9 Deep-sea divers are exposed to high pressures during their dives, and at high pressure, gases become more soluble in the blood and tissue fluids due to Henry's law (Solubility = Constant × Pressure). So when the divers do their thing at great depth, high concentrations of these gases dissolve in the blood. If the divers rise to the surface too quickly at the end of a dive, the solubility of these dissolved gases changes too quickly in response to the diminished pressure. This situation can lead to the formation of tiny gas bubbles in the blood and tissues, which can be deadly.

10 Nothing dramatically fizzy happened when Reefus opened the first soda because that soda was still cold from the refrigerator. As the game progressed, however, the remaining sodas warmed to room temperature as they sat beside Reefus's beanbag chair. Gases (like carbon dioxide) are less soluble in warmer liquids. So when Reefus opened the warm, fourth-quarter soda, a reservoir of undissolved gas burst forth from the can.

11 4.3×10^{-2} mol / L. To solve this problem, use the two-state form of Henry's law:

$$\frac{S_1}{P_1} = \frac{S_2}{P_2}$$

The initial solubility and pressure are 0.15 mol/L and 3.5 atm, respectively. The final pressure is 1.0 atm. Using these known values, solve for the final solubility:

$$\frac{S_1}{P_1} = \frac{S_2}{P_2}$$

$$S_2 = \frac{S_1 P_2}{P_1}$$

$$S_2 = \frac{\left(0.15 \frac{\text{mol}}{\text{L}}\right)(1.0 \text{ atm})}{3.5 \text{ atm}} = 0.043 \text{ mol / L}$$

If you're ready to test your skills a bit more, take the following chapter quiz that incorporates all the chapter topics.

Whaddya Know? Chapter 18 Quiz

Ready for a quiz? The 10 questions in this section will test the skills you learned in this chapter. When you're done, check out the section that follows for answers and explanations.

1 Determine if the following solutions are saturated, unsaturated, or supersaturated. Explain why.

   a. A student adds 50 grams of solute to a solution. The solute is stirred into the solution and completely dissolves.

   b. A student adds 50 grams of solute to a solution. Most of the solute dissolves but some of the solute remains at the bottom of the solution after stirring.

   c. A student adds 50 grams of solute to a solution. Most of the solute dissolves but some remains on the bottom after stirring. The student heats the solution and stirs it more. This causes the remaining solute to dissolve. The solution then sits and cools to room temperature.

2 If you dissolve 4.3 mol of NaCl into 5.0 L of water what is the molarity of the resulting solution?

3 How many grams of $LiNO_3$ are needed to create a 3.0 L solution with a 2.5 M concentration?

4 A student adds 4.5 mol of $CaCl_2$ to 2300 mL of water, what is the molarity of the resulting solution?

5 If you mix 134 grams of Na_2SO_4 with 450 mL of water, what is the molarity of the resulting solution?

6 How many liters of water do you need to create a 2.33 M HCl solution from 67 grams of HCl?

7 You mix 6.7 mol of HCl with some amount of water and determine the molarity of the resulting solution is 0.50 M. What was the volume of water used, in L, to create this solution?

8 You begin with 2.0 liters of a 6 M HNO_3 solution. How much water would need to be added to the solution to lower the concentration to 1.5 M?

9 A student needs to dilute a solution in their lab from a concentration of 3.5 M NaOH to 1.25 M NaOH. They begin with 8.7 L of solution. What will the final volume of the solution be once the dilution is complete?

10 A young chemist is attempting to dissolve ammonia gas into water. The water temperature is currently 50°F. The chemist thinks that by raising the temperature of the water, more ammonia gas is likely to dissolve in the water. Is the chemist correct? If not, what should they do instead?

Answers to Chapter 18 Quiz

(1) a. This describes an unsaturated solution. If a solute is able to be added to a solution and dissolved completely, that means the solution has not reached its saturation point yet.

b. This describes a saturated solution. The solute being added yet not entirely dissolving indicates a saturated solution. This shows the solution was able to dissolve some of the solute added but at some point while the dissolving process was taking place, the solution reached the limit of what could dissolve in it. Once it becomes saturated, no more solute is able to be dissolved. The remaining solute, as described in the problem, falls to the bottom of the container and will not dissolve.

c. A supersaturated solution. When solute will not totally dissolve, a solution is saturated. The only way to go about making that solute dissolve is to somehow increase the solubility of it in the solvent. Generally, the easiest way to do this is to increase the temperature of the solution. Doing so increases the ability of the solvent to dissolve solute. When the solution then cools down to a lower temperature, the solute that dissolved when heated will stay dissolved in the solution, creating a super-saturated solution, meaning there is more dissolved solute in the solution than the temperature of the solution would normally allow.

(2) **0.86 M NaCl**. The formula for molarity is mol/L = M. To solve this problem you simply need to plug your starting values into the equation to solve. There is no need to convert because your values are already in mols and L.

$$M = \frac{mol}{L} = \frac{4.3 \text{ mol}}{5.0 \text{ L}} = 0.86 \text{ M NaCl}$$

(3) **517.13 g LiNO$_3$**. This problem asks for grams of lithium nitrate, so to determine that you'll first need to calculate the moles required. Once you know that, you convert from moles to grams and that's your answer. To solve for moles you'll need to rearrange the molarity equation, plus to cancel out the units you'll need to rewrite the unit molarity (M) as mol/L. Once you've solved for moles, you will multiply your value by the molar mass of LiNO$_3$ from the periodic table.

$$M = \frac{mol}{L}$$
$$M \times L = mol$$
$$mol = \left(2.5 \frac{mol}{L}\right)(3.0 \text{ L})$$
$$mol = 7.5 \text{ mol LiNO}_3$$
$$\left(7.5 \text{ mol LiNO}_3\right)\left(\frac{68.95 \text{ g LiNO}_3}{1 \text{ mol LiNO}_3}\right) = 517.13 \text{ g LiNO}_3$$

(4) **1.97 M CaCl₂**. To correctly solve this problem, you will need to first convert your 2,300 mL to liters. Once you've done this you can plug in your moles and liters into the molarity equation to solve:

$$(2{,}300 \text{ mL})\left(\frac{1 \text{ L}}{1{,}000 \text{ mL}}\right) = 2.30 \text{ L}$$

$$M = \frac{\text{mol}}{\text{L}} = \frac{4.5 \text{ mol CaCl}_2}{2.30 \text{ L}} = 1.97 \text{ M CaCl}_2$$

(5) **2.09 M Na2SO₄**. To correctly solve this problem, you will need to first convert your 450 mL to liters. In addition, you must convert your gram value to moles. To do this you will divide by the molar mass of Na_2SO_4 calculated from the periodic table. Once you've done this you can plug your moles and liters into the molarity equation to solve:

$$(450 \text{ mL})\left(\frac{1 \text{ L}}{1{,}000 \text{ mL}}\right) = 0.450 \text{ L}$$

$$(134 \text{ g Na}_2\text{SO}_4)\left(\frac{1 \text{ mol Na}_2\text{SO}_4}{142.05 \text{ g Na}_2\text{SO}_4}\right) = 0.94 \text{ mol Na}_2\text{SO}_4$$

$$M = \frac{\text{mol}}{\text{L}} = \frac{0.94 \text{ mol Na}_2\text{SO}_4}{0.450 \text{ L}} = 2.09 \text{ M Na}_2\text{SO}_4$$

(6) **0.79 L HCl**. To solve this problem, you need to rearrange the molarity equation to solve for liters (volume) and solve as shown here:

$$(67 \text{ g HCl})\left(\frac{1 \text{ mol HCl}}{36.46 \text{ g HCl}}\right) = 1.84 \text{ mol HCl}$$

$$M = \frac{\text{mol}}{\text{L}}$$

$$L = \frac{\text{mol}}{M}$$

$$L = \frac{(1.84 \text{ mol HCl})}{\left(2.33 \frac{\text{mol HCl}}{\text{L}}\right)}$$

Volume = 0.79 L HCl

Subbing 1.84 mol in above results in a final answer of 0.79 L HCl

(7) **13.4 L**. To solve this problem, you need to rearrange the molarity equation to solve for liters (volume) and solve as shown here. Once you have done that, plug in your values of moles and molarity to determine the volume of water used to make the solution. The unit will be liters, as the moles of HCl will cancel out when they are divided:

$$M = \frac{\text{mol}}{\text{L}}$$

$$L = \frac{\text{mol}}{M}$$

$$L = \frac{(6.7 \text{ mol HCl})}{\left(0.5 \frac{\text{mol HCl}}{\text{L}}\right)}$$

Volume = 13.4 L

(8) **9 liters of water.** To lower the concentration of a solution you must dilute it. You can solve this calculation using the dilution formula. Your initial volume and concentration of your solution are given in the problem. Your final concentration is also given. So your M_1, V_1, and M_2 are given to you. To solve this problem, you'll need to calculate V_2. However, when you calculate V_2 that will not be your final answer. The calculated V_2 is the total volume of the solution, but the problem specifically asks what volume of water must be added to the initial solution to achieve the desired final concentration. So to find the correct final answer you must subtract the initial volume from the final volume to determine the amount that must be added.

$$M_1V_1 = M_2V_2$$

$$\frac{M_1V_1}{M_2} = V_2$$

$$\frac{(6\,M)(2\,L)}{(1.5\,M)} = V_2 = 8\,L$$

Final volume – Initial volume = Volume of H_2O added

$8\,L - 2\,L = 6\,L$ of H_2O added

(9) **24.4 L.** This is a simple dilution problem. In this case you are determining the final volume of the solution after dilution has taken place. You do not need to determine the amount of water added as was done in the previous problem.

$$M_1V_1 = M_2V_2$$

$$\frac{M_1V_1}{M_2} = V_2$$

$$\frac{(3.5\,M)(8.7\,L)}{(1.25\,M)} = V_2 = 24.4\,L$$

(10) The chemist is incorrect. Generally the colder a liquid is, the easier it is for gases to be dissolved in that liquid. This means if the chemist raises the temperature of the liquid, the ammonia gas will be less soluble in the water than it would be at colder temperatures. The best way to make the ammonia gas more soluble would be to lower the temperature of it. Though not mentioned in the problem, another way to help the solubility of the gas in water would be to increase the pressure at which the gas water mixture is held to further increase the solubility of the ammonia in water.

IN THIS CHAPTER

» Knowing the difference between molarity and molality

» Working with boiling point elevation and freezing point depression

» Deducing molecular masses from boiling and freezing point changes

Chapter **19**

Playing Hot and Cold: Colligative Properties

A s a recently minted expert in solubility and molarity (see Chapter 18), you may be ready to write off solutions as another chemistry topic mastered, but you, as a chemist worth your salt, must be aware of one final piece to the puzzle. Some properties of solutions depend on the specific nature of the solute. In other words, a property or characteristic you can record about the solution depends on the specific chemical identity of the solute. For example, salt solutions taste salty, whereas sugar solutions taste sweet. Salt solutions conduct electricity, but sugar solutions don't. Solutions containing the nickel cation are commonly green, and those containing the copper cation are blue.

Some properties of solution don't depend on the specific type of solute (meaning it doesn't matter which chemical is dissolved in a solution) — just the *number* of solute particles. Properties that simply depend on the relative number of solute particles are called *colligative properties*. These chemically important phenomena arise from the presence of solute particles in a given mass of solvent. The presence of extra particles in a formerly pure solvent has a significant impact on some of that solvent's characteristic properties, such as freezing and boiling points.

This chapter walks you through these colligative properties and their consequences, and it introduces you to a new solution property: molality. No, that's not a typo. Molality is a different way to measure concentration that allows you to solve for the key colligative properties later in this chapter.

Portioning Particles: Molality and Mole Fractions

Like the difference in their names, the practical difference between *molarity* and *molality* is subtle. Take a close look at their definitions, expressed next to one another in the following equations:

$$\text{Molarity} = \frac{\text{Moles of solute}}{\text{Liters of solution}} \qquad \text{Molality} = \frac{\text{Moles of solute}}{\text{Kilograms of solvent}}$$

The numerators in molarity and molality calculations are identical, but their denominators differ greatly. Molarity deals with liters of solution, while molality deals with kilograms of solvent. A *solution* is a mixture of solvent and solute; a *solvent* is the medium into which the solute is mixed.

A further complication to the molarity/molality confusion is how to distinguish between their variables and units. The letter m turns out to be overused in chemistry. Instead of picking another variable (or perhaps a less confusing name that started with a nice uncommon letter like z), chemists decided to give molality the lowercase "script" m as its variable. To help you avoid uttering any four-letter words when confronted with this plethora of m-words and their abbreviations, we've provided Table 19-1.

TABLE 19-1 *M* Words Related to Concentration

Name	Variable	Unit Abbreviation
Molarity	*M*	M
Molality	*m*	m
Moles		mol

Occasionally, you may be asked to calculate the *mole fraction* of a solution, which is the ratio of the number of moles of *either* solute or solvent in a solution to the total number of moles of solute *and* solvent in the solution. By the time chemists defined this quantity, however, they had finally acknowledged that they had too many m variables, and they gave it the variable X. Of course, chemists still need to distinguish between the mole fractions of the solute and the solvent, which unfortunately both start with the letter s. To avoid further confusion, they decided to abbreviate solute and solvent as A and B, respectively, in the general formula, although in practice, the chemical formulas of the solute and solvent are usually written as subscripts in place of A and B. For example, the mole fraction of sodium chloride in a solution would be written as X_{NaCl}.

REMEMBER

In general, the mole ratio of the solute in a solution is expressed as

$$X_A = \frac{n_A}{n_A + n_B}$$

where n_A is the number of moles of solute and n_B is the number of moles of solvent. The mole ratio of the solvent is then

$$X_B = \frac{n_B}{n_A + n_B}$$

These mole fractions are useful because they represent the ratio of solute to solution and solvent to solution very well and give you a general understanding of how much of your solution is solute and how much is solvent.

EXAMPLE

Q. How many grams of dihydrogen sulfide (H_2S) must you add to 750 g of water to make a 0.35 m solution?

A. 8.9 g H_2S. This problem gives you molality and the mass of a solvent and asks you to solve for the mass of solute. Because molality involves moles and not grams of solute, you first need to solve for moles of solvent, and then you use the gram formula mass of sodium chloride to solve for the number of grams of solute (see Chapter 12 for details on how to make this conversion). Before plugging the numbers into the molality equation, you must also note that the problem has given you the mass of the solvent in grams, but the formula calls for it to be in kilograms. Moving from grams to kilograms is equivalent to moving the decimal point three places to the left (if you need a refresher on the ins and outs of unit conversions, please refer to Chapter 2). Plugging everything you know into the equation for molality gives you the following:

$$0.35 \text{ m} = \frac{X \text{ mol } H_2S}{0.750 \text{ kg } H_2O}$$

Solving for the unknown gives you 0.26 mol of H_2S in solution. You then need to multiply this mole value by the molecular mass of H_2S to determine the number of grams that need to be added:

$$(0.26 \text{ mol } H_2S)\left(\frac{34.08 \text{ g } H_2S}{1 \text{ mol } H_2S}\right) = 8.86 \text{ g } H_2S$$

Q. What is the molality of the solution you make when you dissolve 15.0 g of NaCl in 50.0 g of water?

A. To calculate the molality, you need to convert the 50.0 g of water to kilograms (0.0500 kg). Then convert the grams of NaCl to moles of NaCl and divide by the kilograms of water, as in the following equation:

$$\frac{15.0 \text{ g NaCl}}{1} \times \frac{1 \text{ mol NaCl}}{58.44 \text{ g NaCl}} \times \frac{1}{0.0500 \text{ kg}} = 5.13 \text{ m}$$

Notice that the equation shows all the calculations on one line. This is usually the best way to solve a problem because it reduces the chance for error and allows you to ensure that the units are canceling correctly. If necessary, you can break the equation down into individual steps instead by first converting from grams of NaCl to moles and then separately plugging that answer into your molality formula to calculate the molality of the solution.

YOUR TURN

1. What is the molality of a solution made by dissolving 36 g of sodium chloride (NaCl) in 6.0 L of water? (Remember that the density of water is 1.0 kg/L.)

2. How many moles of potassium iodide (KI) are required to produce a 3.5 m solution if you begin with 2.7 L of water?

3. Calculate the mole fraction of each component in a solution containing 2.75 mol of ethanol (C_2H_6O) and 6.25 mol of water.

Too Hot to Handle: Elevating and Calculating Boiling Points

Calculating molality is no more or less difficult than calculating molarity, so you may be asking yourself, "Why all the fuss?" Is it even worth adding another quantity and another variable to memorize? Yes! Although molarity is exceptionally convenient for calculating concentrations and working out how to make dilutions in the most efficient way, molality is reserved for the calculation of several important colligative properties, including boiling point elevation. Each individual liquid has a specific temperature at which it boils (at a given atmospheric pressure). This temperature is the liquid's *boiling point*. However, that boiling point isn't set in stone. It can be changed quite easily. *Boiling point elevation* refers to the tendency of a solvent's boiling point to increase when an impurity (a solute) is added to it.

In fact, the more solute that is added, the greater the change in the boiling point. For example, if you had pure distilled water with no impurities at standard atmospheric pressure, then the water should boil at exactly 100°C. However, if you start adding salt to that water, the boiling point will begin to slowly increase. Thankfully, we have a nice formula that allows you to calculate exactly how much the boiling point will increase, depending on the liquid and the amount of solute added.

REMEMBER

Boiling point elevations are directly proportional to the molality of a solution, but chemists have found that some solvents are more susceptible to this change than others. The formula for the change in the boiling point of a solution, therefore, contains a proportionality constant, abbreviated K_b, which is a property determined experimentally and must be read from a table such as Table 19-2. The formula for the boiling point elevation is

$$\Delta T_b = K_b m$$

where m is molality. Note the use of the Greek letter delta (Δ) in the formula to indicate that you're calculating a *change in* the boiling point, not the boiling point itself. You need to add this number to the boiling point of the pure solvent to get the boiling point of the solution. The units of K_b are typically given in degrees Celsius per molality.

TABLE 19-2 Common K_b Values

Solvent	K_b in °C/m	Boiling Point in °C
Acetic acid	3.07	118.1
Benzene	2.53	80.1
Camphor	5.95	204.0
Carbon tetrachloride	4.95	76.7
Cyclohexane	2.79	80.7
Ethanol	1.19	78.4
Phenol	3.56	181.7
Water	0.512	100.0

Boiling point elevations are a result of the attraction between solvent and solute particles in a solution. Colligative properties such as boiling point elevation depend on only the number of particles *in solution*. Adding solute particles increases these intermolecular attractions because more particles are around to attract one another. To boil, solvent particles must therefore achieve a greater kinetic energy to overcome this extra attractive force, which translates into a higher boiling point. (See Chapter 15 for more information on kinetic energy.)

Q. What is the boiling point of a solution containing 45.2 g of menthol ($C_{10}H_{20}O$) dissolved in 350 g of acetic acid?

EXAMPLE

A. 120.6°C. The problem asks for the boiling point of the solution, so you know that first you have to calculate the boiling point elevation. This means you need to know the molality of the solution and the K_b value of the solvent (acetic acid). Table 19-2 tells you that the K_b of acetic acid is 3.07. To calculate the molality, you must convert 45.2 g of menthol to moles:

$$(45.2 \text{ g menthol})\left(\frac{1 \text{ mol menthol}}{156.3 \text{ g menthol}}\right) = 0.29 \text{ mol menthol}$$

You can now calculate the molality of the solution, taking care to convert grams of acetic acid to kilograms:

$$m = \frac{0.29 \text{ mol menthol}}{0.350 \text{ kg acetic acid}} = 0.83 \text{ m}$$

Now that you have molality, you can plug it and your K_b value into the formula to find the change in boiling point:

$$\Delta T_b = \left(3.07 \frac{°C}{m}\right)(0.83 \text{ m}) = 2.5°C$$

You're not quite done, because the problem asks for the boiling point of the solution, not the change in the boiling point. Luckily, the last step is just simple arithmetic. You must add your ΔT_b to the boiling point of pure acetic acid, which, according to Table 19-2, is 118.1°C. This gives you a final boiling point of $118.1°C + 2.5°C = 120.6°C$ for the solution.

Q. What is the boiling point of a 2.0 m aqueous KNO_3 solution?

A. KNO_3 is a salt, a strong electrolyte. In a 2.0 m solution, you have 4.0 m in particles, because KNO_3 dissociates completely into K^+ and NO_3^-. In addition, you know that the solvent in this instance is water because the solution is described as *aqueous.*(You can easily miss the solvent in a problem like this, so make sure you read every word in the problem; rarely is something included in a chemistry problem for no reason.) The K_b of water is 0.512°C/m (see Table 19-2), so

$$\Delta T_b = K_b m$$
$$\Delta T_b = (0.512°C/m)(4.0 \text{ m})$$
$$\Delta T_b = 2.0°C$$

The change in boiling point is 2.0°C. You know that the boiling point of a solution is always higher than the pure solvent, so the solution's boiling point is

$$100.0°C + 2.0°C = 102.0°C$$

---

4 What is the boiling point of a solution containing 158 g of sodium chloride (NaCl) and 1.2 kg of water? What if the same number of moles of calcium chloride ($CaCl_2$) is added to the solvent instead? Explain why there's such a great difference in the boiling point elevation.

YOUR TURN

5 A clumsy chemist topples a bottle of indigo dye ($C_{16}H_{10}N_2O_2$) into a beaker containing 450 g of ethanol. If the boiling point of the resulting solution is 79.2°C, how many grams of dye were in the bottle?

How Low Can You Go? Depressing and Calculating Freezing Points

The second of the important colligative properties that you can calculate from molality is *freezing point depression*. Not only does adding solute to a solvent raise its boiling point, but it also lowers its freezing point. That's why you sprinkle salt on icy sidewalks. The salt mixes with the ice and lowers its freezing point. If this new freezing point is lower than the outside temperature, the ice melts, eliminating the spectacular wipeouts so common on salt-free sidewalks. The colder it is outside, the more salt you need to melt the ice and lower the freezing point to below the ambient temperature.

REMEMBER Freezing point depressions, like boiling point elevations in the preceding section, are calculated using a constant of proportionality, this time abbreviated K_f. The formula therefore becomes $\Delta T_f = K_f m$, where m is molality. To calculate the new freezing point of a compound, you must *subtract* the change in freezing point from the freezing point of the pure solvent. Table 19-3 lists several common K_f values.

TABLE 19-3 Common K_f Values

Solvent	K_f in °C/m	Freezing Point in °C
Acetic acid	3.90	16.6
Benzene	5.12	5.5
Camphor	37.7	179.0
Carbon tetrachloride	30.0	−23.0
Cyclohexane	20.2	6.4
Ethanol	1.99	−114.6
Phenol	7.40	41.0
Water	1.86	0.0

Adding an impurity to a solvent makes its liquid phase more stable through the combined effects of boiling point elevation and freezing point depression. That's why you rarely see bodies of frozen salt water. The salt in the oceans lowers the freezing point of the water, making the liquid phase more stable and able to sustain temperatures slightly below 0°C.

For a visual reference, look at Figure 19-1. It shows the effect of a solute on both the freezing point and the boiling point of a solvent.

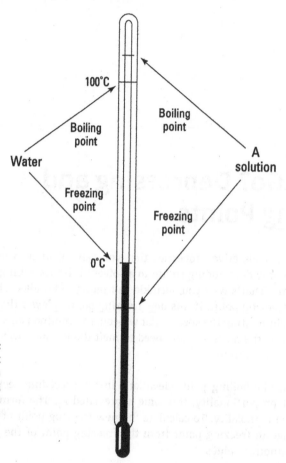

FIGURE 19-1:
Boiling point elevation and freezing point depression of a solution as altered by a solute.

Q. Each kilogram of seawater contains roughly 35 g of dissolved salts. Assuming that all these salts are sodium chloride, what is the freezing point of seawater?

EXAMPLE

A. –2.23°C. Begin by converting grams of salt to moles to figure out the molality. One mole of NaCl is equivalent to 58.4 g, so 35 g is equivalent to 0.60 mol of NaCl. You need to multiply this number by 2 to compensate for the fact that sodium chloride dissociates into twice as many particles in water, so this solution contains 1.20 mol. Next, find the molality of the solution by dividing this number of moles by the mass of the solvent (1 kg), giving you a 1.20 m solution. Last, look up the K_f of water in Table 19-3 and plug all these values into the equation for freezing point depression:

$$\Delta T_f = \left(1.86 \frac{°C}{m}\right)(1.20\ m) = 2.23°C$$

Because this value is merely the freezing point depression, you must subtract it from the freezing point of the pure solute to get $0\,°C - 2.23\,°C = -2.23\,°C$, the freezing point of seawater.

6

Antifreeze takes advantage of freezing point depression to lower the freezing point of the water in your car's engine and keep it from freezing on blistery winter drives. If antifreeze is made primarily of ethylene glycol ($C_2H_6O_2$), how much of it needs to be added to lower the freezing point of 10.0 kg of water by 15.0 °C?

YOUR TURN

7 If 15 g of silver (Ag) is dissolved into 1,500 g of ethanol (C_2H_6O), what is the freezing point of the mixture?

Determining Molecular Masses with Boiling and Freezing Points

Just as a solid understanding of molality helps you to calculate changes in boiling and freezing points, a solid understanding of ΔT_b and ΔT_f can help you determine the molecular mass of a mystery compound that's being added to a known quantity of solvent. When you're asked to solve problems of this type, you'll always be given the mass of the mystery solute, the mass of solvent, and either the change in the freezing or boiling point or the new freezing or boiling point itself. From this information, you then follow a set of simple steps to determine the molecular mass:

1. **Find the boiling point elevation or freezing point depression.**

    If you've been given the boiling point, calculate the ΔT_b by subtracting the boiling point of the pure solvent from the number you were given. If you know the freezing point, add the freezing point of the pure solvent to it to get the ΔT_f.

2. Look up the K_b or K_f of the solvent (refer to Tables 19-2 and 19-3).

3. Solve for the molality of the solution using the equation for ΔT_b or ΔT_f.

4. Calculate the number of moles of solute in the solution by multiplying the molality calculated in Step 3 by the given number of kilograms of solvent.

5. Divide the given mass of solute by the number of moles calculated in Step 4. This is your molecular mass, or number of grams per mole, from which you can often guess the identity of the mystery compound.

Q. 97.30 g of a mystery compound is added to 500.0 g of water, raising its boiling point to 100.78°C. What is the molecular mass of the mystery compound?

EXAMPLE

A. 128.0 g/mol. First subtract the boiling point of water from this new boiling point:

$$\Delta T_b = 100.78\,°C - 100.00\,°C = 0.78\,°C$$

Then plug this value and a K_b of 0.512 into the equation for boiling point elevation and solve for molality:

$$m = \frac{\Delta T_b}{K_b} = \frac{0.78\,°C}{0.512\frac{°C}{m}} = 1.52\ m$$

Next, take this molality value and multiply it by the given mass of the solvent, water, in kilograms:

$$\left(\frac{1.52\ \text{mol solute}}{1\ \text{kg}\,H_2O}\right)(0.5\ \text{kg}\,H_2O) = 0.76\ \text{mol solute}$$

Last, divide the number of grams of the mystery solute by the number of moles, giving you the molecular mass of the compound:

$$\frac{97.30\ g}{0.76\ \text{mol}} = 128.0\ g/\text{mol}$$

YOUR TURN

8 The freezing point of 83.2 g of carbon tetrachloride is lowered by 11.52 °C when 15.0 g of a mystery compound is added to it. What is the molecular mass of this mystery compound?

9 When 8.8 g of a mystery compound is added to 42.1 g of benzene, the boiling point is elevated to 81.9 °C. What is the molecular mass of this mystery compound?

Practice Questions Answers and Explanations

(1) **0.10 m.** This problem gives you the value of your solute (sodium chloride) in grams and the value of your solvent (water) in liters. You must first convert the sodium chloride to moles. To do so, divide by the molar mass (which we discuss in Chapter 7):

$$\left(36 \text{ g NaCl}\right)\left(\frac{1 \text{ mol NaCl}}{58.44 \text{ g NaCl}}\right) = 0.62 \text{ mol NaCl}$$

Then convert liters to kilograms using the given density of water. Thankfully, this is an easy one-to-one conversion: 6.0 L of water becomes 6.0 kg of water. Finally, plug your values into the molality formula to determine the molality of your solution:

$$m = \frac{0.62 \text{ mol NaCl}}{6.0 \text{ kg H}_2\text{O}} = 0.10 \text{ m}$$

(2) **9.5 mol.** This problem gives you your molality value and the liters of solution. You simply have to determine the number of moles of solute (potassium iodide) that you need to add to make this solution. To do so, first convert the liters of solution to kilograms of solution using the density of water (1.0 kg/L). Then plug your values into the molality formula and solve for moles of solute:

$$3.5 \text{ m} = \frac{x \text{ mol solute}}{2.7 \text{ kg solvent}}$$

$$9.5 \text{ mol} = x$$

(3) **The solution has a mole fraction of 0.306 for ethanol and 0.694 for water.** This problem tells you that the molality of ethanol $\left(n_{C_2H_6O}\right)$ is 2.75 mol and the molality of water $\left(n_{H_2O}\right)$ is 6.25 mol. Plug these values into the equations for the mole fraction of solute and solvent:

$$X_{C_2H_6O} = \frac{2.75 \text{ mol}}{2.75 \text{ mol} + 6.25 \text{ mol}} = 0.306$$

$$X_{H_2O} = \frac{6.25 \text{ mol}}{2.75 \text{ mol} + 6.25 \text{ mol}} = 0.694$$

(4) **The solution containing sodium chloride has a boiling point of 102.3°C; the solution containing calcium chloride has a boiling point of 103.5°C.** To solve for the boiling point, you must first solve for the molality. Start by dividing 158 g NaCl by its gram formula mass (58.44 g/mol), which tells you that 2.71 mol of solute is being added to the water. Multiply this value by 2 because each molecule of NaCl splits into two particles in solution, for a total of 5.42 mol. Divide the moles by the mass of solvent (1.2 kg) to give you a molality of 4.5 m. Finally, multiply this molality by the K_b of water, 0.512°C/m, to give you a ΔT_b of 2.3°C. Add this temperature change to the boiling point of pure water (100°C) to give you a new boiling point of 102.3°C.

$$\Delta T_b = K_b m$$

$$\Delta T_b = \left(0.512 \frac{°C}{m}\right)\left(4.5 \text{ m}\right)$$

$$\Delta T_b = 2.3°C$$

$$T_b = 100°C + 2.3°C = 102.3°C$$

If you instead add the same number of moles of calcium chloride (2.71 mol) to the water, the calcium chloride would dissociate into three particles per mole in solution. This gives you $2.71\ mol \times 3 = 8.13\ mol$ of solute in solution. Divide the number of moles by the mass of the solvent (1.2 kg) to get 6.8 m, and multiply by the K_b of water ($0.512°C/m$) to get a ΔT_b of $3.5°C$. This is a difference of more than 1°C! The difference arises because colligative properties such as boiling point elevation depend on only the number of particles *in solution*.

$$\Delta T_b = K_b m$$

$$\Delta T_b = \left(0.512\frac{°C}{m}\right)(6.8\ m)$$

$$\Delta T_b = 3.5°C$$

$$T_b = 100°C + 3.5°C = 103.5°C$$

⑤ **79 g $C_{16}H_{10}N_2O_2$.** This problem requires you to solve a boiling point elevation problem backwards. You're given the solution's boiling point, so the first thing to do is to solve for the ΔT_b of the solution. You do this by subtracting the boiling point of pure ethanol (which you find in Table 19-2) from the given boiling point of the impure solution:

$$\Delta T_b = 79.2°C - 78.4°C = 0.8°C$$

After you look up the K_b of ethanol ($1.19°C/m$), the only unknown in your ΔT_b equation is the molality. Solving for this gives you

$$m = \frac{\Delta T_b}{K_b} = \frac{0.8°C}{1.19\frac{°C}{m}} = 0.67\ m$$

Now that you know the molality, you can find the number of moles of solute. Be sure to convert grams of ethanol to kilograms before you plug everything into the formula.

$$0.67\ m = \frac{x\ mol\ C_{16}H_{10}N_2O_2}{0.450\ kg\ C_2H_6O}$$

$$x = 0.30\ mol\ C_{16}H_{10}N_2O_2$$

Last, translate this mole count into a mass by multiplying by the gram molecular mass of $C_{16}H_{10}N_2O_2$ (262 g/mol), giving you your final answer of 79 g $C_{16}H_{10}N_2O_2$.

⑥ **5.00 kg $C_2H_6O_2$.** Here, you've been given a freezing point depression of 15.0°C, and you're asked to back-solve for the number of grams of antifreeze required to make it happen. Begin by solving for the molality of the solution by plugging all the known quantities into your freezing point depression equation and solving for molality:

$$m = \frac{\Delta T_f}{K_f} = \frac{15.0°C}{1.86\frac{°C}{m}} = 8.06\ m$$

Now solve for the number of moles of the solvent ethylene glycol (antifreeze):

$$8.06\ m = \frac{x\ mol\ C_2H_6O_2}{10.0\ kg\ H_2O}$$

$$x = 80.6\ mol\ C_2H_6O_2$$

Finally, translate this mole count into a mass by multiplying by the gram molecular mass of $C_2H_6O_2$ (62.0 g/mol), giving you 4,997.2 g, which is approximately equal to 5.00 kg $C_2H_6O_2$. Lowering the freezing point of water by such a significant amount requires a solution that is one-third antifreeze by mass!

(7) **−114.8°C.** First calculate the molality of the solution by converting the mass of silver into a mole count ($(15 \text{ g}) \times (1 \text{ mol} / 108 \text{g}) = 0.14$ mol) and dividing it by the mass of ethanol in kilograms (turning 1,500 g into 1.5 kg), giving you 0.093 m. Then multiply this molality by the K_f of ethanol (1.99°C/m), giving you a ΔT_f of 0.19°C. Last, subtract this ΔT_f value from the freezing point of pure ethanol, giving you

$$T_f = -114.6°C - 0.19°C = -114.8°C$$

(8) **470 g/mol.** Step 1 is unnecessary in this case because you already have the ΔT_f, so begin by looking up the K_f of carbon tetrachloride in Table 19-3 (30.0°C/m). Plug both of these values into the equation for freezing point depression and solve for molality:

$$m = \frac{\Delta T_f}{K_f} = \frac{11.52°C}{30.0\frac{°C}{m}} = 0.384 \text{ m}$$

Next, take this molality value and multiply it by the given value for the mass of the solvent (first converting grams to kilograms):

$$\left(\frac{0.384 \text{ mol solute}}{1 \text{ kg CCl}_4} \right)(0.0832 \text{ kg}) = 0.0319 \text{ mol solute}$$

Last, divide the number of grams of the mystery solute by the number of moles, giving you the molecular mass of the compound:

$$\left(\frac{15.0 \text{ g}}{0.0319 \text{ mol}} \right) = 470 \text{ g} / \text{mol}$$

(9) **290 g/mol.** Solve for ΔT_b by subtracting the boiling point of pure benzene (see Table 19-2) from the given boiling point of the solution:

$$81.9°C - 80.1°C = 1.80°C$$

Plug this value and the K_b for benzene (2.53°C/m) into your ΔT_b equation and solve for the molality of the solution:

$$m = \frac{\Delta T_b}{K_b} = \frac{1.80°C}{2.53\frac{°C}{m}} = 0.71 \text{ m}$$

Next, take this molality value and multiply it by the given value for mass of solvent (first converting grams to kilograms):

$$\left(\frac{0.71 \text{ mol solute}}{1 \text{ kg benzene}} \right)(0.0421 \text{ kg}) = 0.030 \text{ mol solute}$$

Last, divide the number of grams of the mystery solute by the number of moles, giving you the molecular mass of the compound:

$$\left(\frac{8.8 \text{ g}}{0.030 \text{ mol}} \right) = 293.3 \text{ g} / \text{mol}$$

The answer needs only two significant figures, so use 290 g/mol.

If you're ready to test your skills a bit more, take the following chapter quiz that incorporates all the chapter topics.

Whaddya Know? Chapter 19 Quiz

Ready for a quiz? The 10 questions in this section will test the skills you learned in this chapter. When you're done, check out the section that follows for answers and explanations.

1. What is the molality of a solution that is made by dissolving 250 grams of potassium nitrate (KNO_3) into 10 L of water?

2. What is the molality of a solution that is made by dissolving 4.0 moles of lithium chloride (LiCl) in 800 g of water?

3. What is the molality of a solution that is composed of 200 grams of calcium bromide ($CaBr_2$) and 60 kg of water?

4. Calculate the amount of solute, in moles, required to create a 5.5 m solution in 10 L of water.

5. How many moles of potassium fluoride are required to be mixed with 15.0 kg of water to make a 10.5 m solution?

6. Calculate the mole fraction of the solvent in a solution that contains 4.5 mol of lithium nitrate and 15 moles of water.

7. What is the mole fraction of solute and solvent in the solution described in problem 1?

8. What is the boiling point of a solution containing 250 grams of potassium nitrate and 0.8 kg of water?

9. The normal boiling point of ethanol is 78.4°C. When 39.2 grams of lithium chloride is added to 890 g of ethanol, what is the change in the boiling point?

10. What is a colligative property? Give two examples.

11. If 40.0 grams of $CaCl_2$ is dissolved into 0.50 kilograms of water, what is the freezing point of the resulting solution?

12. What is the change in freezing point to a 0.93 kg solution of ethanol if 39 grams of carbon dioxide are dissolved into it?

13. Why does adding more solute to a solution generally decrease the freezing point of the liquid? Explain at the molecular level.

14. The freezing point of 131 g of phenol is lowered from 41.0°C to 33.0°C when 30.5 grams of an unknown compound is added to it. What is the molecular mass of this compound?

15. The boiling point of 2.10 kg of ethanol is raised by 4.50°C when 423 grams of a mystery compound is added to it. What is the molecular mass of the mystery compound?

Answers to Chapter 19 Quiz

1. **0.25 m.** This problem gives you the value of your solute (potassium nitrate) in grams and the value of your solvent (water) in liters. You must first convert the potassium nitrate to moles. To do so, divide by the molar mass (which we discuss in Chapter 12):

$$\left(250 \text{ g KNO}_3\right)\left(\frac{1 \text{ mol KNO}_3}{101.10 \text{ g KNO}_3}\right) = 2.47 \text{ mol KNO}_3$$

Then convert liters to kilograms using the given density of water. Thankfully, this is an easy one-to-one conversion: 10 L of water becomes 10 kg of water. Finally, plug your values into the molality formula to determine the molality of your solution:

$$m = \frac{2.47 \text{ mol KNO}_3}{10 \text{ kg H}_2\text{O}} = 0.25 \text{ m}$$

2. **5 m.** This problem gives you the value of your solute in moles so there is no need to do any conversions there. However, your solute value is given to you in grams so you need to convert that to kilograms.

$$\left(800 \text{ g}\right)\left(\frac{1 \text{ kg}}{1,000 \text{ g}}\right) = 0.8 \text{ kg}$$

Once you have the kilograms you then plug your moles and kilograms into the molality formula to determine your answer:

$$m = \frac{4.0 \text{ mol LiCl}}{0.8 \text{ kg H}_2\text{O}} = 5 \text{ m}$$

3. **0.017 m.** This problem is similar to problem 1. Convert your grams of solute to moles of solute using the molar mass and then divide by the kilograms of water given in the problem to determine your molality:

$$\left(200 \text{ g CaBr}_2\right)\left(\frac{1 \text{ mol CaBr}_2}{199.89 \text{ g CaBr}_2}\right) = 1.00 \text{ mol CaBr}_2$$

$$m = \frac{1 \text{ mol KNO}_3}{60 \text{ kg H}_2\text{O}} = 0.017 \text{ m}$$

4. **55 mol.** To solve this problem you are calculating the mol value of the solute. You are given the molality of the solution but you are not given the mass of the solvent. Instead you are given the volume of the solvent, water, in liters. The density of water is 1 L = 1 Kg making the 10 L the water equal to 10 kg of water. Once you have this you just need to plug into your equation and solve for the moles of solute.

$$5.5 \text{ m} = \frac{\text{Mol of solute}}{10 \text{ kg H}_2\text{O}}$$
$$= 55 \text{ mol of solute}$$

5. **157.5 mol KF.** This problem is much the same as number 4. The only difference is that you are already given the kilograms of your solvent:

$$10.5 \text{ m} = \frac{\text{Mol of KF}}{15 \text{ kg H}_2\text{O}}$$
$$= 157.5 \text{ mol KF}$$

6 **0.77 mole fraction of the solvent.** To calculate the mole fraction of the solvent you need to know two things: The number of moles of the solvent and the total number of moles present in the solution of solute and solvent combined. For this solution, the moles of solvent is equal to 15 mol H_2O and the total moles of solution are equal to 19.5 mol (15 mol H_2O + 4.5 mol LiNO3). Once you know this divide and you'll have the mole fraction of the solvent:

$$X_{solvent} = \frac{15 \text{ mol solute}}{19.5 \text{ mol total}}$$

$$X_{solvent} = 0.77$$

7 **Solvent mole fraction is 0.996, and solute mole fraction is 0.004.** To determine the mole fraction of solvent and solute in problem 1, you have to convert the grams of potassium nitrate to moles. This is your moles of solute.

$$\left(250 \text{ g KNO}_3\right)\left(\frac{1 \text{ mol KNO}_3}{101.10 \text{ g KNO}_3}\right) = 2.47 \text{ mol KNO}_3$$

To determine moles of solvent you must first convert liters to kilograms, which is a 1:1 conversion, so you have 10 kg of water. Next, you need to convert kilograms to grams:

$$\left(10 \text{ kg H}_2O\right)\left(\frac{1,000 \text{ g}}{1 \text{ kg}}\right) = 10,000 \text{ g H}_2O$$

Once this is done you have to convert moles of H_2O. This is the moles of your solvent:

$$\left(10,000 \text{ g H}_2O\right)\left(\frac{1 \text{ mol H}_2O}{18.02 \text{ g H}_2O}\right) = 554.93 \text{ mol H}_2O$$

The total moles of the solution is equal to the moles of solute plus solvent: 2.47 mol KNO_3 + 554.93 H_2O = 557.4 total moles. Once you have this calculate the mole fraction of solute and solvent:

$$X_{solvent} = \frac{554.93 \text{ mol solvent}}{557.4 \text{ mol total}} \qquad\qquad X_{solute} = \frac{2.47 \text{ mol solute}}{557.4 \text{ mol total}}$$

$$X_{solvent} = 0.996 \qquad\qquad X_{solute} = 0.004$$

8 **101.59°C.** First convert the grams of potassium nitrate to moles of potassium nitrate, as shown above in problem 7: 2.47 mol KNO_3. Once you have done this determine the molality of the solution:

$$m = \frac{2.47 \text{ mol KNO}_3}{0.8 \text{ kg H}_2O} = 3.1 \text{ m}$$

Once you've determined the molality, plug it into the boiling point elevation formula, determine the K_b value, and solve:

$$\Delta T_b = \left(0.512\frac{°C}{m}\right)\left(3.1 \text{ m}\right) = 1.59°C$$

This shows the boiling point of the solution has been raised by 1.59°C. Since the boiling point of water is 100°C this makes the new boiling point 100°C + 1.59°C = 101.59°C.

9 **1.23°C.** Convert your grams of lithium chloride to moles. Then convert your ethanol value to kilograms. Determine the molality of the solution and then plug in your value to the boiling point elevation formula with the K_b value of ethanol:

$$\left(39.2\text{g LiCl}\right)\left(\frac{1 \text{ mol LiCl}}{42.39 \text{ g LiCl}}\right) = 0.92 \text{ mol LiCl}$$

$$m = \frac{0.92 \text{ mol LiCl}}{0.890 \text{ kg H}_2\text{O}} = 1.03 \text{ m}$$

$$\Delta T_b = \left(1.19\frac{°\text{C}}{\text{m}}\right)\left(1.03 \text{ m}\right) = 1.23°\text{C}$$

10 A colligative property is a property of a solution that depends on the number of solute particles dissolved into that solution. The identity of those particles does not matter at all. The way in which we **track** the number of those particles is usually done with moles. So you would say a colligative property such as freezing point depression or boiling point elevation depends on the number of moles of solute you add to a solution but not the specific identity of that solute.

11 **−1.34°C.** Convert your grams of calcium chloride to moles. Determine the molality of the solution and then plug in your value to the freezing point depression formula with the K_f value of water. Once you know how much the freezing point is going to be depressed, subtract that from the known freezing point of water to determine the new freezing point:

$$\left(40 \text{ g CaCl}_2\right)\left(\frac{1 \text{ mol CaCl}_2}{110.98 \text{ g CaCl}_2}\right) = 0.36 \text{ mol CaCl}_2$$

$$m = \frac{0.36 \text{ mol CaCl}_2}{0.5 \text{ kg H}_2\text{O}} = 0.72 \text{ m}$$

$$\Delta T_f = \left(1.86\frac{°\text{C}}{\text{m}}\right)\left(0.72\text{m}\right) = 1.34°\text{C}$$
$$0°\text{C} - 1.34°\text{C} = -1.34°\text{C}$$

12 **1.91°C.** Convert your grams of carbon dioxide to moles. Determine the molality of the solution and then plug in your value to the freezing point depression formula with the K_f value of ethanol and solve for freezing point:

$$\left(39 \text{ g CO}_2\right)\left(\frac{1 \text{ mol CO}_2}{44.01 \text{ g CO}_2}\right) = 0.89 \text{ mol CO}_2$$

$$m = \frac{0.89 \text{ mol CO}_2}{0.93 \text{ kg C}_2\text{H}_5\text{OH}} = 0.96 \text{ m}$$

$$\Delta T_f = \left(1.99\frac{°\text{C}}{\text{m}}\right)\left(0.96 \text{ m}\right) = 1.91°\text{C}$$

13 When solute is added to a solvent, this lowers the freezing point of that solvent. Freezing occurs when the temperature of a liquid reduces enough to form a solid. The formation of this solid occurs as the molecules slow down. Solids are highly ordered, but dissolved solute makes it harder for this ordered solid to form. Imagine a bunch of molecules trying to get into a nice straight organized line but there are these other molecules that are constantly getting in the way. That is the solute disrupting the freezing of the solvent. Thus the liquid must be cooled down even further to overcome the disruption of the solute.

(14) **217.86 g/mol.** Step one is to determine the change in freezing point, ΔT_f. In this case the change is $41.0°C - 33.0°C = 8.0°C$. Once you know this, look up the K_f of phenol in Table 19-3 (7.40°C/m). Plug both of these values into the equation for boiling point elevation and solve for molality:

$$m = \frac{\Delta T_f}{K_f} = \frac{8.0°C}{7.40\frac{°C}{m}} = 1.08 \text{ m}$$

Next, multiply this molality value by the given value for the mass of the solvent (first converting grams to kilograms):

$$\left(\frac{1.08 \text{ mol solute}}{1 \text{ kg } C_6H_6O}\right)(0.131 \text{ kg}) = 0.14 \text{ mol solute}$$

Last, divide the number of grams of the mystery solute by the number of moles, giving you the molecular mass of the compound:

$$\left(\frac{30.5 \text{ g}}{0.14 \text{ mol}}\right) = 217.86 \text{ g / mol}$$

(15) **53.3 g/mol.** You already know the change in boiling point from the problem, so plug that value and the K_b for ethanol (1.19°C/m) into your ΔT_b equation and solve for the molality of the solution:

$$m = \frac{\Delta T_b}{K_b} = \frac{4.50°C}{1.19\frac{°C}{m}} = 3.78 \text{ m}$$

Next, multiply this molality value by the given value for mass of solvent:

$$\left(\frac{3.78 \text{ mol solute}}{1 \text{ kg ethanol}}\right)(2.10 \text{ kg}) = 7.94 \text{ mol solute}$$

Last, divide the number of grams of the mystery solute by the number of moles, giving you the molecular mass of the compound:

$$\left(\frac{423 \text{ g}}{7.94 \text{ mol}}\right) = 53.3 \text{ g / mol}$$

IN THIS CHAPTER

» Checking out the many definitions of acids and bases

» Calculating acidity and basicity

» Determining the strength of acids and bases with the help of dissociation

Chapter 20

Working with Acids and Bases

Acids and bases are indicators of pH, and if you walk into any kitchen or bathroom, you find a multitude of each. In the refrigerator, you find soft drinks full of carbonic acid. The pantry holds vinegar and baking soda, an acid and a base. Peek under the sink, and you notice the ammonia and other cleaners, most of which are bases. Check out that can of lye-based drain opener — it's highly basic. In the medicine cabinet, you find aspirin, an acid, and antacids, bases. The everyday world is full of acids and bases, and so is the everyday world of the industrial chemist. In this chapter, we cover acids and bases and some good basic chemistry.

Getting to Know Acids and Bases

Before you can fully appreciate acids and bases, you need a basic understanding of what they are. The following chart looks at the properties of acids and bases that you can observe in everyday life.

Acids	Bases
Taste sour (but remember, in the lab, you test, don't taste)	Taste bitter
Produce a painful sensation on the skin	Feel slippery on the skin
React with certain metals (magnesium, zinc, and iron) to produce hydrogen gas	React with oils and greases

Acids	Bases
React with limestone and baking soda to produce carbon dioxide	React with acids to produce a salt and water
React with litmus paper and turn it red	React with litmus paper and turn it blue

Tables 20-1 and 20-2 show some common acids and bases found around the home.

TABLE 20-1 Common Acids Found in the Home

Chemical Name	Formula	Common Name or Use
Hydrochloric acid	HCl	Muriatic acid
Acetic acid	CH_3COOH	Vinegar
Sulfuric acid	H_2SO_4	Auto battery acid
Carbonic acid	H_2CO_3	Carbonated water
Boric acid	H_3BO_3	Antiseptic; eye drops
Acetylsalicylic acid	$C_{16}H_{12}O_6$	Aspirin

TABLE 20-2 Common Bases Found in the Home

Chemical Name	Formula	Common Name or Use
Ammonia	NH_3	Cleaner
Sodium hydroxide	$NaOH$	Lye
Sodium bicarbonate	$NaHCO_3$	Baking soda
Magnesium hydroxide	$Mg(OH)_2$	Milk of magnesia
Calcium carbonate	$CaCO_3$	Antacid
Aluminum hydroxide	$Al(OH)_3$	Antacid

Acids and Bases at the Atomic Level

If you look at Tables 20-1 and 20-2 closely, you may notice that all the acids contain hydrogen, and most of the bases contain the hydroxide ion (OH^-). Two main theories of the behavior of acids and bases use these facts along with one other major theory that looks at electron transfer:

>> Arrhenius theory

>> Brønsted-Lowry theory

>> Lewis theory

The rest of this section takes a closer look at these three theories to help you gain a firmer understanding of acids and bases.

The Arrhenius theory: Must have water

The Arrhenius theory was the first modern acid–base theory developed. In this theory, an acid is a substance that yields H⁺ (hydrogen) ions when dissolved in water, and a base is a substance that yields OH⁻ (hydroxide) ions when dissolved in water. HCl(g) can be considered a typical Arrhenius acid, because when this gas dissolves in water, it *ionizes* (forms ions) to give the H⁺ ion. (Chapter 9) is where you need to go for the riveting details about ions.)

$$HCl(aq) \rightarrow H^+(aq) + Cl^-(aq)$$

According to the Arrhenius theory, sodium hydroxide is classified as a base, because when it dissolves, it yields the hydroxide ion:

$$NaOH(aq) \rightarrow Na^+(aq) + OH^-(aq)$$

Arrhenius also classified the reaction between an acid and a base as a *neutralization* reaction, because if you mix an acidic solution with a basic solution, you end up with a neutral solution composed of water and a salt:

$$HCl(aq) + NaOH(aq) \rightarrow H_2O(l) + NaCl(aq)$$

Look at the ionic form of this equation (the form showing the reaction and production of ions) to see where the water comes from:

$$H^+(aq) + Cl^-(aq) + Na^+(aq) + OH^-(aq) \rightarrow H_2O(l) + Na^+(aq) + Cl^-(aq)$$

As you can see, the water is formed from combining the hydrogen and hydroxide ions. In fact, the net–ionic equation (the equation showing only those chemical substances that are changed during the reaction) is the same for all Arrhenius acid–base reactions:

$$H^+(aq) + OH^-(aq) \rightarrow H_2O(l)$$

The Arrhenius theory is still used quite a bit. But like all theories, it has some limitations. It specifies that the reactions must take place in water and that bases must contain hydroxide ions, but many reactions that do not meet these stipulations resemble acid–base reactions. For example, look at the gas phase reaction between ammonia and hydrogen chloride gases:

$$NH_3(g) + HCl(g) \rightarrow NH_4^+(aq) + Cl^-(aq) \rightarrow NH_4Cl(s)$$

The two clear, colorless gases mix, and a white solid of ammonium chloride forms. We show the intermediate formation of the ions in the equation so you can better see what's actually happening. The HCl transfers one H⁺ to the ammonia. That's basically the same thing that happens in the HCl/NaOH reaction, but the reaction involving the ammonia can't be classified as an acid–base reaction, because it doesn't occur in water and it doesn't involve the hydroxide ion. But again, the same basic process is taking place in both cases. To account for these similarities, a new acid–base theory was developed: the Brønsted-Lowry theory.

The Brønsted-Lowry acid-base theory: Giving and accepting

The Brønsted-Lowry theory attempts to overcome the limitations of the Arrhenius theory by defining an acid as a proton (H^+) donor and a base as a proton (H^+) acceptor. The base accepts the H^+ by furnishing a lone pair of electrons for a *coordinate covalent bond*, which is a covalent bond (shared pair of electrons) in which one atom furnishes both of the electrons for the bond. Normally, one atom furnishes one electron for the bond and the other atom furnishes the second electron. In the coordinate covalent bond, one atom furnishes both bonding electrons.

Figure 20-1 shows the reaction between NH_3 and HCl using the electron-dot structures of the reactants and products.

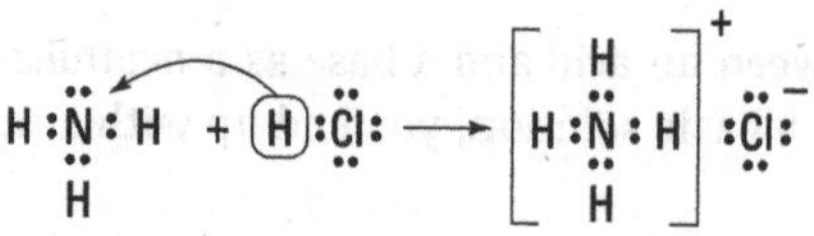

FIGURE 20-1:
The reaction between ammonia and hydrogen chloride gases.

HCl is the acid, so it's the proton donor, and ammonia is the base, the proton acceptor. Ammonia has a lone pair of nonbonding electrons that it can furnish for the coordinate covalent bond.

In many cases, you are going to see a reaction and be asked to identify the Brønsted-Lowry base and acid from the equation. Thankfully, there is a very simple way to determine what is donating a proton and what is accepting a proton. The best way is to keep careful track of hydrogen ions in the equation. Consider, for example, the dissociation of the base sodium carbonate in water. Note that although sodium carbonate is a base, it doesn't contain a hydroxide ion.

$$Na_2CO_3 + 2H_2O \rightarrow H_2CO_3 + 2NaOH$$

This is a simple double replacement reaction (see Chapter 13 for an introduction to these types of reactions). A hydrogen ion from water switches places with the sodium of sodium carbonate to form the products carbonic acid and sodium hydroxide. By the Brønsted-Lowry definition, water is the acid because it donates its hydrogen to Na_2CO_3. This makes Na_2CO_3 the base because it accepts the hydrogen from H_2O.

REMEMBER What about the substances on the right-hand side of the equation? Brønsted-Lowry theory calls the products of an acid-base reaction the *conjugate acid* and *conjugate base*. The conjugate acid (in this case, H_2CO_3) is produced when the base accepts a proton, while the conjugate base (NaOH) is formed when the acid loses its hydrogen. This reaction also brings up a very important point about the strength of each of these acids and bases. Although sodium carbonate is a very strong base, its conjugate acid, carbonic acid, is a very weak acid. Similarly, water is an extremely weak acid, and its conjugate base, sodium hydroxide, is a very strong base.

Lewis relies on electron pairs

In the same year that Brønsted and Lowry proposed their definition of acids and bases, an American chemist named Gilbert Lewis proposed an alternative definition that not only encompassed Brønsted-Lowry theory but also accounted for acid-base reactions in which a hydrogen ion isn't exchanged. Lewis's definition relies on tracking lone pairs of electrons. Under his theory, a *base* is any substance that donates a pair of electrons to form a coordinate covalent bond with another substance, and an *acid* is a substance that accepts that electron pair in such a reaction.

All Brønsted-Lowry acids are Lewis acids, but in practice, the term *Lewis acid* is generally reserved for Lewis acids that don't also fit the Brønsted-Lowry definition. The best way to spot a Lewis acid-base pair is to draw a Lewis (electron dot) structure of the reacting substances, noting the presence of lone pairs of electrons. (We introduce Lewis structures in Chapter 10.) For example, consider the reaction between ammonia (NH_3) and boron trifluoride (BF_3):

$$NH_3 + BF_3 \rightarrow NH_3BF_3$$

At first glance, neither the reactants nor the product appears to be an acid or base, but the reactants are revealed as a Lewis acid-base pair when drawn as Lewis structures as in Figure 20-2. Ammonia donates its lone pair of electrons to the bond with boron trifluoride, making ammonia the Lewis base and boron trifluoride the Lewis acid.

FIGURE 20-2:
The Lewis structures of ammonia and boron trifluoride.

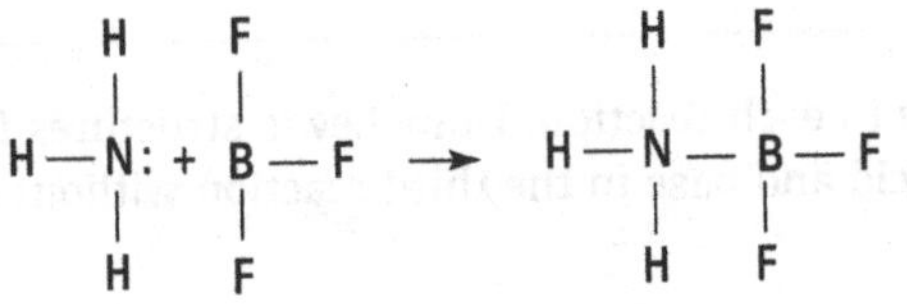

$$
\begin{array}{ccc}
H & F & \\
| & | & \\
H-N: + B-F & \rightarrow & H-N-B-F \\
| & | & \\
H & F &
\end{array}
$$

Sometimes you can identify the Lewis acid and base in a compound without drawing the Lewis structure. You can do this by identifying reactants that are electron rich (bases) or electron poor (acids). A metal cation, for example, is electron poor and tends to act as a Lewis acid in a reaction, accepting a pair of electrons.

TIP

In practice, it's much simpler to use the Arrhenius or Brønsted-Lowry definition of acid and base, but you'll need to use the Lewis definition when hydrogen ions aren't being exchanged. You can pick and choose among the definitions when you're asked to identify the acid and base in a reaction.

EXAMPLE

Q. Identify the acid and base in the following reaction and label their conjugates.

$$NH_3 + H_2O \rightarrow NH_4^+ + OH^-$$

A. This is a classic Brønsted-Lowry acid-base pair. Water (H_2O) loses a proton to ammonia (NH_3), forming a hydroxide anion. This makes water the proton donor, or Brønsted-Lowry acid, and OH⁻ its conjugate base. Ammonia accepts the proton from water, forming ammonium (NH_4^+). This makes ammonia the proton acceptor, or Brønsted-Lowry base, and NH_4^+ its conjugate acid.

1 Consider the following reaction. Label the acid, base, conjugate acid, and conjugate base, and comment on their strengths. How can water act as an acid in one reaction and a base in another?

$$HCl + H_2O \rightarrow H_3O^+ + Cl^-$$

2 Use the Arrhenius definition to identify the acid or base in each reaction and explain how you know.

a. $NaOH(s) + H_2O \rightarrow Na^+(aq) + OH^-(aq) + H_2O$

b. $HF(g) + H_2O \rightarrow H^+(aq) + F^-(aq) + H_2O$

3 Identify the Lewis acid and base in each reaction. Draw Lewis structures for the first two, and determine the Lewis acid and base in the third reaction without a dot structure.

a. $6H_2O + Cr^{3+} \rightarrow Cr(OH_2)_6^{3+}$

b. $2NH_3 + Ag^+ \rightarrow Ag(NH_3)_2^+$

c. $2Cl^- + HgCl_2 \rightarrow HgCl_4^{2-}$

Measuring Acidity and Basicity: p H, p OH, and K_W

A substance's identity as an acid or a base is only one of many things that a chemist may need to know about it. Sulfuric acid and water, for example, can both act as acids, but using sulfuric acid to wash your face in the morning would be a grave error indeed. Sulfuric acid and water differ greatly in *acidity*, a measurement of an acid's strength. The amount of acidity in a solution is related to the concentration of the hydronium ion, H_3O^+, in the solution. The more acidic the solution is, the larger the concentration of the hydronium ion. The hydronium ion is often simply represented as H^+. A similar quantity, called *basicity*, measures a base's strength.

Acidity and basicity are measured in terms of quantities called *pH* and *pOH*, respectively. Both are simple scales ranging from 0 to 14, with low numbers on the pH scale representing a higher acidity and therefore a stronger acid. On both scales, a measurement of 7 indicates a *neutral solution*. On the pH scale, any number lower than 7 indicates that the solution is acidic, with acidity increasing as pH decreases, and any number higher than 7 indicates a basic solution, with basicity increasing as pH increases. In other words, the further the pH gets away from 7, the more acidic or basic a substance gets. The pOH shows exactly the same relationship between distance from 7 and acidity or basicity, only this time low numbers indicate very basic solutions, while high numbers indicate very acidic solutions. Figure 20-3 shows you the entire pH scale.

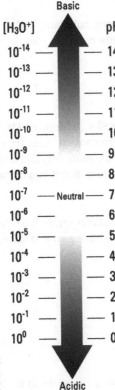

FIGURE 20-3:
The pH scale.

You calculate pH using the formula $pH = -\log[H^+]$, where the brackets around H^+ indicate that it's a measurement of the concentration of hydrogen ions in moles per liter (or molarity; see Chapter 18). You calculate pOH using a similar formula, with OH^- concentration replacing the H^+ concentration: $pOH = -\log[OH^-]$. The word *log* in each formula stands for logarithm.

REMEMBER

Because a substance with high acidity must have low basicity, a low pH indicates a high pOH for a substance and vice versa. In fact, a very convenient relationship between pH and pOH allows you to solve for one when you have the other: $pH + pOH = 14$.

You'll often be given a pH or pOH and be asked to solve for the H⁺ or OH⁻ concentrations instead of the other way around. The logarithms in the pH and pOH equations make it tricky to solve for [H⁺] or [OH⁻], but if you remember that a log is undone by raising 10 to both sides of an equation, you quickly arrive at a convenient formula for [H⁺], namely $\left[H^+\right]=10^{-pH}$. Similarly, [OH⁻] can be calculated using the formula $\left[OH^-\right]=10^{-pOH}$.

Water can act as either an acid or a base, depending on what it's combined with. When an acid reacts with water, water acts as a base, or a proton acceptor. But in reactions with a base (like ammonia; see the preceding section), water acts as an acid, or a proton donor. Substances that can act as either an acid or a base are called *amphoteric*.

But can water react with itself? Yes, it can. When two water molecules react with each other, one donates a proton and the other accepts it:

$$H_2O(l) + H_2O(l) \rightleftarrows H_3O^+(aq) + OH^-(aq)$$

This reaction is an *equilibrium reaction*. A modified equilibrium constant, called the K_w (which stands for *water dissociation constant*), is associated with this reaction. The K_w has a value of 1.0×10^{-14} and has the following form:

$$1.0 \times 10^{-14} = K_w = \left[H^+\right]\left[OH^-\right]$$

In pure water, the [H⁺] equals the [OH⁻] from the balanced equation, so [H₃O⁺] = [OH⁻] = 1.0×10^{-7}. The K_w value is a constant. This value allows you to convert from [H⁺] to [OH⁻], and vice versa, in *any* aqueous solution, not just pure water. In aqueous solutions, the hydronium ion and hydroxide ion concentrations are rarely going to be equal. But if you know one of them, the K_w allows you to figure out the other one.

Take a look at the 2.0 M acetic acid solution problem in the upcoming section "Ionizing partway: Weak acids." You find that the [H⁺] is 6.0×10^{-3}. Now you have a way to calculate the [OH⁻] in the solution by using the K_w relationship:

$$K_w = 1.0 \times 10^{-14} = [H_3O^+][OH^-]$$
$$1.0 \times 10^{-14} = \left[6.0 \times 10^{-3}\right][OH^-]$$
$$1.0 \times 10^{-14} / 6.0 \times 10^{-3} = [OH^-]$$
$$1.7 \times 10^{-12} = [OH^-]$$

We think pH is a great topic in chemistry because you can usually relate to it. At some point, almost everyone has used some form of cleaning solution or had a carbonated beverage. All these household substances cover a wide range of pH values. Table 20-3 lists some common substances and their pH values. You can use it to identify the different acids and bases you encounter in your daily life.

To sustain life, human blood must stay within about ±0.2 pH units of 7.3, a narrow range. Many things, such as foods and hyperventilation, can act to change the pH of your blood. Buffers help regulate blood pH and keep it in the 7.1 to 7.5 range.

TABLE 20-3 Average pH Values of Some Common Substances

Substance	pH
Oven cleaner	13.8
Hair remover	12.8
Household ammonia	11.0
Milk of magnesia	10.5
Chlorine bleach	9.5
Seawater	8.0
Human blood	7.3
Pure water	7.0
Milk	6.5
Black coffee	5.5
Soft drinks	3.5
Aspirin	2.9
Vinegar	2.8
Lemon juice	2.3
Auto battery acid	0.8

Q. Calculate the pH and pOH of a solution with an $[H^+]$ of 1×10^{-8}. Is the solution acidic or basic? Do the same for a solution with an $[OH^-]$ of 2.3×10^{-11}.

EXAMPLE

A. For a solution with an $[H^+]$ of 1×10^{-8}: pH = 8; pOH = 6; the solution is a base. You've been given the H^+ concentration, so first solve for the pH by plugging $[H^+]$ into the formula for pH:

$$pH = -\log\left[1 \times 10^{-8}\right] = 8$$

Plugging this value into the equation relating pH and pOH gives you a value for pOH:

$$8 + pOH = 14$$
$$pOH = 14 - 8$$
$$pOH = 6$$

A pH of 8 indicates that this solution is very slightly basic. It's not just a coincidence that the exponent of the H^+ concentration is equal to the pH. This is true whenever the coefficient of the H^+ concentration is 1.

For a solution with an $[OH^-]$ of 2.3×10^{-11}: pOH = 10.6; pH = 3.4; the solution is an acid. The second portion of the problem gives you an $[OH^-]$, in which case you can use the ion-product constant for water to calculate the $[H^+]$ and then solve for pH and pOH as you did before; or more simply, you can first calculate the pOH and use it to find the pH. Plugging an $[OH^-]$ of 2.3×10^{-11} into the pOH equation yields

$$pOH = -\log\left[2.3 \times 10^{-11}\right] = 10.6$$

Plug this value into the relation between pH and pOH to get the pH:

$$pH + 10.6 = 14$$
$$pH = 14 - 10.6$$
$$pH = 3.4$$

This low pH indicates that the substance is a relatively strong acid.

YOUR TURN

 4 Determine the pH given the following values:

(a) $\left[H^+\right] = 1 \times 10^{-13}$

(b) $\left[H^+\right] = 1.58 \times 10^{-9}$

(c) $\left[OH^-\right] = 2.7 \times 10^{-7}$

(d) $\left[OH^-\right] = 1 \times 10^{-7}$

 5 Determine the pOH given the following values:

(a) $\left[H^+\right] = 2 \times 10^{-8}$

(b) $\left[OH^-\right] = 5.1 \times 10^{-11}$

 6 Determine whether each of the following is acidic, basic, or neutral:

(a) $\left[OH^-\right] = 2.5 \times 10^{-7}$

(b) $\left[OH^-\right] = 3.1 \times 10^{-13}$

(c) $\left[H^+\right] = 4.21 \times 10^{-5}$

(d) $\left[H^+\right] = 8.9 \times 10^{-10}$

 7 Determine the [H⁺] from the following pH or pOH values:

(a) pH = 3.3

(b) pH = 7.69

(c) pOH = 10.21

(d) pOH = 1.26

Dissociating with Strong and Weak Acids

There are two major categories of acids and bases that you can look at when determining acid dissociation: strong and weak. However, remember that acid–base strength is not the same as concentration. *Strength* refers to the amount of ionization or breaking apart that a particular acid or base undergoes. *Concentration* refers to the amount of acid or base that you initially have. You can have a concentrated solution of a weak acid, or a dilute solution of a strong acid, or a concentrated solution of a strong acid, or . . . well, we're sure you get the idea. The following sections point out the main differences between strong and weak acids and bases.

Ionizing completely: Strong acids

Acids that ionize completely are considered strong. If you dissolve hydrogen chloride gas in water, the HCl reacts with the water molecules and donates a proton to them:

$$HCl(g) + H_2O(l) \rightarrow Cl^-(aq) + H_3O^+(aq)$$

The H_3O^+ ion is called the *hydronium ion.* This reaction goes essentially to completion, meaning the reactants keep creating the product until they're all used up. In this case, all the HCl ionizes to H_3O^+ and Cl^-; no more HCl is present. Note that water, in this case, acts as a base, accepting the proton from the hydrogen chloride.

Because strong acids ionize completely, calculating the concentration of the hydronium ion and chloride ion in solution is easy if you know the initial concentration of the strong acid. For example, suppose that you bubble 0.1 mol of HCl gas into a liter of water. You can say that the initial concentration of HCl is 0.1 M (0.1 mol/L). *M* stands for molarity, and *mol/L* stands for moles of solute per liter.

You can represent this 0.1 M concentration for the HCl in this fashion: [HCl] = 0.1. The brackets around the compound indicate molar concentration, or moles per liter. Because the HCl completely ionizes, you see from the balanced equation that for every HCl that ionizes, you get one hydronium ion and one chloride ion. So the concentration of ions in that 0.1 M HCl solution is

$$\left[H_3O^+\right] = 0.1 \text{ M} \quad \text{and} \quad \left[Cl^-\right] = 0.1 \text{ M}$$

This idea is valuable when you calculate the pH of a solution. Table 20-4 lists the most common strong acids — and bases — you're likely to encounter.

TABLE 20-4 Strong Acids and Bases

Acid Name	Chemical Formula	Base Name	Chemical Formula
Hydrochloric acid	HCl	Lithium hydroxide	LiOH
Hydroiodic acid	HI	Sodium hydroxide	NaOH
Hydrobromic acid	HBr	Potassium hydroxide	KOH
Nitric acid	HNO_3	Rubidium hydroxide	RbOH
Sulfuric acid	H_2SO_4	Cesium hydroxide	CsOH
Chloric acid	$HClO_3$	Calcium hydroxide	$Ca(OH)_2$
Perchloric acid	$HClO_4$	Strontium hydroxide	$Sr(OH)_2$
		Barium hydroxide	$Ba(OH)_2$

To measure the amount of dissociation occurring when a weak acid is in aqueous solution, chemists use a constant called the *acid dissociation constant* (K_a). K_a is a special variety of the equilibrium constant. The equilibrium constant of a chemical reaction is the concentration of products over the concentration of reactants, and it indicates the balance between products and reactants in a reaction.

Falling to pieces: Strong bases

A strong base *dissociates* (breaks apart) completely in water. You normally see only one strong base, the hydroxide ion, OH^-. Calculating the hydroxide ion concentration is really straightforward. Suppose that you have a 1.5 M (1.5 mol/L) NaOH solution. The sodium hydroxide, a salt, completely dissociates into ions:

$$NaOH \rightarrow Na^+(aq) + OH^-(aq)$$

If you start with 1.5 mol/L NaOH, then you have the same concentration of ions:

$$\left[Na^+\right] = 1.5 \text{ M} \quad \text{and} \quad \left[OH^-\right] = 1.5 \text{ M}$$

Ionizing partway: Weak acids

Acids that only partially ionize are called *weak acids*. One example is acetic acid (CH_3COOH). If you dissolve acetic acid in water, it reacts with the water molecules, donating a proton and forming hydronium ions. It also establishes an equilibrium in which you have a significant amount of un-ionized acetic acid. (In reactions that go to completion, the reactants are completely used up creating the products. But in equilibrium systems, two exactly opposite chemical reactions — one on each side of the reaction arrow — are occurring at the same place, at the same time, with the same speed of reaction.)

The acetic acid reaction with water looks like this:

$$CH_3COOH(l) + H_2O(l) \rightleftarrows CH_3COO^-(aq) + H_3O^+(aq)$$

The acetic acid that you added to the water is only partially ionized, so it's a weak acid. In the case of acetic acid, about 5 percent ionizes, and 95 percent remains in the molecular form. The amount of hydronium ion that you get in solutions of acids that don't ionize completely is much less than it is with a strong acid.

TIP

Calculating the hydronium ion concentration in weak acid solutions isn't as straightforward as it is in strong solutions, because not all of the weak acid that dissolves initially has ionized. To calculate the hydronium ion concentration, you must use the equilibrium constant expression for the weak acid. For weak acid solutions, you use a mathematical expression called the K_a — the *acid ionization constant*.

Take a look at the generalized ionization of some weak acid HA (hypothetical acid):

$$HA + H_2O \rightleftarrows A^- + H_3O^+$$

The K_a expression for this weak acid is

$$K_a = \frac{\left[H_3O^+\right]\left[A^-\right]}{[HA]}$$

Note that the [HA] represents the molar concentration of HA *at equilibrium*, not initially. Also, note that the concentration of water doesn't appear in the K_a expression, because there's so much water that it actually becomes a constant incorporated into the K_a expression.

Now go back to that acetic acid equilibrium. The K_a for acetic acid is 1.8×10^{-5}. The K_a expression for the acetic acid ionization is

$$K_a = 1.8 \times 10^{-5} = \frac{\left[H_3O^+ \right]\left[CH_3COO^- \right]}{\left[CH_3COOH \right]}$$

Here's an example of how to use K_a to calculate the hydronium ion concentration in a 2.0 M solution of acetic acid:

1. **Start by considering what information you have about the initial concentration and the products.**

   The initial concentration of acetic acid is 2.0 M, and you know that a little bit has ionized, forming a little hydronium ion and acetate ion. You also can see from the balanced reaction that for every hydronium ion that's formed, an acetate ion is also formed, so their concentrations are the same.

2. **Using the information you've gathered from the balanced reaction, represent the amount of $[H_3O^+]$ and $[CH_3COO^-]$ as x.**

   $$\left[H_3O^+ \right] = \left[CH_3COO^- \right] = x$$

3. **In order to produce the x amount of hydronium and acetate ion, the same amount of ionizing acetic acid is required. Represent the amount of acetic acid remaining at equilibrium as the amount you started with.**

   In this example, you started with 2.0 M; subtract the amount that ionizes, x:

   $$\left[CH_3COOH \right] = 2.0 - x$$

4. **For the vast majority of situations, you can say that x is very small in comparison to the initial concentration of the weak acid. Therefore, you can often approximate the equilibrium concentration of the weak acid with its initial concentration.**

   In this case, you can say that $2.0 - x$ is approximately equal to 2.0. The equilibrium constant expression now looks like this:

   $$K_a = 1.8 \times 10^{-5} = \frac{\left[x \right]\left[x \right]}{\left[2.0 \right]} = \frac{\left[x \right]^2}{\left[2.0 \right]}$$

5. Solve for x, which is the $[H_3O^+]$.

$$\left(1.8\times10^{-5}\right)\left[2.0\right]=\left[x\right]^2$$
$$\sqrt{3.6\times10^{-5}}=\left[x\right]=[H_3O^+]$$
$$6.0\times10^{-3}=[H_3O^+]$$

Refer to Table 20-4 to see some common strong acids. Most of the other acids you encounter are weak.

One way to distinguish between strong and weak acids is to look for an acid ionization constant (K_a) value. If the acid has a K_a value, then it's weak.

REMEMBER

Finding equilibrium with water: Weak bases

Weak bases also react with water to establish an equilibrium system. Ammonia is a typical weak base. It reacts with water to form the ammonium ion and the hydroxide ion:

$$NH_3(g)+H_2O(l)\rightleftharpoons NH_4^+ + OH^-$$

Like a weak acid, a weak base is only partially ionized. The modified equilibrium constant expression for weak bases is K_b. You use it exactly the same way you use the K_a (see the preceding section for details), except you solve for the $[OH^-]$.

Q. Write a general expression for the acid dissociation constant of the following reaction, a dissociation of ethanoic acid:

EXAMPLE

$$CH_3COOH+H_2O \rightarrow H_3O^+ + CH_3COO^-$$

Then calculate its actual value if $\left[CH_3COOH\right]=2.34\times10^{-4}$ and $\left[CH_3COO^-\right]=6.51\times10^{-5}$.

A. 1.80×10^{-5}. Your general expression should be

$$K_a=\frac{\left[H_3O^+\right]\left[CH_3COO^-\right]}{\left[CH_3COOH\right]}$$

You're given $\left[CH_3COO^-\right]=6.51\times10^{-5}$, and $[H_3O^+]$ must be the same. Plugging all known values into the expression for K_a yields

$$K_a=\frac{\left[6.51\times10^{-5}\right]\left[6.51\times10^{-5}\right]}{\left[2.34\times10^{-4}\right]}=1.80\times10^{-5}$$

The K_a is very small, so this is a very weak acid.

8 Calculate the K_a for a 0.50 M solution of benzoic acid, C_6H_5COOH, if the $[H^+]$ concentration is 5.6×10^{-3}.

9 The pH of a 0.75 M solution of HCOOH is 1.93. Calculate the acid dissociation constant.

10 If you have a 0.20 M solution of ammonia, NH_3, that has a K_b of 1.80×10^{-5}, what is the pH of that solution?

Practice Questions Answers and Explanations

(1) In this reaction, hydrochloric acid (HCl) donates a proton to water (H_2O), making it the Brønsted–Lowry acid. Water, which accepts the proton, is the Brønsted–Lowry base. This makes hydronium (H_3O^+) the conjugate acid and chloride (Cl^-) the conjugate base. Water can act as the base in this reaction and as an acid in the example problem because it's composed of both a hydrogen ion and a hydroxide ion; therefore, it can either accept or donate a proton.

(2) For Arrhenius acids, remember to track the movement of H^+ and OH^- ions. If the reaction yields an OH^- product, the substance is a base, whereas an H^+ product reveals that the substance is an acid.

   a. **NaOH, Arrhenius base.** Sodium hydroxide (NaOH) dissociates in water to form OH^- ions, making it an Arrhenius base.

   b. **HF, Arrhenius acid.** Hydrofluoric acid (HF) dissociates in aqueous solution to form H^+ ions, making it an Arrhenius acid.

(3) To identify Lewis acids and bases, track the movement of electron pairs. Draw a Lewis structure to locate the atom with a lone pair available to donate. This is the Lewis base.

   a. **Water (H_2O) is the Lewis base, and chromium (Cr^{3+}) is the Lewis acid.** Your Lewis structure should show that the lone pair of electrons is donated to the bond by H_2O. You can tell that the electrons come from H_2O because chromium doesn't have any electrons available for bonding due to its ionic charge of +3.

   b. **Ammonia (NH_3) is the Lewis base, and silver (Ag^+) is the Lewis acid.** Your Lewis structure should show that the lone pair of electrons is donated to the bond by NH_3. The positive metal (silver) doesn't have any electrons available for bonding, so the electrons for forming the bond must come from NH_3.

c. **The negative charge on the chlorine (Cl⁻) indicates that the chlorine is electron rich, making it the Lewis base. Mercury (Hg) is therefore the Lewis acid.** Mercury is a metal and regularly forms a positive ion, which is another hint that mercury will act as the Lewis acid and accept the electrons.

(4) To calculate the pH values in this problem, use the equation $pH = -\log\left[H^+\right]$ and plug in all known values. If you're given [OH⁻] instead of [H⁺], simply calculate the pOH (using the equation $pOH = -\log\left[OH^-\right]$) instead and then subtract that number from 14 to give you the pH.

a. **pH = 13.** Because the coefficient on the H⁺ concentration is 1, the pH is simply the exponent.

b. **pH = 8.8.** Plug the given [H⁺] into the pH equation, giving you $pH = -\log\left[1.58 \times 10^{-9}\right] = 8.8$.

c. **pH = 7.4.** First, use the OH⁻ concentration to calculate the pOH: $pOH = -\log\left[2.7 \times 10^{-7}\right] = 6.6$. Then use the relationship between pH and pOH $(14 - pOH = pH)$ to determine the pH:.
$14 - 6.6 = 7.4$

d. **pH = 7.** Because the coefficient of this OH⁻ concentration is 1, you can read the pOH directly from the exponent to get pOH = 7. Then use the relationship between pH and pOH $(14 - pOH = pH)$ to determine the pH: $14 - 7 = 7$.

(5) To calculate pOH, simply take the negative logarithm of the OH⁻ concentration. If you're given the H⁺ concentration, use it to calculate the pH and then subtract that value from 14 to yield the pOH.

a. **pOH = 6.3.** First, use the H⁺ concentration to calculate the pH: $pH = -\log\left[2 \times 10^{-8}\right] = 7.7$. Then use the relationship between pH and pOH $(14 - pH = pOH)$ to determine the pOH:
$14 - 7.7 = 6.3$.

b. **pOH = 10.3.** Here, you can simply use the OH⁻ concentration to calculate the pOH directly: $pOH = -\log\left[5.1 \times 10^{-11}\right] = 10.3$.

(6) To determine acidity, you need to calculate pH.

a. **Basic.** Use [OH⁻] to determine pOH: $pOH = -\log\left[2.5 \times 10^{-7}\right] = 6.6$. Subtract this value from 14 to get a pH of 7.4. This pH is very slightly greater than 7, so the solution is slightly basic.

b. **Acidic.** Use [OH⁻] to determine pOH: $pOH = -\log\left[3.1 \times 10^{-13}\right] = 12.5$. Subtract this value from 14 to get a pH of 1.5. This pH is significantly smaller than 7, so the solution is quite acidic.

c. **Acidic.** Use the [H⁺] concentration to determine pH: $pH = -\log\left[4.21 \times 10^{-5}\right] = 4.4$. This solution has a pH smaller than 7 and is therefore acidic.

d. **Basic.** Use the [H⁺] concentration to determine pH: $pH = -\log\left[8.9 \times 10^{-10}\right] = 9.1$. This solution has a pH greater than 7 and is therefore basic.

7 In this problem, use the formula $\left[H^+\right]=10^{-pH}$. If you're given the pOH instead of the pH, begin by subtracting the pOH from 14 to give you the pH and then plug that number into the formula.

a. 5.0×10^{-4}. Use the pH to calculate the H+ concentration using the equation $\left[H^+\right]=10^{-pH}$: $\left[H^+\right]=10^{-3.3}=5.0\times10^{-4}$.

b. 2.04×10^{-8}. Use the pH to calculate the H+ concentration using the equation $\left[H^+\right]=10^{-pH}$: $\left[H^+\right]=10^{-7.69}=2.04\times10^{-8}$.

c. 1.62×10^{-4}. Begin by using the pOH to calculate the pH using the equation $14-pOH=pH$: $14-10.21=3.79$. Then use the pH to calculate the H+ concentration using the equation $\left[H^+\right]=10^{-pH}$: $\left[H^+\right]=10^{-3.79}=1.62\times10^{-4}$.

d. 1.82×10^{-13}. Begin by using the pOH to calculate the pH using the equation $14-pOH=pH$: $14-1.26=12.74$. Then use the pH to calculate the H+ concentration using the equation $\left[H^+\right]=10^{-pH}$: $\left[H^+\right]=10^{-12.74}=1.82\times10^{-13}$.

8 6.3×10^{-5}. The molarity in this problem tells you that the concentration of benzoic acid is 5.0×10^{-1}. You also know that the concentration of benzoic acid's conjugate base is the same as the given H+ concentration. All that remains is to write an equation for the acid dissociation constant and plug in these concentrations.

$$K_a=\frac{\left[H_3O^+\right]\left[C_6H_5COO^-\right]}{\left[C_6H_5COOH\right]}=\frac{\left[5.6\times10^{-3}\right]\left[5.6\times10^{-3}\right]}{\left[5.0\times10^{-1}\right]}=6.3\times10^{-5}$$

9 1.9×10^{-4}. You need to use the pH to calculate the [H+] with $\left[H^+\right]=10^{-pH}$:

$$\left[H^+\right]=10^{-1.93}=1.2\times10^{-2}$$

This value must also be equal to the concentration of the conjugate base. All that remains is to write an equation for the acid dissociation constant and plug in these concentrations:

$$K_a=\frac{\left[H_3O^+\right]\left[HCOO^-\right]}{\left[HCOOH\right]}=\frac{\left[1.2\times10^{-2}\right]\left[1.2\times10^{-2}\right]}{\left[7.5\times10^{-1}\right]}=1.9\times10^{-4}$$

10 **pH = 11.3.** Begin by considering the K_b equation:

$$K_b=\frac{\left[OH^-\right]\left[NH_4^+\right]}{\left[NH_3\right]}$$

Because $\left[OH^-\right]=\left[NH_4^+\right]$, you can rewrite this as

$$K_b=\frac{\left[OH^-\right]^2}{\left[NH_3\right]}$$

Solving this equation for [OH$^-$] yields $\left[OH^-\right]=\sqrt{\left(K_b\right)\left(\left[NH_3\right]\right)}$. Plugging the known values of K_b and [NH$_3$] into this equation yields

$$\left[OH^-\right]=\sqrt{\left(1.8\times10^{-5}\right)\left(\left[0.20\right]\right)}=1.9\times10^{-3}$$

Use this to solve for the pOH:

$$pOH = -\log\left[1.9 \times 10^{-3}\right] = 2.7$$

The final step is to solve for pH by subtracting this number from 14:

$$14 - 2.7 = 11.3$$

If you're ready to test your skills a bit more, take the following chapter quiz that incorporates all the chapter topics.

Whaddya Know? Chapter 20 Quiz

Ready for a quiz? The 15 questions in this section will test the skills you learned in this chapter. When you're done, check out the section that follows for answers and explanations.

1 Identify the Arrhenius acid and the Arrhenius base in the following reaction:

$$H_2SO_4 + 2KOH \rightarrow 2H_2O + K_2SO_4$$

2 Identify the acid, base, conjugate acid, and conjugate base in the following reaction:

$$H_2S + NH_3 \rightarrow NH_4^+ + HS^-$$

3 Identify the acid, base, conjugate acid, and conjugate base in the following reaction:

$$S^{2-} + H_2O \rightarrow OH^- + HS^-$$

4 Identify whether or not the following completely dissociate in water:

a. NaOH

b. HCl

c. H_2CO_3

d. HF

e. HNO_2

a. $Cu(OH)_2$

a. KOH

5 What is the pH of a solution containing that has an OH^- concentration of $3.2 \times 10^{-5} M$?

6 What is the pOH of a solution that has an OH^- concentration of 8.5×10^{-12}? Is this solution acidic or basic?

7 What is the hydronium ion concentration $[H^+]$ of a solution that has a pH of 4.3?

8 What is the hydroxide ion concentration $[OH^-]$ of a solution that has a pH of 11.2?

9 What is the pH of a solution when you mix 2.5 mol of HCl into 3.0 liters of water?

10 What is the pH of a solution when you mix 1.8 mol of NaOH into 4.0 liters of water?

11 You mix 59 grams of hydrogen bromide with 3.0 liters of water. What is the resulting pH of the solution? Is this solution acidic or basic?

12 What is the pH of a solution of hydrochloric acid with a concentration of 0.0000030 M?

13 What is the difference between a strong acid and a weak acid in terms of solubility?

14 You have a 0.30 M solution of HCN. The K_a of HCN is 6.2×10^{-10}. What is the pH of the solution?

15 Using the K_w equation, what is the concentration of H^+ found in solution if the OH^- concentration in that solution is 4.3×10^{-6} M?

Answers to Chapter 20 Quiz

1. The Arrhenius acid in the reaction is H_2SO_4. This is the acid because when it is put into to solution the hydrogen will dissociate from the formula and increase the overall H^+ concentration of the solution. The Arrhenius base is KOH. This is the base because when it is put into solution the formula will dissociate and increase the OH ion concentration in solution.

2. The acid in this reaction is H_2S and the base is NH_3. The conjugate acid is NH_4^+ and the conjugate base is HS^-. You can see this because one hydrogen from H_2S is donated (acids donate protons) to NH_3 (bases accept protons) during the reaction. This is observable as the products show that NH_3 became NH_4^+ and H_2S became HS^-. If the reaction were to reverse itself NH_4^+ would donate a proton to HS^- making NH_4^+ the conjugate acid and HS^- the conjugate base for accepting the proton.

3. The acid in this reaction is H_2O (remember water is amphoteric and can act as an acid or base as needed) and the base is S^{2-}. The conjugate acid is HS^- and the conjugate base is OH^-. You can see this because one hydrogen from H_2O is donated (acids donate protons) to S^{2-} (bases accept protons) during the reaction. This is observable as the products show that H_2O became OH^- and S^{2-} became HS^-. If the reaction were to reverse itself HS^- would donate a proton to OH^- making HS^- the conjugate acid and OH^- the conjugate base for accepting the proton.

4. To determine if an acid or base dissolves complete in water you need to know whether they are a strong acid/base, as shown in Table 20-4. If an acid/base is strong it will completely dissociate in water. If it is not a strong acid/base then it will not complete dissociate.

   a. Strong acid, yes will completely dissociate

   b. Strong acid, yes will completely dissociate

   c. Weak acid, no will not completely dissociate

   d. Weak acid, no will not completely dissociate

   e. Weak acid, no will not completely dissociate

   f. Weak base, no will not completely dissociate

   g. Strong base, yes will completely dissociate

5. **pH = 9.51.** If given an OH^- concentration you can take the –log of that concentration to determine pOH. Once you have pOH you subtract from 14 using the pH + pOH = 14 formula to solve for pH.

$$[OH^-] = 3.2 \times 10^{-5}$$
$$-\log[OH^-] = pOH$$
$$-\log[3.2 \times 10^{-5}] = pOH$$
$$pOH = 4.49$$
$$pH = 14 - 4.49$$
$$pH = 9.51$$

(6) pOH = 11.07. **The solution is acidic.** Take the −log of your OH concentration to determine the pOH of the solution. Once you have this the simplest way to determine if the solution is acidic or basic is to convert the pOH value to pH and then use the pH scale of 0–6 being acidic, 7 being neutral, and 8–14 being basic.

$$[OH^-] = 8.5 \times 10^{-12}$$
$$-\log[OH^-] = pOH$$
$$-\log[8.5 \times 10^{-12}] = pOH$$
$$pOH = 11.07$$
$$pH = 14 - 4.49$$
$$pH = 2.93$$

Since the pH value is 2.93 that makes the solution acidic.

(7) $[H^+] = 5.0 \times 10^{-5}$. To convert from pH to $[H^+]$ concentration you use the formula $10^{-pH} = [H^+]$ $10^{-pH} = [H^+]$. In this case you simply plug in your pH value into the formula and solve. Make sure you remember to add the negative in front of the pH value. That is a common mistake.

$$10^{-pH} = [H^+]$$
$$10^{-4.3} = 5.0 \times 10^{-5}$$
$$[H^+] = 5.0 \times 10^{-5}$$

(8) $[OH^-] = 1.6 \times 10^{-3}$. This question asks you to calculate the OH^- concentration of a solution but it only gives you the pH of solution. If you take the antilog of the pH value you get the $[H^+]$ value of that solution but not the pH. You can then use the K_w equation to solve for $[OH^-]$. There is a simpler way to solve this, though, and that is the way that will be shown below. As a reminder, it is usually a good idea to always try to use the simplest solution possible because it will lead to less error. In this case, calculate the pOH of the solution from your pH and then take the antilog of the pOH to determine the $[OH^-]$ concentration.

$$pH + pOH = 14$$
$$11.2 + pOH = 14$$
$$pOH = 2.8$$
$$10^{-pOH} = [OH^-]$$
$$10^{-2.8} = 1.6 \times 10^{-3}$$
$$[OH^-] = 1.6 \times 10^{-3}$$

(9) **pH = 0.081.** You first need to calculate the molarity of the solution by dividing the given moles of 2.5 by 3 L of water.

$$\frac{mol}{L} = Molarity$$

$$\frac{2.5 \text{ mol HCl}}{3.0 \text{ L H}_2\text{O}} = 0.83 \text{ M HCl}$$

Once you have this you can use the molarity value of HCl as your [H+] value because HCl is a strong acid and completely dissociates in water. You then take the −log of the [H+] and calculate the pH.

$$-\log[H^+] = pH$$
$$-\log[0.83] = pH$$
$$pH = 0.081$$

(10) You first need to calculate the molarity of the solution by dividing the given moles of 2.5 by 3 L of water.

$$\frac{mol}{L} = Molarity$$

$$\frac{1.8 \text{ mol NaOH}}{4.0 \text{ L H}_2\text{O}} = 0.45 \text{ M NaOH}$$

The molarity value you calculate is equal to the hydroxide ion concentration [OH−] because NaOH is a strong base and dissociates completely in solution. Once you have this value, you will calculate the pOH and then subtract that value from 14 to determine the pH of the solution.

$$[OH^-] = 0.45 \text{ M}$$
$$-\log[OH^-] = pOH$$
$$-\log[0.45] = pOH$$
$$pOH = 0.35$$
$$pH = 14 - 0.35$$
$$pH = 13.65$$

(11) **pH = 0.62.** You first need to convert the mass value of hydrogen bromide to moles using the molar mass of hydrogen bromide (HBr) which is 80.91 g/mol. Once you have this mole value you can calculate the molarity of the solution. That molarity will be equal to the hydrogen ion concentration [H+] as HBr is a strong acid and completely dissociates in solution. Once you have that [H+] you can calculate pH.

$$\left(59.0 \text{ g HBr}\right)\left(\frac{1 \text{ mol HBr}}{80.91 \text{ g HBr}}\right) = 0.73 \text{ mol HBr}$$

$$\frac{mol}{L} = Molarity$$

$$\frac{0.73 \text{ mol HBr}}{3.0 \text{ L H}_2\text{O}} = 0.24 \text{ M HBr}$$

$$[H^+] = 0.24 \text{ M}$$
$$-\log[H^+] = pH$$
$$-\log[0.24] = pH$$
$$pH = 0.62$$

(12) **pH = 5.5.** Hydrochloric acid is a strong acid so it entirely dissociates in solution. That means the concentration given for HCl is equal to the [H⁺] concertation of the solution, 0.0000030 M or 3.0×10^{-6}. Take the –log to determine the pH.

$$-\log\left[3.0 \times 10^{-6}\right] = \text{pH}$$
$$\text{pH} = 5.5$$

(13) A strong acid is an acid that completely dissociates when dissolved into a solution. A weak acid does not completely dissociate in solution. Instead the amount of dissociation varies depending on the relative strength of the acid, as shown by the weak acids K_a value.

(14) **pH = 4.9.** Begin by considering the K_a equation:

$$K_a = \frac{\left[CN^-\right]\left[H^+\right]}{\left[HCN\right]}$$

Because $\left[CN^-\right] = \left[H^+\right]$, you can rewrite this as

$$K_a = \frac{\left[H^+\right]^2}{\left[HCN\right]}$$

Solving this equation for [H⁺] yields $\left[H^+\right] = \sqrt{(K_a)(HCN)}$. Plugging the known values of K_a and [HCN] into this equation yields

$$\left[H^+\right] = \sqrt{\left(6.2 \times 10^{-10}\right)\left(\left[0.30\right]\right)} = 1.36 \times 10^{-5}$$

Use this to solve for the pH:

$$\text{pH} = -\log\left[1.36 \times 10^{-5}\right] = 4.9$$

(15)
$$K_w = 1 \times 10^{-14}$$
$$K_w = [OH^-][H^+]$$
$$[H^+] = \frac{K_w}{[OH^-]}$$
$$[H^+] = \frac{\left(1 \times 10^{-14}\right)}{\left(4.3 \times 10^{-6}\right)}$$
$$[H^+] = 2.36 \times 10^{-9}$$

Chapter **21**

Achieving Neutrality with Titrations and Buffers

I n the real world of chemistry, acids and bases often meet in solution, and when they do, they're drawn to one another. These unions of acid and base are called *neutralization reactions* because the lower pH of the acid and the higher pH of the base essentially cancel one another out, resulting in a neutral solution. In fact, when a hydroxide-containing base reacts with an acid (containing H^+), the products are simply an innocuous salt and water.

Although strong acids and bases have their uses, the prolonged presence of a strong acid or base in an environment not equipped to handle it can be very damaging. In the laboratory, for example, you need to handle strong acids and bases carefully, deliberately performing neutralization reactions when appropriate. A lazy chemistry student who attempts to dump a concentrated acid or base down the laboratory sink will get a browbeating from her hawk-eyed teacher. Dumping stuff like this can damage the pipes and is generally unsafe. A responsible chemist neutralizes acidic laboratory waste with a base such as baking soda and neutralizes basic waste with an acid. Doing so makes a solution perfectly safe to dump down the drain and often results in the creation of a satisfyingly sizzly solution while the reaction is occurring.

In this chapter, we explore three topics surrounding acid–base chemistry:

>> **Titrations and indicators:** *Titrations* are the process by which chemists add acids to bases (or vice versa) a little bit at a time, gradually using up acid and base equivalents as the two neutralize each other. Indicators are chemicals that allow you to quickly and easily identify whether you're dealing with an acidic or basic solution. They can also do a fairly good job of telling you the pH of the solution you're working with. When combined with a titration, some indicators can help you determine the exact concentration of hydronium and hydroxide ions in a solution.

>> **Buffer solutions:** Buffer solutions are mixtures that contain both acid and base forms of the same compounds and serve to maintain the pH of the solution even when extra acid or base is added.

>> **Solubility product:** Because salts are produced when acids react with bases, we discuss the solubility product, K_{sp}, a number that tells you how soluble a salt is in solution.

TIP

At heart, neutralization reactions in which the base contains a hydroxide ion are simple double-replacement reactions of the form $HA + BOH \rightarrow BA + H_2O$ (in other words, an acid reacts with a base to form a salt and water). You're asked to write a number of such reactions in this chapter, so be sure to review double replacement reactions and balancing equations in Chapter 13 before you delve into the new and exciting world of neutralization.

Using Indicators and Titration to Figure Out Molarity

Indicators are substances (organic dyes) that change color in the presence of an acid or base. You may be familiar with the hydrangea, an acid–base indicator plant. If it's grown in acidic soil, it turns pink; if it's grown in alkaline soil, it turns blue. Another common substance that acts as a good acid–base indicator is red cabbage. You can chop up some red cabbage, boil it (most students really *love* this part), and then use the leftover liquid to test substances. When mixed with an acid, the liquid turns pink; when mixed with a base, it turns green. In fact, if you take some of this liquid, make it slightly basic, and then exhale your breath into it through a straw, the solution eventually turns pink, indicating that the solution has turned slightly acidic. The carbon dioxide in your breath reacts with the water, forming carbonic acid:

$$CO_2(g) + H_2O(l) \rightleftharpoons H_2CO_3(aq)$$

Carbonated beverages are slightly acidic due to this reaction. Carbon dioxide is injected into the liquid to give it fizz. A little of this carbon dioxide reacts with the water to form carbonic acid. This reaction also explains why rainwater is slightly acidic. It absorbs carbon dioxide from the atmosphere as it falls to earth.

In chemistry, indicators are used to indicate the presence of an acid or a base. Chemists have many indicators that change at slightly different pH levels. The two most commonly used indicators, which we discuss in the following sections, are litmus paper and phenolphthalein.

Taking a quick dip with litmus paper

Litmus is a substance that is extracted from a type of lichen and absorbed into porous paper. (In case you're curious, *lichen* is an organism that's made up of an alga and a fungus that live intimately together and mutually benefit from the relationship. Sounds kind of sordid to us.)

There are three types of litmus:

» Red litmus is used to test for bases.

» Blue litmus is used to test for acids.

» Neutral litmus can be used to test for both.

If a solution is acidic, both blue and neutral litmus turn red. If a solution is basic, both red and neutral litmus turn blue. Litmus paper is a good, quick test for acids and bases. And you don't have to put up with the smell of boiling cabbage.

Titrating with phenolphthalein

Phenolphthalein (pronounced fe-nul-*tha*-leen) is another commonly used indicator. Until a few years ago, phenolphthalein was used as the active ingredient in a popular laxative. In fact, you can extract the phenolphthalein from the laxative by soaking it in either rubbing alcohol or gin (being careful not to drink it). You then use this solution as an indicator for all the acid-base related issues you encounter on a daily basis.

Phenolphthalein is clear and colorless in an acid solution and pink in a basic solution. It's commonly used in a procedure called a *titration*, in which the concentration of an acid or base is determined by its reaction with a base or acid of known concentration. Titration is very common in many areas of chemistry, and it's quite handy.

Imagine you're a newly hired laboratory assistant who's been asked to alphabetize the chemicals on the shelves of a chemistry laboratory during a lull in experimenting. As you reach for the bottle of sulfuric acid, your first-day jitters get the better of you, and you knock over the bottle. Some careless chemist failed to screw the cap on tightly! You quickly neutralize the acid with a splash of baking soda and wipe up the now nicely neutral solution. As you pick up the bottle, however, you notice that the spilled acid burned away most of the label. You know it's sulfuric acid, but there are several different concentrations of sulfuric acid on the shelves, and you don't know the molarity of the solution in this bottle. Knowing that your boss will surely blame you if she sees the damaged bottle and not wanting to get sacked on your very first day, you quickly come up with a way to determine the molarity of the solution and save your job.

You know that the bottle contains sulfuric acid of a mystery concentration, and you notice bottles of 1 M sodium hydroxide, a strong base, and phenolphthalein, a pH indicator, among the chemicals on the shelves. You measure a small amount of the mystery acid into a beaker and add a little phenolphthalein. You reason that if you drop small amounts of sodium hydroxide into the solution until the phenolphthalein indicates that the solution is neutral by turning the appropriate color, you'll be able to figure out the acid's concentration.

You can find the concentration by doing a simple calculation of the number of moles of sodium hydroxide you've added and then reasoning that the mystery acid must have an equal number of moles to have been neutralized. This then leads to the number of moles of acid, and that, in turn, can be divided by the volume of acid you added to the beaker to get the molarity. Whew! You've just performed a titration.

In a titration calculation, you generally know the identity of an acid or base of unknown concentration, and you know the identity and molarity of the acid or base that you're going to use to neutralize it. Given this information, you then follow six simple steps:

1. **Place a known volume (say, 25.00 mL measured accurately with a pipette) in an Erlenmeyer flask.**

   An *Erlenmeyer flask* is a flat-bottomed, conical-shaped container.

2. **Add a pH indicator such as phenolphthalein to the mystery acid or base.**

   Because you're adding the indicator to an acidic solution, the solution in the flask remains clear and colorless.

3. **Add small amounts of a standardized base solution of known molarity with a buret (in most chemistry labs, you'll use sodium hydroxide as the base). Keep adding the base until the solution turns the faintest shade of pink.**

   A *buret* is a graduated glass tube with a small opening and a stopcock, which helps you measure precise volumes of a solution.

   When the titration turns a faint shade of pink, you've reached the *endpoint* of the titration. This is the point at which the indicator shows that the acid has been exactly neutralized by the base. Figure 21-1 shows the titration setup.

4. **Calculate the number of moles added.**

   Multiply the number of liters of acid or base added by the molarity of that acid or base to get the number of moles you added.

5. **Calculate the unknown moles.**

   Use the balanced equation and basic stoichiometry to determine how many moles of the mystery substance being neutralized are present (see Chapter 14 for details on stoichiometry).

6. **Solve for molarity.**

   Divide the number of moles of the mystery acid or base by the number of liters measured out in Step 1, giving you the molarity.

People often visualize the titration process using a graph that shows the concentration of base on one axis and the pH on the other, as in Figure 21-2. The interaction of the two concentrations traces out a *titration curve*, which has a characteristic *s* shape. At the equivalence point in Figure 21-2, the amount of base present is equal to the amount of acid present in the solution. If you're using an indicator such as phenolphthalein, the equivalence point marks when the first permanent color change takes place.

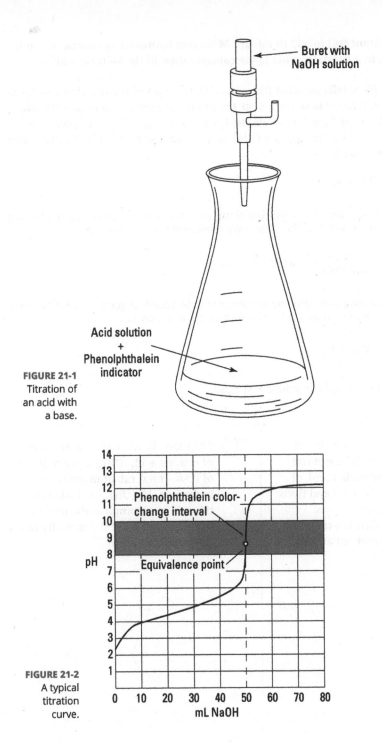

Buret with
NaOH solution
Acid solution
+
Phenolphthalein
indicator

FIGURE 21-1 Titration of an acid with a base.

14
13
12
11
10
9
8
7
6
5
4
3
2
1
pH
Phenolphthalein color-
change interval
Equivalence point
0 10 20 30 40 50 60 70 80
mL NaOH

FIGURE 21-2 A typical titration curve.

Q. If a laboratory assistant had to add 10 mL of 1 M sodium hydroxide to neutralize 5 mL of the sulfuric acid in a titration, what is the concentration of the sulfuric acid?

EXAMPLE **A.** 1 M H_2SO_4. The problem tells you that the volume from Step 1 of the titration process is 5 mL and that the volume of base from Step 3 is 10 mL. In Step 4, you must calculate the number of moles of sodium hydroxide the lab assistant added by multiplying the volume in liters (0.01 L) by the molarity (1 M) to give you 0.01 mol NaOH. The balanced equation for this reaction is

$$H_2SO_4 + 2NaOH \rightarrow 2H_2O + Na_2SO_4$$

You use the coefficients from this equation along with the mole value calculated in Step 4 to determine the mole value of the unknown substance you're titrating:

$$\left(\frac{0.01 \text{ mol NaOH}}{1} \right)\left(\frac{1 \text{ mol } H_2SO_4}{2 \text{ mol NaOH}} \right) = 0.005 \text{ mol } H_2SO_4$$

The final step is to divide this value by the liters of acid added to get the molarity. Your initial acid volume was given as 5 mL, which converts to 0.005 L.

$$\frac{0.005 \text{ mol } H_2SO_4}{0.005 \text{ L}} = 1 \text{ M } H_2SO_4$$

The concentration of the sulfuric acid is 1 M H_2SO_4.

YOUR TURN

 1 In doing a titration of a solution of calcium hydroxide of unknown concentration, a student adds 12 mL of 2 M HCl (hydrochloric acid) and finds that the molarity of the base is 1.25 M. How much $Ca(OH)_2$ must the student have measured out at the start of the titration?

2 Titration shows that a 5.0 mL sample of nitrous acid, HNO_2, has a molarity of 0.50. If 8.0 mL of magnesium hydroxide, $Mg(OH)_2$, was added to the acid to accomplish the neutralization, what must the molarity of the base have been?

3 How much could a chemist find out about a mystery acid or base through titration if neither its identity nor its concentration is known?

Maintaining Your pH with Buffers

You may have noticed that the titration curve shown in Figure 21-2 has a flattened area in the middle where pH doesn't change significantly as base is added. This region is called a *buffer region.*

Certain solutions, called *buffered solutions,* resist changes in pH like a stubborn child resists eating her Brussels sprouts: steadfastly at first but choking them down reluctantly if enough pressure is applied (such as the threat of no dessert). Although buffered solutions maintain their pH very well when relatively small amounts of acid or base are added to them or the solution is diluted, they can withstand the addition of only a certain amount of acid or base before becoming overwhelmed.

Buffers are most often made up of a weak acid and its conjugate base, though they can also be made of a weak base and its conjugate acid. (Conjugate bases and acids are the products in acid-base reactions; see Chapter 20 for details.) A weak acid in aqueous solution will be partially dissociated, and the amount of dissociation depends on its pK_a value (the negative logarithm of its acid dissociation constant). The dissociation will be of the form $HA + H_2O \rightarrow H_3O^+ + A^-$, where A^- is the conjugate base of the acid HA. The acidic proton is taken up by a water molecule, forming hydronium. If HA were a strong acid, approximately 100 percent of the acid would become H_3O^+ and A^-, but because it's a weak acid, only a fraction of the HA dissociates and the rest remains HA.

The K_a of this reaction is defined by

$$K_a = \frac{[H_3O^+][A^-]}{[HA]}$$

Solving this equation for the $[H_3O^+]$ concentration allows you to devise a relationship between the $[H_3O^+]$ and the K_a of a buffer:

$$\left[H_3O^+ \right] = \left(K_a \right) \left(\frac{[HA]}{[A^-]} \right)$$

REMEMBER

Taking the negative logarithm of both sides of the equation and manipulating logarithm rules yields an equation called the *Henderson–Hasselbalch equation*, which relates the pH and the pK_a:

$$pH = pK_a + \log \left(\frac{[A^-]}{[HA]} \right)$$

With logarithm rules, you can manipulate this equation to get $\frac{[A^-]}{[HA]} = 10^{(pH - pK_a)}$, which may be more useful in certain situations.

TIP

The very best buffers and those best able to withstand the addition of both acid and base are those for which [HA] and [A$^-$] are approximately equal. When this occurs, the logarithmic term in the Henderson–Hasselbalch equation disappears, and the equation becomes $pH = pK_a$. When creating a buffered solution, chemists therefore choose an acid that has a pK_a close to the desired pH.

If you add a strong base such as sodium hydroxide (NaOH) to this mixture of dissociated base (A$^-$) and undissociated acid (HA), the base's hydroxide is absorbed by the acidic proton, replacing the exceptionally strong base OH$^-$ with a relatively weak base A$^-$ and minimizing the change in pH:

$$HA + OH^- \rightarrow H_2O + A^-$$

This causes a slight excess of base in the reaction, but it doesn't affect pH significantly. You can think of the undissociated acid as a reservoir of protons that are available to neutralize any strong base that may be introduced to the solution. When a product is added to a reaction, the equilibrium in the reaction changes to favor the reactants or to "undo" the change in conditions. Because this reaction generates A$^-$, the acid dissociation reaction happens less frequently as a result, further stabilizing the pH.

When a strong acid, such as hydrochloric acid (HCl), is added to the mixture, its acidic proton is taken up by the base A$^-$, forming HA:

$$H^+ + A^- \rightarrow HA$$

This causes a slight excess of acid in the reaction but doesn't affect pH significantly. It also shifts the balance in the acid dissociation reaction in favor of the products, causing it to happen more frequently and recreating the base A$^-$.

Figure 21-3 summarizes the addition of acid and base and their effect on the ratio of products and reactants.

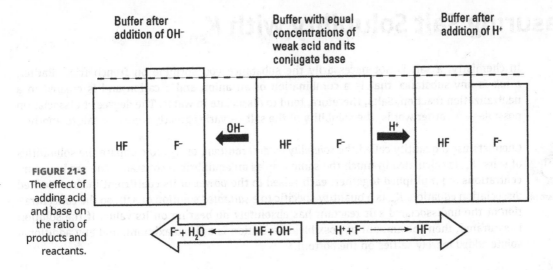

FIGURE 21-3
The effect of adding acid and base on the ratio of products and reactants.

Buffers have their limits, however. The acid's proton reservoir, for example, can compensate for the addition of only a certain amount of base before it runs out of protons that can neutralize free hydroxide. At this point, a buffer has done all it can do, and the titration curve resumes its steep upward slope.

Q. Consider a buffered solution that contains the weak acid ethanoic acid $\left(K_a = 1.8 \times 10^{-5} \right)$ and its conjugate base, ethanoate. If the concentration of the solution is 0.5 M with respect to ethanoic acid and 0.3 M with respect to ethanoate, what is the pH of the solution?

EXAMPLE

A. pH = 4.5. This problem is a simple application of the Henderson–Hasselbalch equation. Remember to take the negative logarithm of K_a to get pK_a. Plugging in known values yields

$$pH = pK_a + \log\left(\frac{[A^-]}{[HA]} \right) = -\log\left(1.8 \times 10^{-5} \right) + \log\left(\frac{[0.3]}{[0.5]} \right) = 4.5$$

YOUR TURN

4 Ethanoic acid would make an ideal buffer to maintain what approximate pH?

 5 Describe the preparation of 1,000 mL of a 0.2 M carbonic acid buffer of pH = 7.0. Assume the pK_a of carbonic acid is 6.8.

Measuring Salt Solubility with K_{sp}

In chemistry, a salt is not necessarily the substance you sprinkle on french fries. Rather, a *salt* is any substance that is a combination of an anion and a cation and is created in a neutralization reaction. Salts, therefore, tend to dissociate in water. The degree of dissociation possible — in other words, the solubility of the salt — varies greatly from one salt to another.

REMEMBER Chemists use a quantity called the *solubility product constant*, or K_{sp}, to compare the solubilities of salts. K_{sp} is calculated in much the same way as an equilibrium constant. The product concentrations are multiplied together, each raised to the power of its coefficient in the balanced dissociation equation. K_{sp} is a quantity specific to a *saturated* solution of salt, so the concentration of the undissociated salt reactant has absolutely no bearing on its value. If the solution is saturated, then the amount of possible dissociation is at its maximum, and any additional solute added merely settles on the bottom.

Q. Write a formula for the solubility product constant of the reaction $CaF_2 \rightarrow Ca^{2+} + 2F^-$.

EXAMPLE

A. $K_{sp} = \left[Ca^{2+}\right]\left[F^-\right]^2$. You construct the solubility product constant by raising the concentrations of the two products to the power of their coefficients, so $K_{sp} = \left[Ca^{2+}\right]\left[F^-\right]^2$.

YOUR TURN

6 Write the solubility product dissociation constants for silver(I) chromate (Ag_2CrO_4) and strontium sulfate ($SrSO_4$).

7 If the K_{sp} of silver(I) chromate is 1.1×10^{-12} and the silver ion concentration in the solution is 0.0005 M, what is the chromate concentration?

Practice Questions Answers and Explanations

(1) **9.6 mL Ca(OH)₂.** Begin by finding the number of moles of HCl by multiplying the molarity (2 M) by the volume (0.012 L) to get 0.024 mol HCl. Next, calculate the number of moles of Ca(OH)₂ needed to neutralize this amount using the balanced chemical equation $Ca(OH)_2 + 2HCl \rightarrow CaCl_2 + 2H_2O$:

$$\left(\frac{0.024 \text{ mol HCl}}{1} \right) \left(\frac{1 \text{ mol Ca(OH)}_2}{2 \text{ mol HCl}} \right) = 0.012 \text{ mol Ca(OH)}_2$$

Divide this number by the molarity of the base to get the volume of base added:

$$\left(\frac{0.01 \text{ mol Ca(OH)}_2}{1.25 \text{ M Ca(OH)}_2} \right) = 0.0096 \text{ L Ca(OH)}_2$$

(2) **0.16 M Mg(OH)₂.** This problem gives you Step 6 in the titration procedure and asks you to back-solve for the molarity of the base. Start by finding the number of moles of acid present in the solution by multiplying the molarity (0.50 M) by the volume (0.005 L), giving you 0.0025 mol. Next, examine the balanced neutralization reaction to determine the number of moles of magnesium hydroxide needed to neutralize 0.0025 mol HNO_2:

$$2HNO_2 + Mg(OH)_2 \rightarrow Mg(NO_2)_2 + 2H_2O$$

From the balanced equation, you can see that for every 2 mol of HNO_2, you need 1 mol of magnesium hydroxide to neutralize it:

$$\left(\frac{0.0025 \text{ mol HNO}_2}{1} \right) \left(\frac{1 \text{ mol Mg(OH)}_2}{2 \text{ mol HNO}_2} \right) = 1.25 \times 10^{-3} \text{ mol Mg(OH)}_2$$

Divide this value by the volume of base added (8.0 mL, or 0.008 L) to get the molarity:

$$\left(\frac{1.25 \times 10^{-3} \text{ mol Mg(OH)}_2}{0.008 \text{ L}} \right) = 0.16 \text{ M Mg(OH)}_2$$

(3) Without an identity or a concentration, a chemist could still determine the number of moles per liter of the mystery acid or base. If he titrates an extra-mysterious acid with a base of known concentration until he achieves neutrality, then he knows the number of moles of acid in the solution (equal to the number of moles of base added). This information may even allow him to guess at its identity.

(4) **pH of 4.7.** Ethanoic acid, also known as *acetic acid* or *vinegar*, would be an ideal buffer for a pH close to its pK_a value. You're given the K_a of ethanoic acid in the example problem (it's $K_a = 1.8 \times 10^{-5}$), so simply take the negative logarithm of that value to get your pK_a value and therefore your pH:

$$-\log(1.8 \times 10^{-5}) = pK_a = 4.7$$

(5) You're given a pH and a pK_a, which suggests that you need to use the Henderson–Hasselbalch equation. You have the total concentration of acid and conjugate base, but you don't know either of the concentrations in the equation individually, so begin by solving for their ratio:

$$\frac{[A^-]}{[HA]} = 10^{(pH-pK_a)} = 10^{(7-6.8)} = 1.6$$

So you have 1.6 mol of base for every 1 mol of acid. Expressing the amount of base as a fraction of the total amount of acid and base therefore gives you

$$\frac{[A^-]}{[HA]+[A^-]} = \frac{1.6}{1+1.6} = 0.62$$

Multiply this number by the molarity of the solution (0.2 M) to give you the molarity of the basic solution, or 0.12 M. The molarity of the acid is therefore 0.2 M − 0.12 M = 0.08 M.

This means that to prepare the buffer, you must take 0.2 mol of carbonic acid and add it to somewhat less than 1,000 mL of water (say 800 mL). To this you must add enough of a strong base such as sodium hydroxide (NaOH) to force the proper proportion of the carbonic acid solution to dissociate into its conjugate base. Because you want to achieve a base concentration of 0.12 M, you should add 120 mL of NaOH. Finally, you should add enough water to achieve your final volume of 1,000 mL.

(6) **For silver(I) chromate:** $K_{sp} = [Ag^+]^2[CrO_4^{2-}]$; **for strontium sulfate:** $K_{sp} = [Sr^{2+}][SO_4^{2-}]$. To determine the solubility product constants of these solutions, you first need to write an equation for their dissociation in water (see Chapter 8 for details):

$$Ag_2CrO_4 \rightarrow 2Ag^+(aq) + CrO_4^{2-}(aq)$$
$$SrSO_4 \rightarrow Sr^{2+}(aq) + SO_4^{2-}(aq)$$

Raising the concentration of the products to the power of their coefficients in these balanced reactions yields the K_{sp} for each:

$$K_{sp} = [Ag^+]^2[CrO_4^{2-}]$$
$$K_{sp} = [Sr^{2+}][SO_4^{2-}]$$

(7) 4×10^{-6}. You write an expression for the K_{sp} of silver(I) chromate in the preceding question. Solve the equation for the chromate concentration by dividing K_{sp} by the silver ion concentration:

$$[CrO_4^{2-}] = \frac{K_{sp}}{[Ag^+]^2} = \frac{(1.1 \times 10^{-12})}{(5 \times 10^{-4})^2} = 4 \times 10^{-6}$$

If you're ready to test your skills a bit more, take the following chapter quiz that incorporates all the chapter topics.

Whaddya Know? Chapter 21 Quiz

Ready for a quiz? The 10 questions in this section will test the skills you learned in this chapter. When you're done, check out the section that follows for answers and explanations.

1 What do buffered solutions resist?

2 Would 3 L of 1 M of NaOH completely neutralize a 1 M HCl solution that has a volume of 3 L?

3 During a titration experiment you determine that 152 mL of 0.50 M HCl is required to neutralize a 340 mL solution of NaOH. What is the concentration of the NaOH?

4 You add 400 ml of 2 M HCl to a 1 L solution of 0.25 M NaOH. You mix the solution and allow it to sit for 5 minutes. You then use litmus paper to perform a pH test. The paper turns red. Is the solution currently acidic or basic? You then add 4 L of 1 M NaOH and test with litmus paper again and this time the paper turns blue. Is the solution now acidic or basic?

5 What is the pH of a buffer solution that is 0.88 M hydrocyanic acid (HCN) and 0.23 M sodium cyanide. The pK_a of HF is 3.17.

6 What is the pH of a buffer solution that is 0.29 M HNO_2 and 0.14 M $NaNO_2$. The pK_a of HNO_2 is 3.39.

7 What is the pH at the equivalence point for the titration of 50.00 mL of a 0.35-M potassium hydroxide (KOH) solution with 0.7500-M nitric acid (HNO_3)?

8 Consider a buffered solution that contains the weak acid, carbonic acid ($K_a = 4.3 \times 10^{-7}$) and its conjugate base, HCO_3^-. If the concentration of the solution is 0.6 M with respect to carbonic acid and 0.4 M with respect to the bicarbonate ion, what is the pH of the solution?

9 In doing a titration of a solution of potassium hydroxide of unknown concentration, a student adds 24 mL of 1.5 M HBr (hydrobromic acid) and finds that the molarity of the base is 1.75 M. How much KOH must the student have measured out at the start of the titration?

10 What is the equivalence point in a titration?

Answers to Chapter 21 Quiz

(1) Buffered solutions resist change in pH to a solution. That means if an acid that contributes H^+ is added to the buffered solution the buffer will work to neutralize that acid as it is added. The same can be said if a base is added that contributes OH^-, the buffer will work to neutralize the addition. There can come a point, however, though when the entire conjugate pair of the buffered solution has reacted and there is nothing left to prevent a pH change. At that point any addition of acid or base to the buffered solution can result in a change in pH.

(2) **Yes.** NaOH is a strong base and HCl is a strong acid. This means they both completely dissolve when put into water. They also react at a 1 to 1 ratio, implying that for every mol of HCl you would need one mole of NaOH to neutralize it. In this case since the solutions are of the same volume and the same molarity, they would completely neutralize one another.

(3) **0.22 M NaOH.** The problem tells you that the volume from Step 1 of the titration process is 340 mL and that the volume of acid from Step 3 is 152 mL. In Step 4, you must calculate the number of moles of hydrochloric acid the lab assistant added by multiplying the volume in liters (0.152 L) by the molarity (0.5 M) to give you 0.076 mol HCl. The balanced equation for this reaction is

$$HCl + NaOH \rightarrow H_2O + NaCl$$

You use the coefficients from this equation along with the mole value calculated in Step 4 to determine the mole value of the unknown substance you're titrating:

$$\left(0.076 \text{ mol HCl}\right)\left(\frac{1 \text{ mol NaOH}}{1 \text{ mol HCl}}\right) = 0.076 \text{ mol NaOH}$$

The final step is to divide this value by the liters of NaOH added to get the molarity. Your initial NaOH volume was given as 340 mL, which converts to 0.340 L.

$$\frac{0.076 \text{ mol NaOH}}{0.340 \text{ L}} = 0.22 \text{ M NaOH}$$

The concentration of the sodium hydroxide is 0.22 M NaOH.

(4) The solution is acidic after the first litmus test. After the second litmus paper test the solution is basic. All of the numbers given here are actually inconsequential to answering the question. All it is looking for is for you to make a determination as to whether a litmus paper reading is acidic or basic. Remember that if litmus paper turns red it indicates an acidic solution and when it turns blue it indicates a basic solution. So the first reading showing red indicates acid and after the second solution was added it indicates that it is basic due to turning blue.

(5) **pH = 2.59.** This is a buffer solution, so you should solve it using the Henderson–Hasselbalch equation.

$$pH = pK_a + \log\left(\frac{\left[A^-\right]}{\left[HA\right]}\right)$$

The acid in this buffer solution is the hydrocyanic acid [0.88 M] and the corresponding conjugate base is sodium cyanide [0.23 M]. Simply plug the concentration for each of these into the equation above and apply the pK_a given in the problem and you will be able to solve for the pH:

$$pH = 3.17 + \log\left(\frac{\left[0.23\right]}{\left[0.88\right]}\right)$$

$$pH = 2.59$$

(6) **pH = 3.07.** This is a buffer solution, so you should solve it using the Henderson–Hasselbalch equation as shown above.

The acid in this buffer solution is the HNO_2 [0.29 M] and corresponding conjugate base is $NaNO_2$ [0.14 M]. Simply plug the concentration for each of these into the equation above and apply the pK_a given in the problem and you will be able to solve for the pH:

$$pH = 3.39 + \log\left(\frac{[0.14]}{[0.29]}\right)$$

$$pH = 3.07$$

(7) The pH at the equivalence point of any strong acid base titration is 7. In this case potassium hydroxide is a strong base and nitric acid is a strong acid. When these are titrated together, regardless of the initial concentration the equivalence point of the titration will be at a pH of 7.

(8) **pH = 6.19.** This problem involves using a buffer solution, so you'll be using the Henderson–Hasselbalch equation as shown above. Plug your values for the acid and its conjugate base into the equation as shown below. Be sure to convert your K_a to pK_a by first taking the negative log (–log) of your K_a value before plugging it into the equation.

$$pK_a = -\log\left(K_a\right)$$

$$pK_a = -\log\left(4.3 \times 10^{-7}\right)$$

$$pK_a = 6.37$$

$$pH = pK_a + \log\left(\frac{[A^-]}{[HA]}\right)$$

$$pH = 6.37 + \log\left(\frac{[0.4]}{[0.6]}\right)$$

$$pH = 6.19$$

(9) **29 ml KOH.** Begin by finding the number of moles of HBr by multiplying the molarity (1.5 M) by the volume (0.024 L) to get 0.036 mol HBr. Next, calculate the number of moles of KOH needed to neutralize this amount using the balanced chemical equation as shown below. In this case all of the coefficients are 1, which makes the conversion very easy!

$$KOH + HBr \rightarrow KBr + H_2O:$$

$$\left(\frac{0.036 \text{ mol HBr}}{1}\right)\left(\frac{1 \text{ mol KOH}}{1 \text{ mol HBr}}\right) = 0.036 \text{ mol KOH}$$

Divide this number by the molarity of the base to get the volume of base added:

$$\left(\frac{0.036 \text{ mol KOH}}{1.25 \text{ M KOH}}\right) = 0.029 \text{ L KOH or 29 ml KOH}$$

(10) The equivalence point of a titration occurs when the amount of titrant added to the solution is enough to completely neutralize the solution. In simpler terms the equivalence point occurs when the number of moles of acid/base added to the solution is equal to the number of moles that were initially present in the solution. When doing a titration the equivalence point occurs when you first see a permanent color change take place in the solution; generally this is a very light pink color if your indicator being used is phenolphthalein.

Glossary

absolute zero The temperature at which all molecular motion stops, 0 kelvin.

acid A compound that is a proton (H^+) donor.

acidic A solution whose pH is less than 7.

acid rain Rain that has a pH in the acid range due to pollutants.

activation energy The minimum amount of energy that must be supplied in order to start a chemical reaction.

activity series A list of the metals in order of decreasing ease of oxidation.

actual yield The amount of product actually formed in a chemical reaction.

alkali metals The elements in the IA (1) family on the periodic table.

alkaline earth metals The elements in the IIA (2) family on the periodic table.

alpha particle Essentially a helium nucleus (two protons and two neutrons).

amorphous solid A solid that lacks extensive ordering of the particles.

amphoteric A substance that acts either as an acid or a base depending on what it's combined with.

amplitude The height of a wave.

amu An atomic mass unit, $\frac{1}{12}$ of the mass of a C-12 nucleus.

angular momentum quantum number (*l*) This number describes the shape of an orbital.

anions Ions that have a negative charge.

aqueous solution A mixture in which the solvent is water.

atom The smallest particle of matter that represents a particular element.

atomic number (*Z*) The number of protons in the nucleus.

atomic orbital The volume of space in which you're most likely to find a specific electron in an atom.

Aufbau principle States that the electrons in an atom fill the lowest energy levels first.

Avogadro's law States that the volume of a gas and the number of moles of gas are directly related if the temperature and pressure are held constant.

Avogadro's number The number of particles (atoms, ion, molecules) in a mole; it is numerically equal to 6.022×10^{23}.

barometer An instrument that measures atmospheric pressure.

base A compound that is a proton (H⁺) acceptor.

basic A solution with a pH greater than 7.

beta particle Essentially an electron.

binary compound A compound composed of only two elements.

biological oxygen demand (BOD) The amount of dissolved oxygen needed to oxidize the biological material in water.

boiling The process of going from a liquid state to a gaseous state.

boiling point (bp) The temperature at which a liquid boils; also the temperature at which the vapor pressure of the liquid equals atmospheric pressure.

bond order Relates the bonding and antibonding electrons in a molecular orbital. It's equal to

$$\frac{\left(\begin{array}{c}\text{\# electrons in} \\ \text{bonding molecular orbitals}\end{array}\right) - \left(\begin{array}{c}\text{\# electrons in} \\ \text{antibonding molecular orbitals}\end{array}\right)}{2}$$

Boyle's law States that an inverse relationship exists between the volume and pressure of a gas if the temperature and amount are held constant.

buffers Solutions that resist a change in pH when either an acid or a base is added to them.

calorie The amount of energy needed to raise the temperature of 1 gram of water 1 degree Celsius.

calorimetry A laboratory technique used to measure the amount of heat released or absorbed during a chemical or physical process.

capillary action The spontaneous rising of a liquid in a narrow tube against the force of gravity.

catalyst A substance that speeds up a reaction and is (at least theoretically) recoverable at the end of the reaction in an unchanged form.

cations Ions that have a positive charge.

Charles's law States that a direct relationship exists between the volume and Kelvin temperature if the pressure and amount of the gas are held constant.

chemical equilibrium Established when two exactly opposite reactions are taking place at the same time in the same place and with equal rates of reaction.

colligative properties Properties of solutions that depend only on the number of solute particles present and not on the type of solute.

colloids Homogeneous mixtures in which the solute diameters are between those of solutions and suspensions.

combined gas equation Relates the temperature, pressure, and volume of a gas, assuming the amount is held constant.

combustion reaction A reaction in which a chemical species combines rapidly with oxygen and usually emits heat and/or light.

compounds Pure substances composed of two or more different elements.

concentrated A qualitative term that describes a solution with a relatively large amount of solute in comparison to the amount of solvent.

concentration The measure of the amount of solute dissolved in a solution.

conjugate acid-base pair A pair of compounds (one an acid and one a base) that differ by only a single H^+.

continuous spectrum A spectrum of light in which all wavelengths of light are present.

coordinate covalent bond A covalent bond between two atoms in which one atom has furnished both electrons for the bond.

covalent bond A bond in which one or more electron pairs are shared between two atoms.

critical point The point on the phase diagram beyond which the gas and liquid phases of a substance are indistinguishable from each other.

crystal lattice A three-dimensional structure that crystalline solids occupy.

crystalline solids A solid in which the particles are arranged in a very regular ordering called a crystal-line lattice.

Dalton's law States that in a mixture of gases, the total pressure is the sum of the pressures of the individual gases.

decomposition reaction A reaction in which a compound breaks down into two or more simpler substances.

dilute A qualitative term describing a solution that has a relatively small amount of solute compared to the amount of solvent.

dipole A molecule in which one end is negative and the other end is positive.

dipole-dipole interaction An intermolecular force that occurs between polar molecules.

double displacement reaction (metathesis) A reaction in which at least one insoluble product is formed when mixing two solutions.

effective nuclear charge The net attraction to the nucleus that an electron experiences, taking into account the shielding effect that the other electrons contribute.

electrolyte A substance that conducts an electrical current when melted or dissolved in water.

electromagnetic spectrum The range of radiant energy composed of gamma rays, X-rays, and so on.

electron The subatomic particle that has a negative charge and very little mass.

electron affinity The energy change that results from adding an electron to a gaseous atom or ion.

electron capture A radioactive decay mode in which an electron from the 1s orbital is captured by the nucleus.

electron cloud A volume of space in which the probability of finding an electron is high. (Also called *electron density*.)

electron configuration A condensed method of representing the pattern of electrons in an atom.

electronegativity A measure of the attractive force that an atom has on a pair of bonding electrons.

empirical formula A chemical formula of a compound that indicates which atoms are present and the simplest whole-number ratio of the elements.

endothermic A type of reaction that absorbs energy from its surroundings.

endpoint The point of a titration at which an indicator signals that an equivalent amount of titrant as substance being titrated has been added.

enthalpy change (ΔH) The heat gained or lost by a system during constant pressure conditions.

exothermic A type of reaction that gives off heat to its surroundings.

families The vertical columns on the periodic table. (Also called *groups*.)

frequency (v) The number of waves that pass by a reference point per second.

gamma emission A radioactive decay mode in which high-energy, short-wavelength photons are emitted from the nucleus.

gas A state of matter that has no definite shape or volume.

Gay-Lussac's law Describes the direct relationship between the pressure of a gas and its Kelvin temperature if the volume and amount of the gas are held constant.

Graham's law Shows that the speed of gas diffusion or effusion is inversely proportional to the square root of the gas's molar mass.

greenhouse effect The warming of the atmosphere due to the absorption of radiant energy by certain gases.

groups The vertical columns on the periodic table. (Also called *families*.)

half-life The amount of time that it takes for a substance to decay to exactly one-half of its initial concentration.

halogens The elements in the VIIA (17) family on the periodic table.

heat capacity The amount of heat needed to change the temperature of a substance 1 kelvin.

heat of vaporization The amount of heat needed to change a liquid into a gas.

Henry's law States that the solubility of a gas increases with the increasing partial pressure of the gas.

Hess's law States that if a reaction occurs in a series of steps, then the enthalpy change for the overall reaction is simply the sum of the enthalpy changes of the individual steps.

Hund's rule States that electrons add to orbitals of the same energy, half-filling them all before the electron spins pair.

hybrid orbitals Atomic orbitals formed as a result of the mixing of the atomic orbitals of the atoms involved in a covalent bond.

hydrogen bonding A strong dipole-dipole intermolecular force that results when a hydrogen atom bonded to an oxygen, nitrogen, or fluorine atom on one molecule is attracted to an oxygen, nitrogen, or fluorine atom on another molecule.

ideal gas A gas that obeys the five postulates of the kinetic molecular theory of gases.

ideal gas equation Relates the Kelvin temperature, the pressure, the volume, and the amount of a gas; it has the mathematical form $PV = nRT$.

indicators Compounds added to the substance being titrated that change color to signal the endpoint.

inner transition elements The two horizontal groups that are pulled out of the body of the periodic table.

intermolecular forces Attractive or repulsive forces between molecules.

ion-dipole interaction Intermolecular forces between an ion and a polar molecule.

ionic bond A bond resulting from a metal reacting with a nonmetal; the metal loses electrons, forming a cation, while the nonmetal gains electrons, forming an anion. The attractive force between the unlike ions is the ionic bond.

ionic equation Shows the soluble reactants and products in the form of ions.

ionic solids Solids with crystal lattices composed of ions held together by the attraction of the charges of the ions.

ionization energy The energy needed to completely remove an electron from a gaseous atom.

isoelectronic Having the same electronic configuration.

isotopes Atoms of the same element that have different numbers of neutrons.

joule The SI unit of energy.

kinetic energy Energy of motion.

kinetic molecular theory A model that attempts to describe the properties of gases at the microscopic level.

law of conservation of matter Says that matter is neither created nor destroyed in ordinary chemical reactions.

Lewis structure A structural formula that represents the elements and their valence electrons.

limiting reactant The reactant that is first totally consumed in a chemical reaction.

line spectrum A series of fine lines representing the wavelengths of photons characteristic of a particular element.

liquid A state of matter that has a definite volume but no definite shape.

London smog A gaseous atmospheric mixture of fog, soot, ash, sulfuric acid, and sulfur dioxide.

magnetic quantum number (m_l) Describes the orientation of the orbital around the nucleus.

main-group elements The groups on the periodic table that are labeled with an A.

manometer An instrument used to measure the pressure of a confined gas.

mass number For an element, the sum of its protons and neutrons.

mass (weight) percent The mass of the solute divided by the mass of the solution and then multiplied by 100.

mass (weight)-volume percent The mass of the solute divided by the volume of the solution and then multiplied by 100.

melting point (mp) The temperature at which the solid state of a substance converts to the liquid state at atmospheric pressure.

metallic bonding A type of bonding in metals in which the electrons of each atom are delocalized and free to move throughout the entire solid.

metalloid A group of elements that have properties of both metals and nonmetals.

metals Elements that are malleable, ductile, and good conductors. They tend to lose electrons in chemical reactions. Mercury is the only liquid metal; the rest are solids.

metathesis reaction A reaction in which at least one insoluble product is formed from the mixing of two solutions.

molality (m) A solution concentration unit that is defined as the moles of solute per kilogram of solvent.

molar heat capacity (S) The amount of heat needed to change the temperature of 1 mole of a substance 1 kelvin.

molarity (M) A solution concentration unit that is defined as the number of moles of the solute per liters of solution.

mole (mol) The number of particles in exactly 12 grams of carbon-12. At the microscopic level, there are 6.022×10^{23} particles/mole, and at the macroscopic level, a mole is the number of grams in the molar mass of a substance.

molecular equation An equation that shows all reactants and products in the undissociated form.

molecular formula Shows what elements are in the compound and the actual number of each. (Also referred to as the *true* or *actual formula*.)

molecular orbital (MO) theory Describes covalent bonding as the combination of atomic orbitals to form molecular orbitals that encompass the entire molecule.

molecule A covalently bonded compound.

net-ionic equation An equation in which the spectator ions are not shown. Only the chemical species involved in the chemical reaction are shown.

network solid Covalently bonded substances whose crystal lattice is extremely large.

neutral A solution with a pH of 7.

neutralization reaction An acid-base reaction in which an acid reacts with a base to give a salt and water.

noble gases The VIIIA (18) group on the periodic table. They tend to be unreactive due to their filled valence shells.

nonelectrolytes Substances that don't conduct electricity when melted or dissolved in water.

nonmetals The elements that generally have properties the opposite of metals. They tend to gain electrons during chemical reactions.

nonpolar covalent bond A bond in which the bonding electrons are shared equally between two atoms.

nucleus The dense central core of an atom holding the protons and neutrons.

octet rule States that during chemical reactions, atoms lose, gain, or share electrons in order to achieve a filled valence shell, to complete their octet of eight electrons.

orbital (wave function) A quantum mechanical description of the location of an electron in an atom.

osmosis The passing of solvent molecules through a semipermeable membrane.

osmotic pressure The amount of pressure that must be exerted on a solution in order to stop osmosis.

particulates Small, solid particles suspended in the air.

pascal The SI unit of pressure.

periods Horizontal groupings on the periodic table.

phase changes Changes of state.

phase diagram A graphical representation of the relationship of the states of matter of a substance to temperature and pressure.

photochemical smog An atmospheric gas produced when sunlight initiates certain chemical reactions involving unburned hydrocarbons and oxides of nitrogen.

pi (π) bonds Bonds resulting from the overlap of atomic orbitals above and below a line connecting the two nuclei.

polar covalent bond A bond in which the bonding electron pairs are unequally shared.

positron Essentially an electron that has a positive charge.

potential energy Stored energy.

precipitate An insoluble material that forms in a solution from ions present.

precipitation reaction A reaction involving the formation of an insoluble precipitate from the mixing of two soluble compounds.

pressure Force per unit of surface area.

principal quantum number (n) Describes the size of the orbital and relative distance from the nucleus.

proof Twice the volume percent of an aqueous ethyl alcohol solution.

quantized A term for an atom with only certain distinct energies associated with its state.

quantum numbers Describe each electron in an atom; quantum numbers tell the orbital size, shape, orientation in space, and spin of each electron.

radioactivity The spontaneous decay of an unstable isotope to a more stable one.

reactants The starting material in a chemical reaction.

reaction intermediate A substance that is formed but then consumed during the reaction.

resonance A way of describing a molecular structure that cannot be represented by a single Lewis structure. Several different Lewis structures are used, each differing in the position of the electron pairs.

reverse osmosis A process that takes place when the pressure on the solution side of a semipermeable membrane exceeds the osmotic pressure and solvent molecules are forced back through the membrane.

saturated solution A solution in which the maximum amount of solute per given amount of solvent at a certain temperature has been dissolved.

semipermeable membrane A thin porous film that allows the passage of solvent molecules but not solute particles.

shells The various energy levels at different distances from the nucleus in which electrons in an atom are located.

SI system The system of units used in science; it is related to the metric system.

sigma bonds Bonds with the orbital overlap on a line connecting the two nuclei.

single displacement (replacement) reaction Reaction in which atoms of an element replace the atoms of another element in a compound.

solid A state of matter that has both a definite shape and volume.

solute The component of a solution that is present in the smallest amount.

solution A homogeneous mixture composed of a solvent and one or more solutes.

solvation The formation of a layer of solvent molecules surrounding a solute particle.

solvent The component of a solution that is present in the largest amount.

specific heat capacity (_Cp_) The quantity of heat needed to raise the temperature of 1 gram of a substance 1 kelvin.

spectator ion An ion directly involved in a chemical reaction that maintains electrical neutrality of the solution.

speed of light The speed that light travels in a vacuum, 3.00×10^8 meters/second.

spin quantum number (ms) Indicates the direction the electron is spinning.

standard enthalpy of formation ($\Delta H_f°$) The change in enthalpy when 1 mole of the compound is formed from its elements when all substances are in their standard states.

stoichiometry The calculation of amount (mass, moles, particles) of a substance in a chemical reaction.

strong acid A proton donor that completely ionizes in water.

strong base A proton acceptor that contains the hydroxide ion and completely ionizes in water.

strong electrolytes Electrolytes that completely dissociate or ionize in water.

structural isomers Compounds that have the same molecular formula but differ in how the groups are attached to each other.

sublimation A change of state in which a substance goes directly from the solid state to the gaseous state without becoming a liquid.

subshells Within the electron shells, orbitals of slightly different energies in which electrons are grouped.

supersaturated solution A solution in which more than the maximum amount of solute has been dissolved in the solvent.

surface tension The amount of force required to break through the molecular layer at the surface of a liquid.

surroundings In thermodynamics, a term that represents the rest of the universe that is being affected by some type of change.

suspension A heterogeneous mixture whose particles are large (greater than 1,000 nanometers).

system The thermodynamics term for the part of the universe that you are studying.

theoretical yield The maximum amount of product that can be formed during a chemical reaction.

thermochemistry The part of thermodynamics that deals with changes in heat that take place during chemical reactions.

thermodynamics The study of heat and its changes.

titrant The solution in a titration that has a known concentration.

titration A laboratory technique in which a solution of known concentration is used to determine the concentration of an unknown solution.

transition elements The elements categorized as B groups on the periodic table.

transmutation A nuclear reaction in which an element is created by another one.

triple point The combination of temperature and pressure on a phase diagram at which all three states of matter of a substance can exist.

Tyndall effect Exhibited when a light is shown through a colloid and the light beam is visible due to the reflection of light from the large colloid particles.

unit cells The repeating units in a crystal lattice.

unsaturated solution A solution with less than the maximum amount of solute dissolved in a given amount of solvent at a given temperature.

valence bond theory Describes covalent bonding as the overlap of atomic orbitals to form a new type of orbital: a hybrid orbital.

valence electrons The electrons (normally only in s and p orbitals) that are in the outermost energy level.

van der Waals equation A modification of the ideal gas equation to compensate for the behavior of real gases.

van't Hoff factor The ratio of moles of solute particles formed to moles of solute particles dissolved in solution.

vapor pressure The pressure exerted by the gaseous molecules that are in contact with a liquid in a closed container.

viscosity The resistance to flow of liquids.

volume percent For a solution, the volume of the solute divided by the volume of the solution, with the result multiplied by 100.

VSEPR (valence-shell electron-pair repulsion) theory Predicts molecular geometry by considering that the valence electron pairs around a central atom try to maximize the distance from each other in order to minimize repulsive forces.

wave function A mathematical description of the electron's motion.

wavelength (λ) The distance between two identical points on a wave.

weak acid A proton donor that only partially ionizes in water.

weak base A proton acceptor that only partially ionizes in water.

weak electrolyte An electrolyte that only partially ionizes in water.

Index

Symbols

Δ (delta symbol), 250
+ (plus sign), 250
→ (yields symbol), 250

A

absolute zero, 32, 355, 461
accuracy
  defined, 14
  error, 14
  example questions and answers, 14–15
  percent error, 14
  practice questions, 15
  practice questions answers and explanations, 24
  vs. precision, 13–15
acetates, 203, 263
acetic acid
  boiling point of, 405
  formula/common name or use, 207, 420
  freezing point of, 407
acetylsalicylic acid, 420
acid dissociation constant, 429, 430
acid rain, 461
acidic, 461
acidity measurement, 424–428
acids. *See also* bases
  Arrhenius definition of, 421
  binary, 207
  Brønsted-Lowry definition of, 422
  common, 207, 420
  concentration, 428
  conjugate, 422
  defined, 461
  example questions and answers, 423, 427–428, 432
  Lewis definition of, 423
  measuring acidity, 424–428
  naming, 206
  neutralization reactions, 421
  oxy-acids, 207
  practice questions, 424, 428, 433

practice questions answers and explanations, 433
  properties of, 419–420
  quiz answers, 438–439
  quiz questions, 438–439
  strength, 428–433
  strong, 429
  weak, 430–432
actinides, 85
actinium, 66
activation energy, 134, 251, 461
activity series, 261–262, 461
actual value, 14
actual yield, 461
addition
  in scientific notation, 11–12
  significant figures, 18
alkali metals
  cations, 146–147
  defined, 81, 461
  electron configurations, 94
  solubility, 263
alkaline earth metals
  cations, 146–147
  defined, 81, 461
  electron configurations, 94
alkanes, 210
allotropes, 325
alloys, 156, 380
alpha decay, 126–127
alpha particles, 126–127, 461
aluminum, 66, 147, 262, 279, 325
aluminum hydroxide, 220, 420
aluminum sulfate, 279
americium, 66
ammonia, 182, 278–279, 420
ammonium chloride, 220
amorphous solid, 461
amphoteric, 426, 461
amplitude, 102–103, 461
angular momentum quantum number *l*, 97
anions, 69, 149, 461

D

d orbitals, 92, 98

Dalton's law, 362–363, 463

dec- prefix, 211

deca- prefix, 30, 205

decane, 211

decay. *See also* radioisotopes
  alpha, 126–127
  beta, 127
  example questions and answers, 128
  gamma, 127–128
  half-lives, 125, 129–132
  practice questions, 128–129
  practice questions answers and explanations, 137–138
  quiz answers, 140–142
  quiz questions, 139

deci- prefix, 30

decomposition reactions, 261, 463

deka- prefix, 30

delta symbol (Δ), 250

Democritus, 59

denominator, 10

density
  calculating requirements for, 34–36
  example questions and answers, 35–36
  formula, 34
  practice questions, 36
  practice questions answers and explanations, 44
  quiz answers, 47
  quiz questions, 46
  of typical solids and liquids, 35
  units, 34

deposition, 306

derived units, 33

deuterium, 134

di- prefix, 205

diamond, 325

diatomic molecules, 155, 156

dicarbon heptahydride, 209

dicarbon tetrahydride, 206

dichlorine, 189

dichlorine octoxide, 206

diffusion, 364

dihydrogen phosphate, 203

dilute, defined, 382, 463

dilutions, 388–390

dimensional analysis. *See* factor label method

dinitrogen monoxide, 204, 206

diphosphorous hexahydride, 219

dipole, 177, 179, 463

dipole moment, 179

dipole-dipole interaction, 187, 463

dissociation constant, 426

division
  in scientific notation, 9–11
  significant figures, 18

double displacement (double replacement), 262

double displacement reaction (metathesis), 463

ductility, 83

E

effective nuclear charge, 463

effusion, 364

electrolysis, 279

electrolytes, 159, 463

electromagnetic radiation, 102

electromagnetic spectrum, 102–103, 463

electron affinity, 463

electron capture, 127, 463

electron clouds, 96, 463

electron configuration, 464

electron dot formula
  defined, 156
  for water, 166–167

electron dot structures, 145–146

electron shells, 97

electronegativity, 118–120, 149, 175, 464

electronic clouds, 155

electronic con, 83

electron-pair geometry, 181

electrons
  Aufbau filling diagram, 92–93
  charge, 60
  configurations, 91–95
  defined, 60, 463
  example questions and answers, 95, 99, 100–101, 104–105
  excited state, 104
  ground state, 104
  mass, 60
  practice questions, 95, 100, 101, 105
  practice questions answers and explanations, 106–107

R

radiant energy, 102
radio waves, 103, 103–104
radioactive decay, defined, 125
radioactivity, 12, 64, 467
radioisotopes
  alpha decay, 126–127
  beta decay, 127
  defined, 125
  example questions and answers, 128, 131–132
  half-lives, 129–132
  neutron-rich, 126
  practice questions, 128–129, 132
  practice questions answers and explanations, 137–138
  quiz answers, 140–142
  quiz questions, 139
radium, 67
radon, 67
radon-222, 131
reactants, 250, 467
reaction intermediate, 467
reaction mechanisms, 251
reactions
  activation energy of, 251
  collision theory, 251
  combination (synthesis), 259–260, 260
  combustion, 264, 462
  decomposition, 261, 463
  endergonic, 254
  endothermic, 251, 253–254, 319, 330–331, 464
  equilibrium, 426
  example questions and answers, 255, 264
  exothermic, 251, 253, 319, 330–331
  limiting reactants, 283–285
  metathesis, 466
  neutralization
    buffers, 451–452
    defined, 263, 445, 466
    example questions and answers, 449, 453, 454
    indicators, 446
    practice questions, 449–450, 453, 454
    practice questions answers and explanations, 455–456
    quiz answers, 458–459

quiz questions, 457
    solubility product constant, 454
    titration process, 447–449
  one-step, 251
  overview, 259–260
  practice questions, 255, 264–265
  practice questions answers and explanations, 268–270
  precipitation, 262–263, 467
  quiz answers, 273–275
  quiz questions, 271–272
  single displacement (single replacement), 261–262
  single displacement (single replacement) reactions, 468
  synthesis, 260
reagents, 281
red cabbage, 446
red litmus, 447
registration, 3
relative abundance, 71
resonance, 467
reverse osmosis, 467
rhenium, 67
rhodium, 67
rounding off, 19
rubidium, 67
rubidium hydroxide, 429
ruthenium, 67
Rutherford, Ernest, 63–64
rutherfordium, 67

S

s orbital, 92, 98
(s) symbol, 250
saltpeter, 386
salts, ionic, 153
samarium, 67
saturated solution, 381, 468
scandium, 67
scientific notation
  adding in, 11–12
  converting decimal form into, 8
  converting into decimal form, 8
  dividing in, 9–11
  example questions and answers, 11–12

About the Authors

Christopher R. Hren is a high school chemistry teacher. He has been happily married for a while now and loves spending time with his family. Chris is a proud graduate of Michigan State University and spends every Saturday in the fall watching his Spartans on the field. Chris has coached football and track and is a pretty big sports fan in general. Chris has spent a good deal of his time perfecting his forehand on the tennis court and his jump-shot on the basketball court. In the minimal free time he has, Chris loves working with computers and all things technology along with reading a healthy dose of science fiction. His family is lucky enough to have a small cottage on a lake in Northern Michigan, where he enjoys many outdoor activities as well.

John T. Moore, EdD, grew up in the foothills of western North Carolina. He attended the University of North Carolina–Asheville, where he received his bachelor's degree in chemistry. He earned his master's degree in chemistry from Furman University in Greenville, South Carolina. After a stint in the United States Army, he decided to try his hand at teaching. In 1971, he joined the chemistry faculty of Stephen F. Austin State University in Nacogdoches, Texas, where he still teaches chemistry. In 1985, he started back to school part time, and in 1991 he received his doctorate in education from Texas A&M University.

John's area of specialty is chemical education. He has developed several courses for students who are planning to teach chemistry at the high school level. In the early 1990s, he shifted his emphasis to training elementary education majors and in-service elementary teachers in hands-on chemical activities. He has received four Eisenhower grants for professional development of elementary teachers, and along with one of his former students, he has served as coeditor of the "Chemistry for Kids" feature of The Journal of Chemical Education. He is the author of several books on chemistry and is coauthor of several more, including *Chemistry For Dummies, Biochemistry For Dummies,* and *Organic Chemistry II For Dummies.*

Peter J. Mikulecky, PhD, grew up in Milwaukee, an area of Wisconsin unique for its high human-to-cow ratio. After a breezy four-year tour in the Army, Peter earned a bachelor of science degree in biochemistry and molecular biology from the University of Wisconsin–Eau Claire and a PhD in biological chemistry from Indiana University. With science seething in his DNA, he sought to infect others with a sense of molecular wonderment. Having taught, tutored, and mentored in classroom and laboratory environments, Peter was happy to find a home at Fusion Learning Center and Fusion Academy. There, he enjoys convincing students that biology and chemistry are, in fact, fascinating journeys, not entirely designed to inflict pain on hapless teenagers. His military training occasionally aids him in this effort.

Dedication

From Chris: This book is dedicated to my family. To my wife, Laurie, and my children, Madeline and Luke, who are amazing. Without them none of this happens. And to my parents, who have always been so supportive of me in everything I have done throughout my entire life. Thank you.

Authors' Acknowledgments

From Chris: I would like to acknowledge the incredible work of everyone at Wiley on this book. Executive Editor Lindsay Lefevere has been a constant source of support and guidance over the years for me with every project I have done. Without her I would never have had the opportunity to write this book or any other, so I thank her immensely for her confidence in me. Chrissy Guthrie has been incredible every step of the way in providing support and guidance. Every time we work together, she seems to understand exactly what I need, and it shows in the quality of the work. She is an amazing editor. In addition, I'd like to thank everyone else who worked on this book including copy editor Marylouise Wiack and technical editor Caleb Zelencik. Finally, I would like to thank all my students over the past years. The intellectual challenge you have provided me h every day is wonderful. No two days at school are ever the same, and that is thanks to you. I appreciate so very much the many conversations we've had about so many different things. Seeing you grow into fully functional adults over the years has been a true joy.

Publisher's Acknowledgments

Executive Editor: Lindsay Sandman Lefevere

Project Manager and Development Editor: Christina N. Guthrie

Managing Editors: Michelle Hacker and Kelsey Baird

Copy Editor: Marylouise Wiack

Technical Editor: Caleb Zelencik

Project Editor: Pradesh Kumar

Project Manager: Chrissy Guthrie

Cover Image: © CasPhotography/Getty Images